BIOREMEDIATION AND PHYTOREMEDIATION

Technologies for Toxic Pollution

BIOREMEDIATION AND PHYTOREMEDIATION

Technologies for Toxic Pollution

Ashwini A. Waoo, PhD

AAP | APPLE ACADEMIC PRESS

First edition published 2024

Apple Academic Press Inc.
1265 Goldenrod Circle, NE,
Palm Bay, FL 32905 USA

760 Laurentian Drive, Unit 19,
Burlington, ON L7N 0A4, CANADA

CRC Press
2385 NW Executive Center Drive,
Suite 320, Boca Raton FL 33431

4 Park Square, Milton Park,
Abingdon, Oxon, OX14 4RN UK

Library and Archives Canada Cataloguing in Publication

Title: Bioremediation and phytoremediation : technologies for toxic pollution / Ashwini A. Waoo, PhD.
Names: Waoo, Ashwini A., 1980- author.
Description: First edition. | Includes bibliographical references and index.
Identifiers: Canadiana (print) 20230532144 | Canadiana (ebook) 20230532209 | ISBN 9781774913840 (hardcover) | ISBN 9781774913857 (softcover) | ISBN 9781003409595 (ebook)
Subjects: LCSH: Bioremediation. | LCSH: Phytoremediation. | LCSH: Pollution. | LCSH: Biotechnology.
Classification: LCC TD192.5 .W36 2024 | DDC 628.5—dc23

Library of Congress Cataloging-in-Publication Data

CIP data on file with US Library of Congress

ISBN: 978-1-77491-384-0 (hbk)
ISBN: 978-1-77491-385-7 (pbk)
ISBN: 978-1-00340-959-5 (ebk)

About the Author

Ashwini A. Waoo, PhD
Professor, Department of Biotechnology,
Faculty of Life Science and Technology,
AKS University, Satna, Madhya Pradesh, India

Ashwini A. Waoo, PhD, is currently working as a Professor in the Department of Biotechnology in the Faculty of Life Science and Technology at AKS University in Satna, Madhya Pradesh, India. She has over 21 years of academic and research experience and has published more than 30 research papers in international journals. She has also contributed to many international journals as a reviewer. Dr. Waoo was honored with a Women Researcher Award at the 1st International Scientist Awards in Engineering, Science and Medicine, organized by the VD Good Association in Chennai, India. Dr. Waoo has made research contributions, including a major University Grants Commission research project. She has supervised several PhD students and has overseen over 100 dissertations at the under- and postgraduate levels. She has undertaken many responsibilities, such as academic coordinator of PG and PhD biotechnology courses, coordinator of the academic committee, and member of the training and placement cell. She was recently appointed a Bentham Ambassador to India for Bentham Science journals. She is also an academic coordinator of postgraduate biotechnology courses, coordinator on academic committees, and a member of training and placement cells. Dr. Waoo's credentials include a doctorate, GATE, CSIR-NET Life Science, MPhil (Biotechnology), and MSc (Biotechnology) along with IIT-Bombay and SWAYAM certifications. She earned a young scientist's scholarship from the MP Council of Science and Technology.

Contents

Abbreviations

AAS	atomic absorption spectrometry
ACC	1-aminocyclopropane-1-carboxylate
AES	atomic emission spectrometry
AFM	atomic force microscopy
AI	autoinducers
AM	arbuscular mycorrhizal
AMD	acid mine drainage
AMTW	acid mine tailing waste
APX	ascorbate peroxidase
As	arsenic
ASV	anodic stripping voltammetry
ATP	adennosine tri phosphate
AtPCS	arabidopsis thaliana phytochelatin synthase
BAC	bacterial artificial chromosome/bioaccumulation coefficient
BAP	6-benzyl amino purine
BCF	biological concentration factor
BOD	biological oxygen demand
BTEX	benzene, toluene, ethylene, and xylene
CAT	catalase
CCl4	carbon tetrachloride
Cd	cadmium
CEC	cation exchange capacity
CI	color index
COD	chemical oxygen demand
Cr	chrome
CRISPR	clustered regularly interspaced palindromic repeats
CV-AAS	cold vapor AAS technique
DAPG	2,4-diacetylphloroglucinol
DNA	deoxyribonucleic acid
DW	dry weight
EDB	ethylene dibromide
EDL	electrodeless discharge lamp
EDTA	ethylene diamine tetra acetic acid
EMB	eosin methylene blue
EPA	Environmental Protection Agency
EPS	extracellular polymeric substances
EPS	extracellular polysaccharide

ET-AAS	electrothermal atomizer
F-AAS	flame atomization
FAO	Food and Agriculture Organization
FDA	fluorescein diacetate
Fe	iron
GFP	green fluorescent protein
GMOs	genetically modified organisms
GPX	guaiacol peroxidase
GSH	glutathione
HCD	hollow cathode lamp
HCI	hydrochloric acid
HEPA	high efficiency particulate air
HSP	heat shock protein
IAA	indoleacetic acid
ICP	inductively coupled plasma
ICPMS	inductively coupled plasma-mass spectrometry
ISR	Induced systemic resistance
KI	potassium iodide
KOH	potassium hydroxide
LOD	limits of detection
LPS	lipopolysaccharides
MAPK	mitogen-activated protein kinase
MEMS	micro-electro-mechanical systems
MRVP	methyl red and Voges-Proskauer
MSA	mannitol salt agar/multi-sequence alignments
MT	metallothionein
MTM	macro and trace metals
NAA	neutron activation analysis
NAM	nutrient agar media
NGS	next-generation sequencing
Ni	nickel
NO2	nitrogen dioxide
NPK	nitrogen, phosphorus, and potassium
OPH	organophosphate hydrolase
Ops	organophosphates
PAHs	polycyclic aromatic hydrocarbons
Pb	lead
PC	phytochelatin
PCB	polychlorinated biphenyls
PCR	polymerase chain reaction
PCs	phytochelatin
PDA	potato dextrose agar
PDOC	produced dissolved organic carbon

PGPR	plant growth-promoting rhizobacteria
PGRs	plant growth regulators
PM	particulate matter
PNP	p-nitrophenols
POX	peroxidase
PPCP	pharmaceutical and personal care products
PUF	polyurethane foam
QA	quality assurance
QS	quorum sensing
RBC	rotating biological contactors
RDX	hexahydro-1,3,5-trinitro-1,3,5-triazine
RNA	ribonucleic acid
ROS	reactive oxygen species
RT-PCR	real-time PCR
Se	selenium
SEM	scanning electron microscopy
SO2	sulfur dioxide
SOD	superoxide dismutase
SRB	sulfate-reducing bacteria
TCE	trichloroethylene
TF	translocation factor
TFS	transfer factors from soil
TNT	2,4,6-trinitrotoluene
UV	ultraviolet
VAM	vesicular-arbuscular mycorrhizal fungi
WHC	water holding capacity
WHO	world Health Organization
XRF	X-ray fluorescence
Zn	zinc
γ-EC	γ-glutamyl cysteine synthetase

Foreword 1

With great pride and anticipation, I introduce this exceptional book on biotechnology, mainly focused on bioremediation and phytoremediation, authored by our university's finest faculty member, Prof. Ashwini A. Waoo. The dynamic synergy of her expertise and dedication to advancing knowledge in these critical fields shines brilliantly in the pages that follow.

In an era where environmental challenges continue to grow, biotechnology and bioremediation represent promising solutions. As the Pro-Chancellor of AKS University, I applaud our distinguished faculty for her unwavering commitment to research and education. This book exemplifies our university's enduring pursuit of excellence, fostering innovation, and cultivating a new generation of scholars and problem solvers. I do not doubt that this book will inspire, inform, and drive positive change in biotechnology and bioremediation, contributing to a more sustainable and prosperous world.

With a commitment to excellence and a shared vision for a brighter tomorrow, I am confident that this book will inspire and empower generations of scholars, researchers, and visionaries. Together, let us forge a path toward a sustainable future.

—Er. Anant Kumar Soni
Pro-Chancellor,
AKS University, Satna, MP, India

Foreword 2

Few biotechnology book that come along address all environmental toxic pollution concerns. This book on bioremediation and phytoremediation, by Dr. Ashwini A. Waoo, is one of those books. This book is both concise and comprehensive at the same time. There is a wealth of valuable information in its 18 chapters. It is therefore with great pleasure that I am contributing this foreword.

Biotechnology will continue to flourish and expand into new horizons as one of the most significant subjects that has shaped the human future. Different branches of biotechnology have emerged from a small set of disciplinary micro-niches to well-defined, dynamic, academic, and industrial research fields that, shortly, will have both substantial academic and industrial impact. Industrial activity, agriculture, and waste disposal practices have caused the contamination of soil and water with organic compounds and heavy metals, causing harm to ecosystems and humans.

Bioremediation, or the use of biological processes for the cleanup of polluted sites, is one of the most rapidly growing areas of environmental biotechnology. There has been extensive research on microorganisms in bioremediation, with most of the research focused on organic pollutants. However, microorganisms can also remove inorganic pollutants, such as toxic metals and metalloids. The bioremediation of contaminated sites is more likely to be achieved by microbial consortia than by a single strain. A plant-microbe-associated bioremediation technique is a promising and cost-efficient method of removing pollutants from polluted sites.

Phytoremediation is also a low-cost, elegant approach for decontaminating polluted sites with the use of plants and has received widespread approval from the public. Therefore, phytoremediation technology is being examined as a means of treating large areas of land and water that are currently polluted. An overview of environmental phytoremediation strategies is provided in this book. Several chapters of this book discuss phytoremediation by inoculating plants with bacteria from their rhizospheres that can remove heavy metals from soil.

The utilization of our natural resources in the past decade has shown carelessness and negligence to varying degrees. Natural resource contamination is a major issue in many countries. The environment is contaminated by hazardous substances produced, used, and disposed of as a result of industrial activities. Globally, there are many contaminated sites. There is now widespread awareness that contaminated environments can pose a health threat, and efforts have been made at numerous sites around the world to remediate these contaminated areas, so they can be redeveloped. This book can bring awareness to all sectors of society and introduce this eco-friendly concept worldwide in implementation mode.

Clean-up of the existing environmental contamination is becoming increasingly dependent on innovative treatment technologies. In addition, industries are being pressured to reduce their production or generation of hazardous pollutants. The use of bioremediation is an economical and viable option for both field and enclosed remediation. Regulatory requirements dictate the use of innovative treatment technologies. To control pollutants released into the air, water, and land, various environmental statutes have been enacted and hundreds of regulations have been implemented in the last two decades.

Biotechnology helps in the recycling or degradation of hazardous pollutants during industrial activities. Numerous environmental and industrial benefits are possible with these applications, including increased on-site treatment; reduced regulatory effect for the government; reduced liability for industries; and ultimately improved process optimization.

A series of captivating chapters highlighting different techniques of bioremediation and phytoremediation has been crafted to showcase such a wide and dynamic topic. Featuring attractive illustrations of complex theoretical and experimental details and the overall mechanism, this book is sure to win over enthusiastic readers. My favorite thing about the book is the simplicity of the language. In my opinion, this book is very well organized, and at the same time, the book is very provocative. I feel privileged to write the foreword for this prestigious, peer-reviewed publication on a topic that I am passionate about. I wish the author, Dr. Ashwini A. Waoo a great success and expect to see a future edition culminating from this book.

—Professor B. A. Chopade
Chairman,
PAC, SEED Division, DST, Govt. of India, New Delhi, India
and
Former Vice-Chancellor,
Dr. Babasaheb Ambedkar Marathwada University,
Aurangabad, Maharashtra, India

Overview of the Book

The book is organized into eighteen chapters. A brief description of each of the chapters follows:

- Chapter 1 identifies the existing challenges of toxic pollution in the biosphere, its sources, and adverse effects on the environment. It gives an account of heavy metal pollution. It also focuses on the importance of establishing a remediation plan for the environment.
- Chapter 2 introduces the concept of bioremediation technology for the reclamation of the environment, it includes ex-situ and *in situ* techniques of bioremediation, the role of microorganisms in bioremediation, and critical factors affecting the rate of contaminant degradation. It concludes with the advantages and limitations of bioremediation.
- Chapter 3 illustrates the significance of the immobilization technique for bioremediation. It involves detailing immobilization methods, carrier materials, and immobilized cell bioreactors.
- Chapter 4 introduces pollution due to chemical dyes and plastics, their adverse effects on the environment, and the need for bioremediation of these dyes and plastics using microorganisms.
- Chapter 5 reviews the concept of biochar and its application in bioremediation while discussing points such as factors affecting biochar quality and applications of biochar in agro-environment development.
- Chapter 6 focuses attention on the use of cyanobacteria in the bioremediation of wastewater. It also discusses the detoxification mechanism with some examples of cyanobacteria species and the concepts such as metal biosorption, and genetic engineering approach for cyanobacteria.
- Chapter 7 addresses the issue of mine wastes and their implications on the environment. The problem of acid mine drainage and the strategy for its remediation with plants forms the body of the chapter. It includes a recent concept of the use of nanoremediation for mine waste reclamation and the use of algae in mines bioremediation.

- Chapter 8 discusses the concept of biosensors and bioindicators. The biosensor section includes components of biosensors, molecular biosensors, and cell microarrays whereas the bioindicator section includes the species used and their advantages.
- Chapter 9 reviews phytoremediation concepts, their mechanisms, advantages, and the genetic engineering approach for phytoremediation.
- Chapter 10 discusses the rhizosphere, its dynamic interaction, and PGPR mediated remediation of pollutants. It also includes the use of metagenomics in rhizosphere studies. It reviews quorum sensing for pollutant degradation and the concept of biofilm.
- Chapter 11 focuses on rhizoremediation-specific research for heavy metal pollution, biosurfactant ad organic acid production.
- Chapter 12 indicates the need for soil analysis for further environmental remediation technologies.
- Chapter 13 illustrates the overall strategies of phytoremediation of heavy metal contaminated sites. It includes the heavy metal uptake mechanisms in plants.
- Chapter 14 comprises phytoremediation of wastewater, the concept of vegetated caps, buffer strips, etc.
- Chapter 15 introduces the basic protocol, instruments, and composition of media of plant tissue culture and its immense use in phytoremediation.
- Chapter 16 is a review of the historical developments in the field of phytoremediation, key findings, and novel insights into phytoremediation research.
- Chapter 17 presents a case study on phytoremediation of contaminated sites, detailed methodology, observation, results, and its interpretation.
- Chapter 18 is the last chapter of this book; it illustrates the overall strategies for the re-extraction of heavy metals from hyperaccumulator biomass with a sustainable approach.

Acknowledgments

Achievement is not always a success. It is an honest endeavor, a persistent effort to do the best possible under any circumstances. During my book writing process, I have been supported and accompanied by many people and this is a coveted opportunity to express my gratitude for them all.

I would like to express my gratitude to Dr. Akhilesh A. Waoo, Professor, Department of Computer Science, AKS University Satna, my beloved husband who saw me through this book; provided support, talked things over, read, wrote, offered comments, allowed me to quote his remarks, and assisted in the editing, proofreading, and design. I dedicate this book to him.

I would have found it more difficult to pursue this project without the support of my family. As role models, they are beyond comparison. My son, Master Aaradhya, who provides unending inspiration to me, deserves the greatest thanks.

Above all, I want to thank my parents and in-laws, and all family members who supported and encouraged me despite all the time it took me away from them. It was a long and difficult journey for them. I would like to thank my parents for their love and guidance during all of my endeavors.

Also, thanks to my research scholar, Charu Vyas, for supporting work in soil heavy metal analysis.

A very special thanks to the honorable pro-chancellor of AKS University, Er. Anant Soni and the entire team of AKS. I'm honored to be a part of the AKS family. Thank you for letting me serve, and giving me inspiration for turning my ideas into reality.

Thanks to Apple Academic Press (AAP), my publisher, who encouraged me.

Last and not least, I beg forgiveness of all those who have been with me over the years and whose names I have failed to mention.

Preface

To protect our environment from chemical contamination, we need to make sure pollution levels don't reach unsustainable levels. Science and technology can provide solutions to global environmental challenges, such as persistent organic pollutants, climate change, and contaminant bioaccumulation.

An important aim of the book is to make the key concepts and issues that underlie the environmental awareness that has resulted in regulatory measures aimed at rectifying past mistakes and protecting the environment from future contamination and exploitation. One of the alternative technologies for the removal of pollutants from the environment, restoring contaminated sites, and preventing further pollution is bioremediation. The bioremediation process utilizes living organisms, primarily microorganisms, and plants, to degrade and reduce or detoxify waste products and pollutants. Providing the challenges of bioremediation, especially those associated with *in situ* techniques, can be overcome, bioremediation holds the potential to offer a low-cost, non-intrusive, natural way of reducing the adverse effects of toxins in soil over time.

The book provides an overview of issues related to bioremediation and phytoremediation. Human health and wildlife are both affected by environmental contaminants. Plant-based bioremediation offers a cost-effective alternative by using plants and associated soil microbes to help reduce contaminants' content or effects on the environment. A variety of phytoremediation technologies are available, including (a) phytostabilization, which involves retaining contaminants in soil; (b) phytodegradation, which involves converting organic contaminants into less harmful substances; (c) phytovolatilization. As contaminants are converted to a gaseous state inside plants, they are released into the atmosphere through evapotranspiration; and (d) as they accumulate aboveground, harvestable biomass through phytoextraction. It is crucial to understand how heavy metals accumulate in plants and how they tolerate them to improve phytoremediation.

PART I

Bioremediation

CHAPTER 1

Consequences of Toxic Pollution

1.1 INTRODUCTION

Environmental pollution affects the entire biosphere, directly or indirectly. A boom in population and industrialization needs many natural resources and sources of energy in developing as well as underdeveloped countries. Industrial effluents are released directly or indirectly in the soil or natural water resources without any pre-treatment, leading to serious hazards to the ecosystem. In the biosphere, air, water, and soil are the main natural components which badly affected by contaminants. This chapter will focus on the concerns arising due to their pollution. Ultimately, soil pollution will be discussed in detail by attracting the research interest towards remediation of heavy metal polluted sites.

1.2 AIR POLLUTION

Pollutants released into the air can harm human health and the environment. According to the World Health Organization (WHO), seven million people die due to air pollution every year. As per the WHO's guidelines for pollutants, nine out of 10 people breathe air that exceeds those standards, and those in low-income countries breathe air that exceeds those standards the most. From smog obscuring cities to indoor smoking, air pollution is a significant threat to health and the climate. The risks of strokes, heart attacks, chronic obstructive pulmonary disease, lung cancer, and acute respiratory infections are higher due to external air pollution (outdoor air pollution) and household air pollution. The following categories are used to classify pollutants:

- ***Suspended Particulate Matter (PM)*** – There are different types of total suspended particles such as PM10, which can reach lung alveoli

Bioremediation and Phytoremediation: Technologies for Toxic Pollution
Ashwini A. Waoo, PhD (Author)

and is the most hazardous. PM2.5 is produced by the condensation of gaseous pollutants such as sulfur dioxide (SO_2) and nitrogen dioxide (NO_2). PM can consist of mineral dust, coal fly ash, smoke, wood, limestone, asbestos, cement, metal dust, and pesticide mists.

- ***Gaseous Pollutants*** – Compounds included in this category include sulfur compounds, nitrogen compounds, carbon monoxide, and organic compounds such as volatile polycyclic aromatic hydrocarbons, hydrocarbons, and halogen derivatives such as aldehydes and odors. Various household and work products emit volatile organic compounds when they burn. Oxygen oxides and hydrocarbons are converted to ozone at ground level by the reaction of sunlight and nitrogen oxides. This level of ozone is dangerous for human health.
- **Sources of Outdoor Air Pollution**

Industry, vehicles, and power plants contribute to outdoor air pollution. Agricultural waste and wood combustion also cause pollution. Dust from industrial processes, such as cement factories, gas releases, and metal smelters, contributes to this to a greater extent. PM, NO_2, nitric oxide, carbon monoxide, lead, and organic compounds are all released into the air by motor vehicles. Radiation pollution from a malfunctioning nuclear power station around Chernobyl in 1986 is a major disaster for the biosphere. A catastrophic release of organic chemicals, such as in Bhopal, India, in 1984, can also have serious health consequences.

1.2.1 EXPOSURE TO AIR POLLUTANTS

Air pollution causes health effects based on exposure levels to pollutants. People's activity patterns and time schedules determine their total daily exposure. Children and the elderly may be exposed to lower levels of air pollution during the day, and their exposure may be correlated with air pollution levels in their homes. Air pollution adversely affects children's development.

Meteorological factors, such as wind speed and direction, influence air pollution. With topography and temperature alterations, they are usually the strongest determinants of pollution hazards.

1.2.2 EFFECT OF AIR POLLUTION ON HEALTH

As a result of burning toxic substances, respiratory and cardiovascular indisposition and mortality have increased. An air pollution crisis caused 4,000 premature deaths in only a week in London in 1952, and another 8,000 died over the next few months. During the 1970s/80s, new statistical methods and computer-based technologies were used to study mortality at much lower levels of pollution.

1.3 WATER POLLUTION

Contamination of surface water with hazardous chemicals can create health hazards, as these water sources are often linked with drinking water reservoirs. Toxic chemicals like arsenic and fluoride released from industrial dumping sites can be dissolved into groundwater from the soil and rock layers causing direct contamination. The disastrous Minamata disease outbreak that occurred in 1956 in Japan was the result of seawater pollution near the coast. At extremely low concentrations, toxins like polychlorinated biphenyls (PCBs) and dioxins can pose serious health hazards in seawater.

1.3.1 SOURCES OF WATER POLLUTION

Point sources or nonpoint sources of pollution can pollute water. An industrial site, for example, is one source of point-source pollution. In nonpoint-source, pollution is caused due to many small sources, such as rainwater carrying pollutants like insecticides, fertilizers, herbicides, and dropped into rivers, lakes, oceans, or groundwater.

Paper and pulp mills discharge several thousand tons of waste products into the environment each year. The liquid waste contains a high biological oxygen demand, suspended solids, and chlorinated compounds such as dioxins. The storage and transportation of solid waste may also contaminate surface waters. Sugar mill effluents are also characterized by high levels of ammonia and biological oxygen demand. There is a possibility that sugarcane rinsed liquid may contain toxic pesticide residues. Leather tanneries generate large amounts of hair, hides, and solid waste. There is also contamination from textile and dye industries, which emit hazardous pollutants, such as sulfides, chlorides, chromium, and acids. Due to insufficient care taken during mining activities, toxic metals were released into water sources by grinding ores. Most lead and zinc ores contain more toxic cadmium as a minor component. Water pipes can also be corroded by copper in water. Mining and industrial effluents can also carry mercury into waterways.

Mercury is released into the environment through the incineration process of medical waste and medical equipment. As a result, sulfate-reducing bacteria react with mercury and methylate it, increasing mercury's toxicity. Long-term exposure to methylmercury causes serious neurological damage.

Discharged runoff from farmland, carrying sediments contribute to high turbidity, and the extra addition of nitrogen and phosphates from manure or fertilizers. As a result of these chemicals, the water resources become drier, resulting in eutrophication.

Nitrogen fertilizers tend to cause higher concentrations of nitrates in groundwater, thereby causing high levels of nitrates in drinking water, which can cause methemoglobinemia, including the "blue baby" syndrome in young children. From surface waters, pesticides can seep into groundwater. A set of specific rules and regulations applies to the use of most toxic pesticides in developed countries, and they are enforced appropriately. It is possible that rules of application are unknown or are not enforced in developing countries. Therefore, pesticide pollution posed health risks.

1.3.2 HEALTH EFFECTS OF WATER POLLUTION

Water pollutants can trigger acute irritation, rash, or inflammation of the skin, nose, and gastrointestinal system, but the most significant health effects often result from chronic exposure like liver toxicity due to copper, arsenic, or chromium in drinking water. Among the toxins that are toxic to the kidney are copper, cadmium, chlorobenzene, mercury, and cadmium, which are excreted by the kidney.

Pesticides and other pollutants can persist for long periods in the environment and can be transported over long distances by water or air. They disrupt the endocrine system, causing developmental, reproductive, and behavioral problems, reduced fertility, and an increase in birth defects. As a result, it caused cancers such as those of the testes, breasts, and prostates.

Also, there are effects such as impairments to mental and psychomotor development, cognitive impairment, and behavior abnormalities (WHO, 2002). Chemical pollutants, such as alkylphenols, organochlorines, phytestrogens, and PCBs, as well as pharmaceuticals, such as antiseptics and contraceptives, disrupt the endocrine system. Toxins present in drinking water can also cause cancer. (International Agency for Research on Cancer, 2004).

1.4 SOIL POLLUTION

Earth's total surface was covered with a mixture of organic and inorganic compounds known as soil. Among these, the organic portion constitutes dead remains of plants and animals which are situated in the uppermost topsoil layer. The inorganic portion of the compound is derived from the weathering of rocks over thousands of years. This interface is composed of rock, air, and water, and constitutes the soil. The term soil pollution refers to the long-term presence of radioactive materials, toxic components like heavy metals, salts, and chemicals, or disease-causing agents, which adversely affect plant and animal health. Pollution deteriorates the fertility, quality, texture, mineral content, and chemical composition of the soil, as well as disrupt the biological balance of the ecosystem.

Soil pollution can be defined as, contamination of soil by humans and other natural activities which may cause hazardous effects on living organisms. The composition of the soil is shown in Figure 1.1.

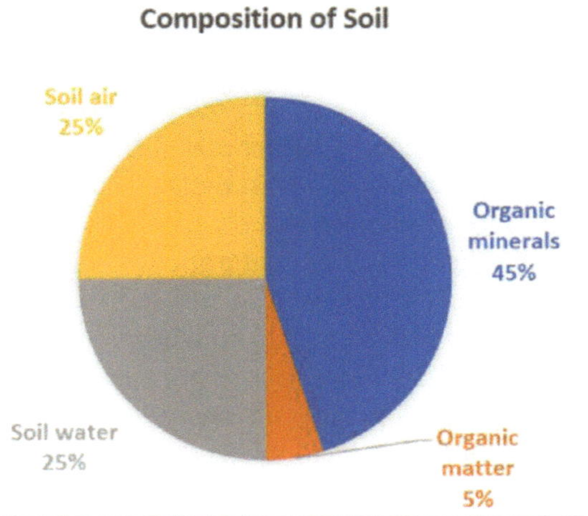

FIGURE 1.1 Composition of soil.

Soil is subjected to various contaminants due to anthropogenic activities like industrial, agricultural, transport. The entire composition of the soil is environmentally important, particularly its heavy metal concentration because heavy metal concentration can directly affect the soil fertility and can also accumulate in the food chain, which ultimately can pose a threat to human health (Facchinelli et al., 2001).

Contamination of soil by industrial effluents and mining activities is a major concern around the world. Because the soil acts as a buffer zone between the atmosphere and groundwater, its contamination has a detrimental effect on the entire ecosystem. The purpose of the present study is to conduct an in-depth investigation of the remediation of contaminated sites from various metal contaminants by using specific plants.

Across India, there are over 125 mostly contaminated sites, more than 40% of chemical fertilizers leach into the soil, over 65% of India's villages are exposed to residual pesticide risks, and heavy metal concentrations are above permissible levels in around 13 states.

There are various sources of soil pollution such as urbanization, untreated wastes, haphazard use of agrochemicals and agricultural, illegal, and unscientific mining, direct discharge of industrial wastes, accidental leakages.

Nature's most valuable resource is soil. It is essential for human and animal life. All human beings, creatures, and vegetation living on Earth depend on it for food. All fruits, vegetables, and flowers produced from the soil will be clean as well if their womb is vibrant and pollute-free. Approximately 329 million acres of land cover India.

It has diverse topographic and climate conditions. This diversity makes it a suitable place, where plants of a wide variety can grow across its length and breadth. In the Western Ghats and the North-Eastern Himalayas, there are more than 17,000 species of higher plants, 7,500 of which are medicinal (Shiva, 1996). Among the world's existing flora, this is the largest proportion of plants used for medicinal purposes.

The soil is considered the thin, natural covering that covers the surface of the earth, in which life thrives. The soil forms the environment for the complex root system of plants and other underground parts such as rhizomes, corn, and bulbs as well as for several soil organisms. Soil is the most crucial factor in the growth, development, and survival of vegetation. It is a natural product of many soil-forming factors. The soil provides the medium in which plant roots can grow and have anchorage. It also provides the plants with steady supplies of water and minerals salt. The soils on the earth's surface are undergoing continual change, which escapes a causal study of the soil. Each soil has life in terms of geologic time This dynamic and evolutionary nature is reflected in the definition of soil as a natural body of the earth's surface with properties resulting from the integrated action of climate, and living matter (plants, animals) acting upon parent material over time, as determined by relief (slope).

Soil consists of products of weathering of rocks, intermixed with living organisms and the products of their decay, the moisture and air filling the interstitial space. It, therefore, has a microclimate of its own, controlled by water, air, and organisms. Different types of soil have different proportions of mineral matter, organic matter, water, and air.

The plant growth depends on the soil for essential parameters such as water and nutrient elements. Beyond this, the soil must provide the environment in which roots can function. The genesis of soil is mainly governed by topography (relief), climate condition, groundwater, and organic activity (primary vegetal). The soil often acts upon a much more extensive portion of the plant body than does the atmosphere.

Plants grow naturally in soil; therefore, both their physical and biological characteristics should be examined. There was significant ecological significance to soil moisture, pH, organic matter, and organic carbon (edaphic factors), altitude, slope, and a degree from sea level (topographic factors) in the region investigated. Industrial emissions disturbed soil coarse sand, water-holding capacity, organic matter, calcium carbonate, pH, as well as available sulfate and heavy metals (Zn, Cu, and Cr). Researchers have studied the effects of soil on plants on soils in industrial areas.

Natural and human-born sources of soil pollution have an impact on the quality of the soil. Soil production capacity depends on its fundamental properties. Land becomes barren if the fundamental properties of soil are lost as a result of excessive pollution, thus reducing crops and vegetation. On December 5, King Bhumibol Adulyadej celebrates 'World Soil Day' to promote soil science and soil conservation. Food and Agriculture Organization of the United Nations (FAO), also recognizes soil's importance as a natural resource and a major contributor to human and other animal health.

1.4.1 SOURCES OF SOIL POLLUTION

Sources of Soil Pollution

- **Industrial Effluents**
- **Urban Garbage**
- **Agricultural Practices**
- **Radioactive Pollutants**
- **Bio Agents**

In today's scenario, there is a thrust for "Going Green," as the world is becoming unhealthy in all aspects. Soil is said to be the farmer's gold.

When soil quality deteriorates, the economy of the country will suffer, thus affecting our food supply. It is necessary to appropriately handle and dispose of liquid and solid residues from factories, and it is possible to plant trees to remedy soils and water resources, two of the essential elements of life. This section explains soil pollution sources.

- ***Industrial Wastes*** – Industrial pollutants discharged from several origins such as mining industries, drugs, cement, glass, petroleum, and industries as well as contaminants are added from the pulp, paper mills, sugar factories, chemical fertilizers, oil refineries, tanneries, distileries, textiles, coal, steel, fertilizers, pesticides, etc. The disposal of these industrial wastes is a major concern regarding soil pollution.
- ***Urban Wastes*** – Sludge and sewage collected from the urban environment are added from commercial and domestic sources. Plastic, paper, bottles, metal cans, rubber, street sweepings, fibers, residues from fuel, leaves, abandoned vehicles, and other discarded materials make up this garbage. Nondegradable urban wastes can pose a threat to public health.
- ***Agricultural Practices*** – The soil is contaminated to a greater degree by modern agro-technology practices. Increasing crop yields required the addition of large quantities of fertilizers, herbicides, pesticides, and weedicides. Aside from these items, manure, slurry, debris, and soil erosion that contains organic chemicals also pollute the soil.
- ***Radioactive Pollutants*** – Radioactive substances from nuclear explosions and industrial purposes resulted in radioactive wastes that penetrate the soil and accumulate causing soil contamination.
- ***Biological Agents*** – Biological agents include any living organisms like bacteria, viruses, or any part such as animal dander and cat saliva, mites, house dust, and pollen. By controlling growth parameters these biological agents can be minimized.

1.4.2 IMPACTS OF SOIL POLLUTION

The main cause of soil pollution is man-made activities that disturb the natural soil environment. A major source of soil pollution is the discharge of industrial wastes directly into the soil, the rupture of underground storage links for chemicals or oils, use of pesticides, and percolation of contaminated surface water into the ground. It's mostly pesticides, hydrocarbons from oil, lead, and other heavy metals that are involved. Depending on the level of industrialization and quantities of chemical usage, they may occur. The use of contaminated soil in agriculture or as a source of underground water should be avoided. The chemicals were leaching into the food and water, and they were extremely harmful to human health. Following are the main issues of concern regarding soil pollution (Figure 1.2).

Consequences of Toxic Pollution	**Endangering Human Health**
	Economic Losses
	Air and Water Contamination
	Effect on Plant Life
	Acidification
	Acidification
	Diminished Soil Fertility
	Changes in the Structure of Soil
	Increase in Soil Salinity

FIGURE 1.2 Consequences of toxic pollution.

- ***Endangering Human Health*** – Human health can be affected indirectly by bioaccumulation or food poisoning through any type of contact with polluted soil. More than 70 percent of the total soil pollutants are carcinogenic, causing cancer in exposed individuals. Benzene and polychlorinated biphenyls (PCBs), for example, are linked to leukemia and liver cancer respectively with long-term exposure. Soil pollution badly affects human health causing the muscular blockage, skin diseases, and central nervous system disorders. All these resulted in acute illnesses leading to death.
- ***Economic Losses*** – Most soil pollutants accumulated in the crops grown on polluted soil and make them poisonous for human consumption, which leads to tremendous economic losses. Heavily polluted soil with metals and chemicals is considered unsuitable for crop production.
- ***Air and Water Contamination*** – The discharge of soil pollution into the atmosphere results in air pollution. A high concentration of toxic pollutants in the soil causes a high level of toxic particles and foul gasses to be released into the atmosphere. The spread of toxic

chemicals and materials into groundwater and runoff on the surface of the soil results in water pollution when toxic chemicals and heavy metals contaminate groundwater, stormwater runoff, and ends up in lakes, streams, rivers, or oceans.

- ***Effect on Plant Life*** – Plants cannot grow in highly polluted soil. Phytochemical, biological, and structural properties of soil are affected by soil pollutants. In a polluted environment, the soil's beneficial flora, nutrients, and chemicals are altered and begin to degrade, decreasing soil fertility. Due to pollutant contamination, a complete loss of ecological balance is caused, i.e., loss of vegetation, loss of animals that depend on vegetation, ultimately leading to the extinction of wildlife. Plant metabolic pathways are altered by soil contaminants, lowering crop yield. Contaminants in the soil can also enter the food chain and cause problems for humans and animals.
- ***Acidification*** – In soil, nitrogen is emitted through denitrification, ammonia volatilization, and organic degradation. Acid rain is caused by the release of sulfur compounds and sulfur dioxides into the atmosphere. Acid rain long-term damages soil chemistry and nutrients, further contributing to ecological imbalances and soil erosion. If the soil becomes acidic, it will alter soil properties, such as soil pH, which will kill all organisms within the soil.
- ***Diminished Soil Fertility*** – Soil contaminated with various chemicals and heavy metals due to human activities like mining have low soil fertility or might even be completely lost. The harmful pollutants and heavy metals in the soil decrease soil microbial and chemical activity.

  The chemical elements can also denature active enzymes that revitalize healthy soil activities. Minerals like magnesium and calcium are also leached away by soil acidification.
- ***Changes in the Structure of Soil*** – Soil microorganisms help in breaking down organic matter that in majority promotes soil structure regarding vitality and water penetration as well as retention but it was all disturbed in contaminated soil.
- ***Increase in Soil Salinity*** – Increased levels of salt accumulations are always linked to soil pollution. Some agricultural processes like irrigation, discharge nitrate, and phosphate deposits in the soil are the major cause of soil salinity. Due to increased soil salinity, plants

face difficulties in absorbing soil moisture. Crops and plants grown in these regions accumulate other soil pollutants and become highly poisonous causing serious health disorders after consumption.

All these disastrous effects are the direct or indirect results of industrial pollution. It's a serious threat to the planet and must be managed within time by focusing on some remediation technologies.

1.5 INDUSTRIAL POLLUTION

Industrial pollution causes the entry of hazardous chemicals into the food chain of humans from the soil and water source; disturbs the biochemical process by biomagnification, ultimately leading to deadly effects on the biosphere. In the case of soil, these pollutants alter the chemical composition and biological nature of the soil and make it barren.

Following are the main factors concerned with industries in developing countries like India.

- ***Lack of Policies*:** Several industries were able to bypass pollution control board laws due to a lack of effective policies and this resulted in large-scale pollution that adversely affected the health of many people.
- ***Unplanned Industrial Growth*:** A polluted environment is often a consequence of unplanned development in industrial townships.
- ***Use of Conventional Technologies*:** Conventional technologies generate an enormous amount of waste but are still operated in some industries for financial adjustments. These technologies cause a large quantity of pollution.
- ***Presence of Many Small-Scale Industries:*** Many small industries and small factories escape from environmental rules and regulations and result in the release of enormous amounts of harmful gases in the atmosphere generating pollution.
- ***Inefficient Waste Disposal:*** Pollution of soil and water is frequently a result of industrial and urban waste disposal. This type of industrial pollution leads to severe health problems. For Example, air pollution in surrounding areas causes various respiratory disorders in humans.
- ***Leaching of Resources:*** To meet the high demand for raw materials from nature, more minerals need to be extracted from beneath the

earth. If spilled on the upper soil layers, these minerals can also cause soil pollution. The leakage of a vessel may cause oil spills that may prove harmful to marine life.

Among the major industrial pollutants, toxic heavy metals are the most hazardous ones for the ecosystem due to their long-term existence in soil. A further section focused on Heavy metal chemistry, their anthropogenic sources, and their toxicity health concerns.

1.6 HEAVY METALS IN SOIL

The natural metal concentrations in soil vary enormously because of mining, electroplating, gas exhaust, as well as fertilizers and pesticides. (Kabata-Pendias and Pendias, 1989). Solvents and industrial wastes are major sources of hazardous compounds or chemicals. Foul compounds or chemicals from agriculture are also major soil polluters. Their deposition changes the properties of the soil in the neighboring area by changing its chemical and biological composition. Metals and metalloids with an atomic weight between 63.54 and 200.59 are considered heavy metals, having densities greater than 5 g cm^{-3}, and having atomic numbers greater than 20. They are mainly associated with pollution and toxicity. Trace quantity of heavy metals is required by living organisms. Heavy metals such as Mn, Fe, Ni, Zn, and Cu, are vital nutrients for higher plants. Heavy metals in standard concentration play a vital role in plant growth and development but surplus quantities of such metals can be hazardous to them. The compounds can cause oxidative stress, non-enzymatic antioxidative reactions, lipid peroxidation in plants, and enzymatic reactions (Lasat, 2000). Heavy metals may exist in a colloidal, ionic, particulate, or dissolved phase. Soluble forms include ions, unionized organometallic complexes, or chelates. The metal solubility in groundwater is dependent upon pH, the organic matter content, cation exchange capacity, and the oxidation state of the metal.

Heavy metals have a longer-lasting presence in soil than in air or water, so they are more persistent in soil than in other biosphere compartments. The main sources of toxic metals in soils are smelting, metalliferous mining, sewage sludge treatment, waste disposal sites, agricultural fertilizer warfare, military training, and electronic industries. Heavy metal concentrations may be 100- to 1000-fold higher in the environment due to these industrial activities than in the Earth's outer crust (Carral et al., 1995).

A contaminated site may contain lead (Pb), arsenic (As), chromium (Cr), zinc (Zn), cadmium (Cd), mercury (Hg), copper (Cu), or nickel (Ni). Rapid urbanization, mining, dumping of heavy metals waste, paints, fertilizers, leaded gasoline, pesticides, coal combustion residues, livestock manures, atmospheric deposition, and petrochemical spills contributed to soil pollution. By disrupting and accelerating nature's geochemical cycle of metals, human activities result in a build-up of heavy metals at levels above those proven to prove harmful to ecosystems and human health in the long run. A heavy metal pollutant plays a significant role in soil because

- Their higher rates of production through industrial activities than natural ones
- Direct exposure occurs from mines to any environmental regions,
- Their higher concentrations in discarded products, and
- Upon receipt, their chemical form becomes more bioavailable.

Due to the phenomenon of biomagnification and bioaccumulation (Manohar et al., 2006), these toxic heavy metals are incorporated into the environment and can cause various toxic effects on humans and all ecosystems. Especially in industrial areas, heavy metals are the chief environmental pollutants. It has been traditional to dispose of heavy metal waste in the soil before processing it. Afterward, traditional remediation methods for soil pollution with toxic heavy metals are both expensive and not environmentally friendly. Environmental and health problems can result from high levels of heavy metals in the environment. By contaminating food chains, these elements became a hazard to humans as well as the environment. Living organisms accumulate in their bodies due to their presence in ecosystems. Metal-contaminated wastewater is produced by industrial activities. Therefore, environmental clean-up through the exclusion of metals from industrial effluents was a thrust area in today's research. The toxic effect of heavy metals on plants is a necessary aspect of the management of the earth's plants because plants are the elixir for the planet.

1.7 HEAVY METAL TOXICITY ON PLANTS

The phytochemistry and morphology of plants are disrupted by heavy metals and plant growth as well as biomass production are negatively affected by these factors. Metal-rich soils, such as those around mines and industrial areas, can support certain plant species. Heavy metal contamination from these plants can be cleaned up using these plants. Plants are adversely affected by heavy metals for the following reasons:

- Cadmium lowers seed germination, plant growth, and lipid content but can elicit phytochelatins production, which is a metal-binding peptide having a vital role in the cadmium detoxification mechanism in plants.
- Chromium decreases enzyme activity as well as plant growth, also damages the membrane, causes chlorosis and root damage.
- Copper interrupts photosynthesis, plant growth, and reproduction, it also decreases the thylakoid surface area.
- Mercury decreases photosynthetic- activity, antioxidant enzymes, and water uptake.
- Nickel inhibits seed germination, protein synthesis, chlorophyll synthesis, and enzyme synthesis, as well as accumulation of dry mass.
- Lead reduces the production of chlorophyll and plant growth.
- The addition of zinc to moderate concentrations of soil sap increases plant growth and ATP/chlorophyll ratios, although seed germination is decreased.

1.8 UPTAKE OF HEAVY METALS BY PLANTS

Plants require macronutrients such as Nitrogen, Phosphorous, Potassium, Sulphur, Calcium, Magnesium, as well as essential micronutrients such as Zn, Fe, Ni, Mn, Mo, and Cu for their growth and reproduction. The evolution of plants has developed various mechanisms for the uptake, translocation, and storage of nutrients. Transporter proteins in the cell membrane play a significant role in these mechanisms and their specificity is related to their structural and functional characteristics (Pollard and Baker, 1997). Depending on the characteristics of the metal, they are found in the soil in different fractions:

1. In soil solution as free metal ions and soluble metal complexes;
2. As adsorbed compounds by inorganic soil constituents;
3. Associated with soil organic matter;
4. Solid precipitates, either pure or mixed;
5. It may be embedded in the structure of the silicate minerals.

1.9 HEAVY METAL POLLUTION

The industrialization has significantly increased the number of heavy metals in the soil so that people are dependent on them for proper functioning.

Heavy metal compounds are produced, mined, burned, and refined by industrial processes for a variety of reasons. Nature is bombarded with heavy metal burdens in soil, water, and air today. Modern consumer products include them in almost every area, such as construction materials, medicines, processed foods, agents of destruction, and fuel sources. Our environment has become very populated with them.

When toxic heavy metals were detected in high concentrations in the soil of an industrially contaminated area, it is cause for concern. In the following sections, we discuss their mental chemistry and health concerns.

1.9.1 NICKEL TOXICITY

Nickel is a white metallic hard alloy that is used with copper, chrome, iron, and zinc. Widely used for the manufacture of fuels, electro platings, pigments, ceramics, household appliances, food production, jewelry, heat exchangers, magnets, coins, medical prostheses, batteries, etc. Nickel is of transition series of the periodic table, which includes iron, cobalt, palladium, platinum, and five other elements. It is not corrosive to air, water, or alkali, but it dissolves easily. Nickel exists naturally in many mineral forms. About 3% of the earth's surface is composed of this element, which ranks 24th in abundance in the earth's crust. While Ni (II) can exist, nickel is also present as a +2-valence state. Other valances (–1, +1, +3, and +4) are also observed, but less frequently (YouNg R. A., 1995).

Anthropogenic emissions account for more than 80 percent of Ni emissions, and wastewater sludge and fly ash land disposal are major sources of Ni in soil. A higher concentration of Ni in plants generally reflects soil concentrations. Nickel is somewhat mobile in the soil and chemistry is mainly determined by the divalent ion Ni^{2+}.

There is a health risk associated with nickel. As a result of acute exposure to Nickel sulfate or Nickel carbonyl, death is caused by cardiac arrest, pneumonia, brain hemorrhage, allergies (most common health effects), adverse pregnancy outcome (occupational exposure), cancer (occupational exposure). Nickel toxicity may present as a skin rash called nickel dermatitis, diarrhea, nausea, vomiting, dizziness, headache, weakness, chest pain, and coughing. When exposed to nickel vapor, the liver and brain can swell. Also, it may irritate the eyes, throat, and nose, as well as cancer. Hence, nickel is one of the most hazardous heavy metals.

1.9.2 CADMIUM TOXICITY

In general, cadmium plays no biological role, and it is highly toxic to flora and fauna. Normally, the soil contains 0.06–1.1 mg kg^{-1} of cadmium, but this concentration is increased by industrial exhaust and waste incineration. It can be accumulated in higher concentrations at the surface horizons of soils with organic matter and can move downward according to soil properties. There are both organic and inorganic complexes of the Cd^{2+} free ion present in the soil solution. Through increased adsorption density, hydrolysis, and pH-dependent negative charge, soil pH decreases solution Cd concentrations.

Cadmium exposure can cause lung damage. Cadmium in food or water causes stomach irritation, which results in diarrhea and vomiting. When exposed to cadmium at low levels over a long period, cadmium accumulates in the kidneys, resulting in kidney damage. Long-term effects result in lung damage, high blood pressure, weak bones, and high blood pressure. According to the U.S. EPA, cadmium probably causes cancer in humans and is linked to lung cancer.

1.9.3 CHROMIUM TOXICITY

Chrome is a hard, brittle, and lustrous metal. Silver-gray in color, it can be highly polished, and chromic oxide is formed when it is heated. Oxygen makes chromium unstable, so it forms a thin oxide coating that is impermeable to oxygen's rays, protecting the metal beneath.

Textile and leather manufacturing are two of the key human activities responsible for the increase in chromium (III) concentrations. The most common human activities that cause a rise in chromium (VI) concentrations are leather, chemical, electro painting, and textile manufacturing. The concentration of chromium in water will increase primarily as a result of these applications. Through coal combustion, chromium is out into the air, and when waste is disposed of, chromium ends up in the soil. The three common forms of chrome are metallic (Cr^0) – not naturally occurring, trivalent (Cr^{+3}) – nutrient, water-soluble, and hexavalent (Cr^{+6}) – highly toxic. Besides industrial applications, its anticorrosive properties are also widely used in tanning leather, cleaning glassware, and textile production. Chromium is widely used in alloys, like stainless steel, metal ceramics, and chromium plating.

By disposing of fly ash onto land or by using sewage sludge as a fertilizer, significant amounts of chromium can be introduced into soils. Chromite (III) and phytotoxic chromium (VI) are the most stable and common forms of chromium. For reduction in the phytotoxicity of chromium to plants, the strongly oxidizing chromium (VI) species are reduced to chromium (III). Root and aerial chromium uptake are low at or near-neutral pH. The soil chromium concentration shows little relationship with the soil chromium concentration.

In addition to acute renal failure, nausea, vomiting, eczema, allergies, and reproduction toxicity, chrome poisoning increases the risk of contact dermatitis, eczema, allergies, and eczema. When metal-contaminated water is consumed, dangerous metals such as chromium (IV) and hexavalent chromium can enter the body. Chromium (III) occurs in many vegetables, fruits, yeasts, meats, and grains, so most of the chromium is ingested through food. The preparation and packaging of food can affect its chromium content. It is also possible to raise the chromium concentration of food by storing it in steel tanks and cans. Health effects like skin rashes can be caused by high concentrations of chromium (III). Chrome (VI) may cause allergic reactions such as ulcerations, dermatitis, and allergic skin reactions, as well as nose irritations and nosebleeds, bronchospasms, asthmatic bronchitis, etc., and edema with inhalation. Aside from that, it is also carcinogenic. Aside from lung cancer, chromium (VI) has also been linked to respiratory ailments, kidney failure, liver damage, weakened immune systems, alterations to genetic material, and kidney and liver damage.

1.9.4 LEAD TOXICITY

This bluish-white metal is soft, elastic, and shiny. Despite its poor electrical conductivity, it is very corrosion resistant. Lead has a half-life of over 1000 years, making it one of the long-lasting heavy metals. Most natural soil concentrations fall below 20 mg kg^{-1}; low levels of contamination result in 30–100 mg kg^{-1}. Mines and smelting, manure, pesticides, sewage sludge, and vehicle exhausts are the major causes of lead contamination (Davies, 1995). Soil naturally contains lead. Most lead concentrations in the environment are the result of human activities. Gasoline contains lead, which results in a lead cycle that is not natural. Car engines burned lead and lead salts such as bromines, chlorines, and oxides are formed. A lead-acid battery is used mostly in vehicle batteries, and it is also used in ceramic glaze and some candles as projectiles and a coloring element. Organ pipes and electrodes made of it are both used in electrolysis. Glass for television screens and monitors is one of its major uses. In addition to being used in cables, solders, sheeting, and bearings, it is also used as a weight in sporting equipment. The organic fraction is naturally found in the surface soil horizons. The soil solution contains cationic, neutral inorganic species, as well as some organic complexes.

Several ill effects are caused by lead toxicity. It causes anemia, a rise in blood pressure, kidney damage, nerve damage, miscarriages, and brain damage. Lead can be ingested by wild animals and domestic animals while they are grazing. People who are exposed to lead also experience the same effects. Lead can take entry into a fetus via the placenta and harm its nervous system and brain.

1.10 SUMMARY

Globally, environmental pollution is a problem due to toxins, metals, and xenobiotics. However, most traditional remediation technologies are costly and reduce soil fertility, which results in negative ecological impacts. Pollution in the world is becoming increasingly severe, and this subject has been extensively researched. Long-term use is necessary for clean-up and bioremediation. Plants and microorganisms play a critical role in the remediation of the present environment. Contaminant removal efficiency would also be improved by interactions with the surrounding soil.

1.11 EXERCISES

Objective Response Questions

1. **For the first time, GM received a patent in the US through Ananda Chakraborty. It was ____________________.**
   a) The Glo Fish
   b) a GM mouse with the growth hormone gene
   c) Cloned E. Coli
   d) Pseudomonas engineered to degrade petroleum

2. **The process of converting toxic waste to non- or less toxic compounds by microbes is called ____________________.**
   a) Precipitation
   b) Complement fixation
   c) Bioconversion
   d) Bioremediation

3. The bacterium can withstand a dose of radiation that is several folds more than the human cell can handle ___________________.

a) *Escherichia coli*
b) *Conus magus*
c) *Deinococcus radiodurans*
d) *Staphylococcus aureus*

4. Bioremediation involves the use of plants for degradation of pollutants, called ___________________.

a) Composting
b) Biopile
c) Phytoremediation
d) Land farming

5. Following is not the native microbe used for bioremediation.

a) *Pseudomonas aeruginosa*
b) *Piscirickettsia salmonis*
c) *Phanerochaete sordida*
d) *E. coli*

6. Biotech researchers have devised new methods for cleaning up pollution using living organisms. Bioremediation is one of these new processes. Using organisms such as microorganisms, fungi, or green plants, bioremediation is a process that removes hazardous substances from a polluted environment. Using bioremediation, specific contaminants can be reduced. Fern plants are used to remove soil contaminated with arsenic. The bioremediation technique is likely to be successful at cleaning up other sites that are contaminated with the same compound after having cleaned up one site. Some environments cannot be remedied as quickly or as cost-effectively with bioremediation as they can with physical or chemical methods. Bioremediation, however, may result in a lesser level of destruction. What might be the reasons for choosing bioremediation for a specific site?

a) Bioremediation is used to clean up any kind of Pollutant.
b) It is less destructive than any other method.
c) Using living organisms is the quickest method of cleanup.
d) Using living organisms is preferred by all local citizens.

7. Thousands of above- or below-ground storage tanks in the US contain gasoline, oil, chemicals, and other types of contaminants. Corrosion, cracks, and leaks can occur over time in these tanks. These types of tanks have been dumped in landfills for many years, contaminating the ground. These sites have sometimes been surrounded by new housing developments. The Environmental Protection Agency determines a cleanup plan once a contamination site is located within a populated area. Is there any way that prevents contamination from storage tanks near homes?

a) Builders test soil pollution at sites before building.
b) Stop building new homes.
c) Repair leaking storage tanks and dispose of them safely.
d) Move houses to safer locations.
e) An integrated waste management plan

8. Under proper operating conditions and temperatures, municipal solid waste is burned in furnaces, this process is called __________.

a) Vermicomposting
b) Recycling
c) Landfill
d) Incineration

9. Decomposed organic matter, rich in minerals such as N, P, and K, generated by earthworms is called ___________________.

a) Vermicompost
b) Compost bedding
c) Worm casting
d) Humus

10. Global pollution is caused by which of the following reason?

a) Soil pollution
b) Industrial pollution
c) Water pollution
d) Radioactive pollution

Practice Work

Qu 1. What are the consequences of anthropogenic pollution on the biosphere?

Qu 2. What is toxic pollution? Explain the health hazards of toxic pollution on humans and plant health?

Qu 3. What is Pollution? Explain types of pollution in detail.

Qu 4. What are heavy metals? Why they are called hazardous pollutants?

Qu 5. Explain the role of bioremediation in the management of industrial pollution.

Qu 6. Define soil. Explain the causes of soil pollution and their hazardous effects on the ecosystem.

Qu 7. Explain air pollution in detail. List out causes and impacts of air pollution?

Qu 8. Explain the toxic consequences of air pollution on plants and animals.

Qu 9. Write a short note on (i) Sustainable strategies, (ii) Cleaner production.

Qu 10. What are the various industrial wastewater treatment units?

REFERENCES

Carral, E., Puente, X., Villares, R., & Carballeira, A., (1995). Background heavy metal levels in estuarine and organisms in Galicia (Northwest Spain) as determined by modal analysis. *Sci. Total Sediments Environ., 172*, 175–188.

Davies, B. E., (1995). 'Lead.' In: *Heavy Metals in Soils* (2nd edn., 206–223), Alloway, B. J., (ed.), Blackie Academic & Professional, London.

Facchinelli, A., Sacchi, E., & Luca, M., (2001). Multivariate statistical and GIS-based approach to identify heavy metal sources in soils. *Environmental Pollution, 114*(3), 313–324. doi: 10.1016/S0269-7491(00)00243-8.

France, L., (2004). International Agency for Research and Cancer (IARC). *IARC Monographs on the Evaluation of Carcinogenic Risks to Humans.: Tobacco Smoke and Involuntary Smoking*, 83.

Lasat, M. M., (2000). Phytoextraction of metals from contaminated soil: A review of plant/soil/metal interaction and assessment of pertinent agronomic issues. *Journal of Hazardous Substance Research, 2*(1), 1–25. doi: 10.4148/1090-7025.1015.

Pendias, K. A., & Pendias, H., (1989). *Trace Elements in the Soil and Plants*. Boca Raton, FL: CRC Press.

Shiva, M. P., (1996). *Inventory of Forest Resources for Sustainable Management and Biodiversity Conservation*. Indus publishing company, New Delhi.

CHAPTER 2

Bioremediation

2.1 INTRODUCTION

Biotechnology has the potential to assess the state and well-being of ecosystems by transforming pollutants into harmless substances or biodegradable materials and developing environmentally safe technology to remediate them. Environmental biotechnology offers potential strains of microorganisms and applied them to reduce the burden of toxic substances. One of the significant fields in biotechnology is bioremediation. It provides a useful tool that promotes the benefit of the environment by recovering contaminated environments.

Bioremediation term is coined by fusing two words: "bios" means life or living organisms and "to remediate" means to recover or clean up or to solve a problem. Thus, "bioremediation" means to use living organisms to solve an environmental problem of contamination (Sasikumar and Papinazath, 2003).

The emerging, eco-friendly technology of bioremediation is the best way to detoxify the harmful contaminants from the biosphere. Bioremediation has been demonstrated as a bioprocess used to degrade, transform, remove, or break down contaminants. It is a natural process that employs bacteria, fungi, and plants to alter pollutants with their peculiar enzymatic pathways. The metabolism of these organisms can use these contaminants as an energy source and can alter them into a less toxic form.

2.2 EVALUATION OF BIOREMEDIATION

Evaluation of the bioremediation plan is very important before starting the planning for bioremediation. Some critical factors should be considered at the time of evaluation of the bioremediation plan for the clean-up of any contaminated site.

Bioremediation and Phytoremediation: Technologies for Toxic Pollution
Ashwini A. Waoo, PhD (Author)

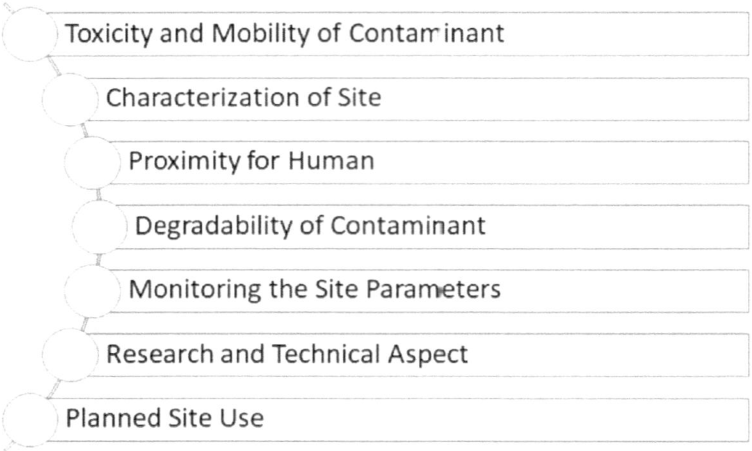

FIGURE 2.1 Criteria for evaluation of bioremediation plan for site clean-up.

Bioremediation deals with controlling environmental pollution through biological agents. It is an eco-friendly technology that is a boon for remediation without disturbing the ecosystem Bioremediation aims to alter hazardous contaminants into less hazardous matter by biological agents. In this scenario, microorganisms are very are easy to handle which exhibit diversified bioprocesses. The bioremediation plan can be evaluated with the criteria shown in Figure 2.1.

- ***Toxicity and Mobility of Contaminants***

Before starting with the bioremediation plan detailed and proper investigation and characterization of contaminated sites must be done examining these points, Horizontal and vertical amount of contamination at the site, Nature, and type of contaminants present, and the mobility or bioavailability of contaminants in the future. This can be useful in getting information about the level of contamination and applying future sustainable strategies for a remediation plan.

- ***Characterization of the Contaminated Site***

The soil layer contains different textures with varying compositions of sand, silt, and clay. Granular suitable-structured soil can allow effective and uniform passage of water, air, and all nutrients to the microbes for *in situ* bioremediation technique. The moisture content is the basic factor that ranges between 25 to 28%. The optimum pH ranges from 5.5–8.0 for the growth of microbes and degradation of contaminants, Biochemical reaction depends on temperature, it ranges from 15–45°C. The existed amount of oxygen in the soil will determine the aerobic or anaerobic mode of bioremediation.

Microbial diversity at the contaminated site had a significant effect on bioremediation, if suitable strains of biodegrading, native bacterial strains absent or at low density then bioaugmentation can be planned with some effective strains of microorganisms such as *Pseudomonas, Flavobacteria, Aeromonas, Chlorobacteria, Acinetobacter, Bacilli, Mycobacteria, Streptomyces, Cyanobacteria*, etc.

- ***Proximity of Human***

Location of a contaminated site, the toxicity of contaminant, and human exposure to that site determine that whether a bioremediation plan is suitable for that environment or not. There must be fewer risks associated with human as well as environmental receptors.

- ***Degradability of Contaminants***

The degradability rate of contaminants through bioremediation must be estimated first. The biological degradation rate is high in the case of natural pollutants. But in case of synthetic compounds having a high molecular weight and complex chemical structures have a slow degradation rate.

- ***Monitoring the Site parameters***

The environmental factors including various chemical and physical factors at the contaminated site determine the bioavailability of contaminants, nutrients, and the activity of biological enzymes. Thus, the specific nature of the contaminants and their possible interactions with the geochemical parameters of the site decides the success of the bioremediation plan.

- ***Research and Technical Aspects***

Conventional treatment technologies have poor performance in bioremediation plans. Thus engineering a specific plan for each polluted site is the major requirement for bioremediation.

- ***Human Resources***

Bioremediation is the latest technology, trained human resources are scarce in this sector. An effective and successful bioremediation plan requires an interdisciplinary approach, comprising multiple fields such as microbiology, engineering, geology, and expert persons from respective fields.

2.3 BIOREMEDIATION STRATEGIES

The major objective of this chapter is to highlight the key factors of the bioremediation process, its types, and principles. Also, discuss their advantages and limitations in the field-level remediation plan. Bioremediation involves 3 strategies to enhance the degradation or remediation process, biostimulation, bioaugmentation, and Intrinsic Bioremediation

2.3.1 BIOSTIMULATION

In this process microorganisms like bacteria are stimulated to initiate the biodegradation process. For this firstly contaminated soil is mixed with beneficial nutrients substances such as liquid form or gas. These substances enhance the growth of microbes and stimulate their metabolic activities, ultimately resulting in efficient and faster removal of pollutants by microorganisms.

2.3.2 BIOAUGMENTATION

Sometimes indigenous microorganisms at the contaminated site are not enough to carry out bioremediation of all contaminants present there. Therefore, to enhance the rate of remediation by incorporating new strains into the contaminated sites that are having the efficiency to biodegrade the specific contaminant referred to as *bioaugmentation*. The process of bioaugmentation is illustrated in Figure 2.2.

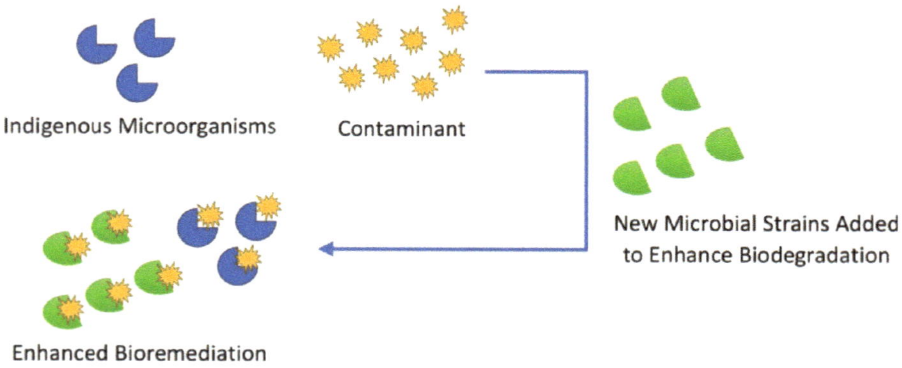

FIGURE 2.2 Process of bioaugmentation.

In this method of application of new wild-type or genetically modified microbial strains are added to contaminated sites to enhance the degradation of compounds (Mrozik and Piotrowska, 2010). The addition of cultured microorganisms for the biodegrading of the contaminant of soil and water is the general practice in such a bioremediation plan.

2.3.3 INTRINSIC BIOREMEDIATION

Soil and water are the most suitable biomes for intrinsic bioremediation because they are the reservoirs of pollutants and toxins. Inherent bioremediation usually takes place in underground tanks and other underground spaces. The difficulty in detecting leaks in such an area allows contaminants and toxins to enter and contaminate the petrol. The toxins are therefore removed from the tanks using microorganisms.

2.4 MECHANISM OF BIOREMEDIATION

Bioremediation represents the thrust area of multidisciplinary research from various fields of microbiology, biochemistry, molecular biology, analytical chemistry, environmental engineering, and others.

The transformation of organic contaminants by microorganisms occurs naturally because the organisms can use these contaminants to obtain their energy requirements. Energy is used for their growth and reproduction. Organic contaminants are useful for microbes in two ways: as a source of carbon, for the process of new cell constituent formation, and they provide electrons, that are extracted to obtain energy (refer to Figure 2.3).

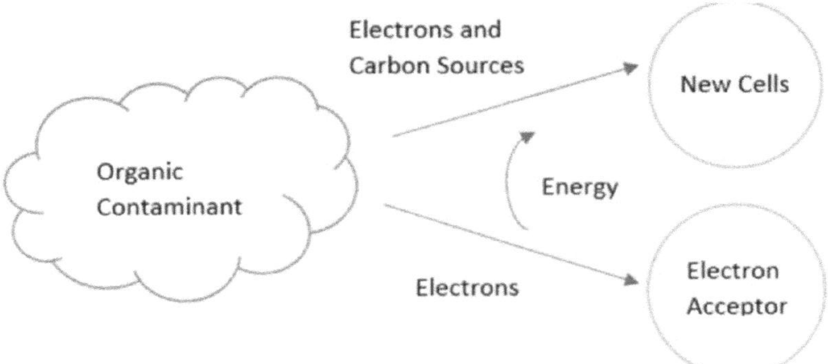

FIGURE 2.3 Basics of microbial metabolism.

Microbes degrade contaminants to gain energy for their growth and reproduction or multiplication. They obtain energy by breaking chemical bonds in the contaminant molecules and transferring electrons to an electron acceptor, e.g., oxygen.

The primary requisites for bioremediation are the presence of a contaminant, an electron acceptor, and the presence of microorganisms having the capacity to degrade it, as illustrated in Figure 2.4. The rate of degradation of contaminant is faster if that contaminant resembles any naturally occurring compound, e.g., petroleum hydrocarbons in the environment.

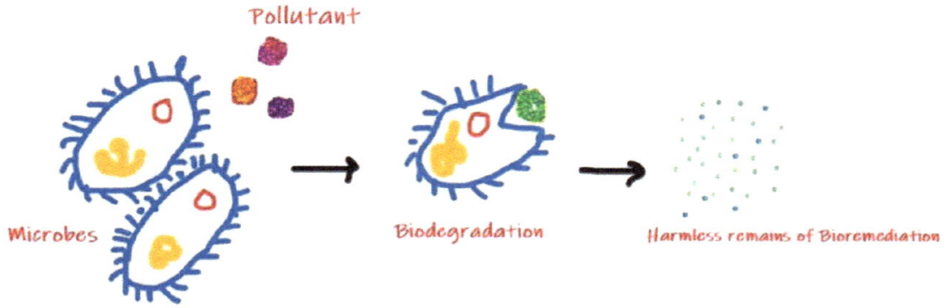

FIGURE 2.4 Mechanism of bioremediation.

2.5 ROLE OF MICROORGANISMS IN BIOREMEDIATION

Isolation of native micro-organisms from any contaminated environmental conditions, and can be used for bioremediation. Microbes representing natural cell tiny bioreactors can operate at aerobic as well as anaerobic conditions. Energy sources and carbon sources are the major requirements for microbial bioremediation. Various parameters can have an impact on the microbial population and distribution at a particular site. Organisms that can degrade some specific contaminants are found predominantly at a particular contaminated site. The rate of degradation of contaminant is directly proportional to the number of microorganisms having the ability to metabolize the contaminant present in a sample.

2.6 CRITICAL FACTORS FOR CONTAMINANT DEGRADABILITY

Contaminants are degraded at a maximum rate when there are optimum conditions for microbial growth with all the favorable environmental factors. It is essential to analyze the characteristics of the contaminated site before beginning the bioremediation plan. The basic factors such as the natural or synthetic nature of the contaminant, effect of halogenation, contaminant mixtures are discussed below.

2.6.1 NATURAL OR SYNTHETIC NATURE OF CONTAMINANTS

Most organic contaminants can be readily metabolized by microorganisms as their carbon or energy source. Adaptation to changing environment is the natural evolutionary process responsible for transformation in the microorganisms so that they can degrade contaminants because most of the natural sites are now becoming industrially contaminated sites due to anthropogenic activities. Thus, there is a development of new metabolic pathways and enzymes to deal with contaminants.

Few natural chemicals have halogen substituents. Thus, halogenated contaminants are the most prominent contaminants of soils and aquatic environments, for example, chlorinated solvents, polychlorinated biphenyls, and chlorinated aromatic compounds. They require more complex pathways for metabolism and involve microbial consortia rather than individual strain.

2.6.2 EFFECT OF HALOGENATION

There are few naturally occurring chemicals having halogen substituents. Synthetic or artificially produced chemicals having carbon-fluorine or carbon-chlorine bonds are slowly metabolized, resisting biodegradation. This class of compounds includes some of the most toxic contaminants of soils and water, like chlorinated solvents, chlorinated aromatic compounds, and polychlorinated biphenyls. These compounds are metabolized by complex pathways involving microbial consortia.

2.6.3 CONTAMINANT MIXTURES

Most of the contaminated sites have complex mixtures of organic compounds that are difficult to metabolize at the same rate. Instead, this rate of contaminant metabolization will differ and be based on several parameters. High concentration results in toxic effects, only low-level contamination can be better suitable for biodegradation.

2.7 TYPES OF BIOREMEDIATIONS

According to the mode of application, bioremediation can be classified as an in-situ or ex-situ process (Figure 2.5). The ex-situ bioremediation processes are mostly costlier due to excavation and transport charges, but they can be useful to remove contaminants under controlled conditions with better and faster results. Apart from this, the equipment installation on-site, and its effective performance and control at a contaminated area, is a major problem in in-situ bioremediation methods. The main factor for determining the suitable method of bioremediation to use is the type of contaminant that will be remediated.

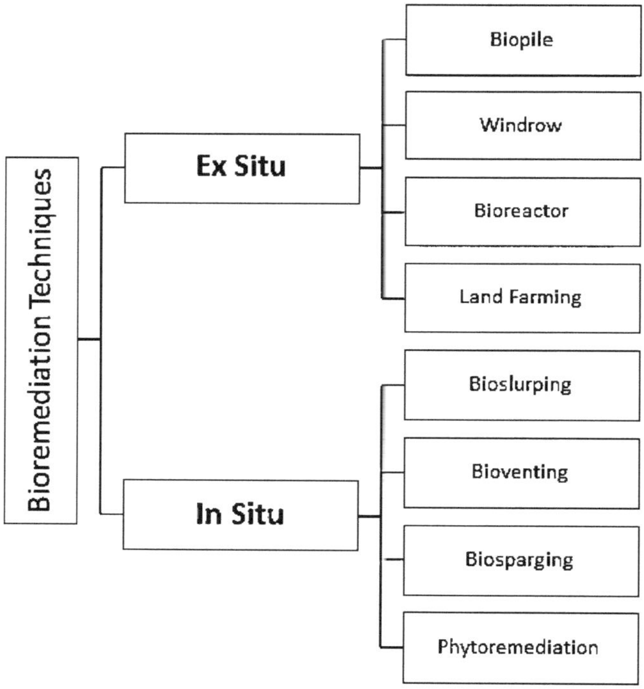

FIGURE 2.5 Types of bioremediations.

2.7.1 EX-SITU BIOREMEDIATION TECHNIQUES

This type of technique involves excavation of pollutants away from contaminated sites and then transporting them to another site for remediation treatment and biodegradation. Ex-situ techniques are based on several factors such as the extent of pollution at depth, type of pollutant, degree of pollution, cost of treatment, geographical location, of the contaminated site (Philp and Atlas, 2005).

- ***Biopile***

In this technique, piling of excavated polluted soil can be carried out above the ground, nutrient amendment, and aeration can also be done to increase the bioremediation rate by promoting microbial growth and interactions (Azubuike et al., 2016). Various components required in this method are: irrigation, aeration, nutrient, leachate collection arrangements, and treatment bed.

Biopile system is used to treat a large amount of polluted soil in a small space. Generally, the biopile plans are easily upgraded from lab scale to field scale (Chemlal et al., 2013). Some organic waste can be used as bulking agents in biopile constructs such as sawdust, straw, bark, or wood chips have been added to enhance the remediation. Uniform aeration, excessive heating of air can be limiting factors of biopiles affecting microbial activities, and retarding the biodegradation phenomenon.

- ***Windrows***

It is an ex-situ bioremediation technique, based on the periodic turning of piled contaminated soil, for enhancing the rate of bioremediation with increased metabolic activities of indigenous microorganisms at the contaminated site. It can be supplemented with the addition of water for moisture content and uniform distribution of pollutants, and nutrients. Microbial activities like biotransformation, assimilation, and mineralization occurred smoothly in the system (Barr, 2002). Windrow treatment showed a higher rate of pollutant removal; however, CH_4 (greenhouse gas) was released during this treatment due to the development of an anaerobic zone within piled contaminated soil.

- ***Land farming***

It is the basic and easy technique of bioremediation, being a low cost and requiring less equipment for operation. It can also be regarded as *in situ* bioremediation technique. Depth of pollutants is a very significant factor in land farming and polluted soils are excavated or tilled.

Land farming is generally used for remediation of hydrocarbon-polluted sites and biodegradation as well as volatilization are the two mechanisms of bioremediation involved in contaminant removal. Land farming has some limitations such as the additional cost of transportation and excavation, the requirement of large operating space, retarded microbial activities, and limited potential of removal of inorganic contaminants (Khan et al., 2004).

- ***Composting***

Composting in bioremediation is the use of a micro-organism in sequestering or in the breakdown of contaminants in contaminated water or soil. These contaminants are biotransformed, biodegraded or metabolized, into humus and simpler by-products, like water, carbon dioxide, and some salts. It is an effective technique for the remediation of chlorinated and nonchlorinated hydrocarbons, solvents, toxic heavy metals, pesticides, petroleum products, and explosives. A special "designed" compost is used to treat pollutants at specific sites. Overall Composting is an aerobic as well as a thermophilic process that may be occurred in static piles or aerated piles.

- ***Bioreactor***

Bioreactors are vessels in which biochemical reactions take place. There are various operating modes of a bioreactor, such as batch, fed-batch, continuous and multistage mode. The selection of mode depends on capital expenditure and raw material used. Optimum growth conditions were provided to microbial cells by the bioreactor. Polluted samples can be poured into a bioreactor as a dry matter or in the form of slurry. This technique has several advantages like controlling the bioprocess parameters like pH, temperature, agitation, and aeration. Appropriate bioaugmentation process increased the pollutant bioavailability, and nutrient addition can be precisely controlled in a bioreactor (Zangi-Kotler et al., 2015).

Bioreactor-based bioremediation is not feasible for field-level practice due to the high volume of polluted soil and its transportation problem. It is difficult to manage all bioprocess parameters of a bioreactor at the optimum level, any parameter can be a limiting factor for the treatment.

- ***Biofilters***

This technology is used to remediate contaminated air produced from toxic volatile organic and inorganic compounds by using microorganisms. Being a low-cost technology, it gradually became popular for waste-removal purposes. Biofilters are made up of packed-bed units covered by an active biofilm and gas is blown through it. A schematic diagram of the biofilter is shown in Figure 2.6.

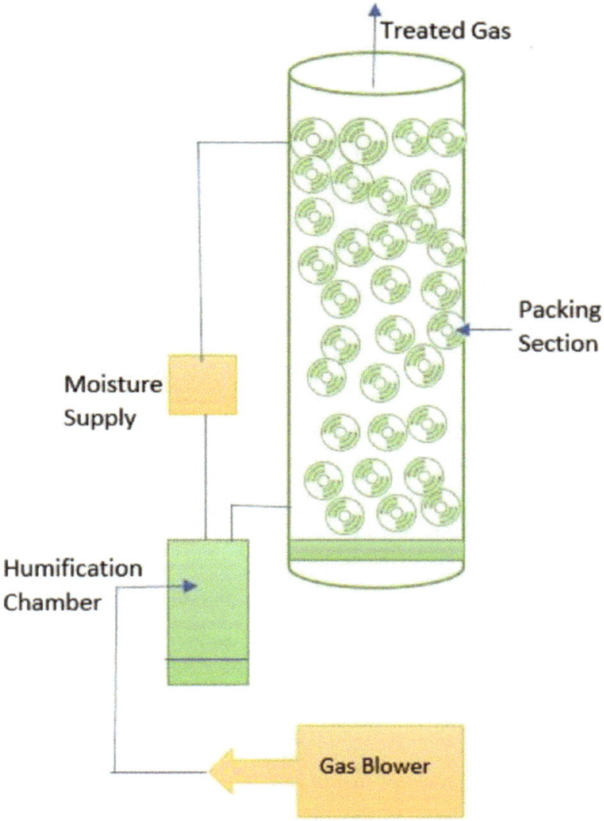

FIGURE 2.6 Schematic diagram of biofilter.

2.7.2 IN-SITU BIOREMEDIATION TECHNIQUES

Treatment of polluted substances at the same site is involved in the in-situ bioremediation technique. Here no need for any type of excavation, thus, these techniques are less expensive than ex-situ bioremediation techniques. Some examples of *in situ* bioremediation techniques are biosparging, bioventing, and phytoremediation. These techniques have been successfully used for remediation of dyes, chlorinated solvents, hydrocarbons, and heavy metals polluted sites (Frascari et al., 2015). All optimum parameters like pH, electron acceptor, moisture content, nutrient availability, and temperature can be managed for a successful *in situ* bioremediation plan. Soil porosity and soil texture strongly affect the use of *in situ* bioremediation to a specific contaminated site.

- ***Bioventing***

This process involves regulated stimulation of airflow with the delivery of oxygen into the unsaturated zone for enhanced bioremediation. In this way, there is an increase in the activities of native microbes accelerating the rate of bioremediation. Nutrients and moisture can also be supplemented to enhance bioremediation. This technique is becoming popular for restoring polluted sites with petroleum products (Höhener and Ponsin, 2014).

- ***Bioslurping***

This technique involves vacuum-enhanced pumping of air and soil vapor extraction in bioventing for soil and groundwater remediation. It is applied for remediation of the contaminated site with volatile as well as semi-volatile organic contaminants. The system uses a "slurp," i.e., free products and soil gas.

During bioslurping excessive soil moisture always limits air permeability and can decrease oxygen transfer rate. As a result, microbial activities are reduced, as well as cost and storage, treatment, and disposal costs are minimized (Philp and Atlas, 2005). Vacuuming a deep, highly permeable site and regulating fluctuating levels of soil and water are concerns associated with this technique.

- ***Biosparging***

Biosparging relies on high airflow rates for pollutant volatilization, it also promotes biodegradation. The use of this technique in treating aquifers has been widespread and used for water contaminated with petroleum pollutants, like diesel and kerosene. The major limiting factor is one can't predict the airflow direction.

2.8 FACTORS AFFECTING BIOREMEDIATION

The regulation of bioremediation processes is a complex phenomenon depending on many factors. These factors may be the microbial population having bioremediation potential, the availability of pollutants, and various environmental factors such as pH, temperature, type of soil, the presence of oxygen, availability of nutrients as shown in Figure 2.7.

Useful microorganisms for bioremediation and the distribution of toxic pollutants in the environment are not uniform. Due to this, the control and optimization of bioremediation processes is a complex task. Various factors affect the process of bioremediation

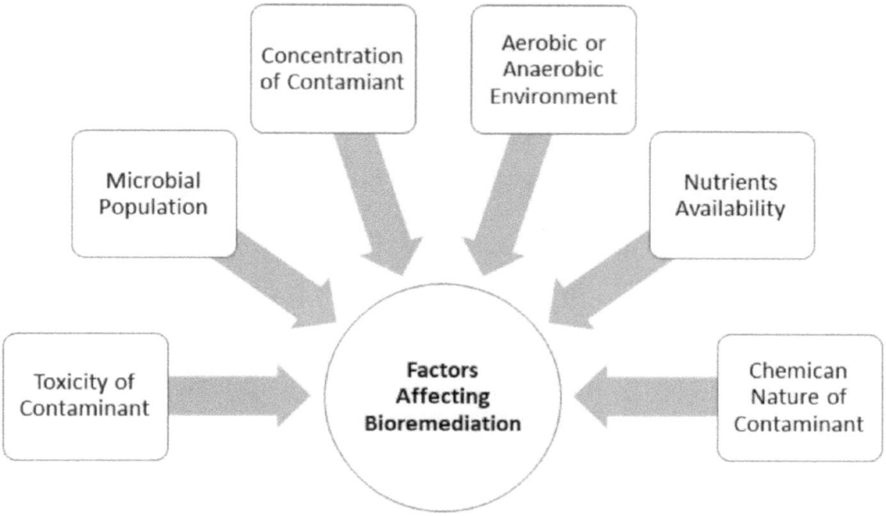

FIGURE 2.7 Factors affecting bioremediation.

- ***Biotic Factors***

Biological factors affect the degradation of contaminants through competition, antagonistic interactions, or predation between microorganisms for limited nutrient sources. Depending on the rate of contaminant degradation, cells express specific biodegradation enzymes at different levels.

The major biological factors are such as mutation, enzyme activity, horizontal gene transfer can affect the Bioremediation process.

- ***Microbial Population***

Microbes having pollutant degradation capacity can grow in the presence of toxic pollutants or hazardous wastes. Their main requirements for growth are energy and carbon source. Aerobic microbes express their degradative abilities with oxygen. *Pseudomonas*, *Sphingomonas*, *Alcaligenes*, *Rhodococcus*, Mycobacterium are some aerobes having biodegrading capacity. Anaerobic microbes grow in the absence of oxygen. Anaerobes can degrade Polychlorinated biphenyls (PCBs).

- ***Environmental Factors***

Microorganisms are the key elements in the bioremediation plan and the growth and activity of Microorganisms are affected by some environmental factors like temperature, pH, moisture, solubility in water, soil structure, nutrients, redox potential, and oxygen content. It also depends upon the physico-chemical bioavailability of the contaminant, its concentration, type, and solubility (Madhavi and Mohini, 2012; Adams et al., 2015). The bioremediation process is generally most efficient at a pH between 6.5 and 8.5.

The amount of moisture in the environment influences the rate of contaminant degradation by affecting the availability of contaminants.

- ***Electron Acceptor***

The bioremediation process needs the electrons used to reduce other compounds. A variety of organic and inorganic compounds may be used as electron acceptors by biological agents. In an aerobic environment, oxygen acts as the electron acceptor. For oil-contaminated matter to biodegrade, oxygen is necessary.

- ***Human Resource Factor***

Bioremediation is the latest technique thus, there is a scarcity of skilled persons in this field. A successful bioremediation plan requires a multi-disciplinary approach by integrating fields such as microbiology, geology, engineering, hydrogeology, soil science as well as project management.

- ***Economic Factor***

The bioremediation process does not offer any type of product. Thus, investment in it as a marketable strategy for research and development is not sufficient in India. More restrictions, the popularity of conventional methods, and performance standards are imposed on bioremediation plans. It cannot adapt at a large-scale level in an industrial area.

2.9 ADVANTAGES OF BIOREMEDIATION

Bioremediation is a biotechnological process involving microorganisms as a bioreactor for solving and minimizing the dangers of toxic pollutants from the environment. It is a natural biological process having the enormous potential to remediate the environment. Microbes having the ability to degrade the contaminant can increase in amount when the pollutant is present in a higher concentration in the contaminated site. When the pollutant is biodegraded, the population becomes less spontaneously. The by-products or the remains from this process are usually harmless in the form of carbon dioxide, cell biomass, and water.

Bioremediation helps in the destruction of toxic pollutants. Hazardous compounds are transformed to harmless form in nature, and there is no need for any future technique required to dispose of the residues. Some more advantages of bioremediation are itemized in Figure 2.8.

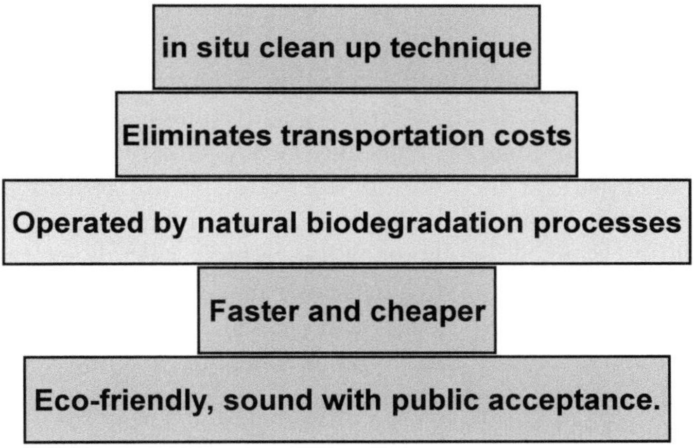

FIGURE 2.8 Advantages of bioremediation.

Any type of toxic chemicals or additives is not required in this process. Biofertilizers can be added to make this process more active and faster (Sharma, 2012). The diverse mode of nutrition of microbes can be discovered for biodegradation of contaminants, i.e., the process of Bioremediation. It is a well-organized procedural activity in microorganisms that are used to transform pollutants into somewhat less toxic or completely non-toxic forms.

2.10 DISADVANTAGES OF BIOREMEDIATION

Although bioremediation has many advantages there are some limiting factors of this process. Bioprocesses are more specific. Important environmental factors govern this process and the presence and number of contaminants limit the microbial populations. When the right nutrients and contaminants are present in each area, optimized bioremediation can occur. Bioremediation is limited to biodegradable compounds thus; non-degradable compounds are still the problem.

According to research, residual biodegradation products may be toxic and harmful than the main toxic pollutant. It is difficult to execute a bioremediation plan lab or field scale. Toxic contaminants may be present in different forms and can be in complex mixture forms. Thus, it is the thrust area of research in bioremediation. Research is needed to improve the actual ground-level bioremediation plan for all types of harmful pollutants.

2.11 ENHANCED BIOREMEDIATION THROUGH PATHWAY ENGINEERING

Anthropogenic pollutant pollution is leading to adverse effects on health and the ecosystem and has led to stricter regulatory limits on heavy metals and radionuclides. Traditional methods may be adequate for treating high levels of contamination, but they are not cost-effective for reducing the levels below regulatory limits.

A pathway engineering approach has recently been developed to selectively and efficiently bioremediate heavy metals and radionuclides. Most microorganisms, when exposed to heavy metals, release phytochelatin (PC) and metallothionein (MT), a metal-binding peptide, and find refuge from the harmful effects of the metals. Several heavy metals are reduced in toxicity by binding to thiol-rich peptides.

In various subcellular locations, these peptides are extensively used to enhance the metal accumulation ability of microorganisms. One such effort was to overexpress the Arabidopsis thaliana phytochelatin synthase (AtPCS), an enzyme essential to the synthesis of PC in *E coli* that accumulates 20-fold more heavy metals. Natural resistance pathways for heavy metals have been developed by many microorganisms, which are tightly controlled by metalloregulatory proteins. A specific accumulation of mercury and arsenic has been demonstrated by exploiting the high affinity of metalloregulatory proteins.

- ***Radionuclide contamination***

Nuclear weapons or leaks from nuclear power plants pose a major issue for the environment. Bacteria naturally resistant to radiation are excellent candidates for metabolic engineering for enhanced radionuclide cleanup. Brim et al., used a radiation-resistant and thermophilic bacterium *Deinococcus geothemalis* and expressed the mer operon from *E. coli* coding for reduction of Hg^{2+}. It was able to reduce mercury at higher temperatures and with ionizing radiation. Additionally, the engineered bacteria were capable of reducing Fe(III), U(VI), and Cr(VI). These results show the possibility of using such engineered microorganisms for a mixture of radioactive wastes at high temperatures.

- ***Organic pollutant removal***

Organophosphates (OPs) have been widely used as pesticides in agriculture, leading to severe environmental pollution. Paraoxin, parathion, and methyl parathion are among the most widely used OPs, as they are easter forms of phosphoric acid. Organisms that live in soil have evolved the ability to degrade OPs with the aid of an enzyme called organophosphate hydrolase (OPH or phosphotriesterase). P-Nitrophene is released as a leaving group as a result of OPH's hydrolysis of P-O linkages. Many researchers have studied the initial hydrolysis by OPH since OP toxicity is significantly reduced when phosphoester bonds are hydrolyzed.

OPH's catalytic activity has been improved using several methods and substrate specificity, including DNA shuffling, site-directed mutagenesis, and error-prone PCR. By shuffled DNA, mutants were obtained with increasing hydrolytic activity for both methyl parathion and chlorpyrifos, which are poorly hydrolyzable OPs.

- ***Nitroaromatic compounds***

Nitrobenzene is widely employed as a dye, pesticide, polymer, or explosive, and is regarded as a priority pollutant. Redox enzymes convert nitroaromatic compounds into amines in most microorganisms. Because of the toxicity of the nitro groups and the electron-removing effect of the nitro groups, biodegradation is slow. The active sites of enzymes that degrade nitrobenzene, nitrotoluene, and dinitrotoluene have been manipulated for an enhanced degradation rate.

- ***Toxic explosive pollutant***

Currently, new toxic-chemical degrading enzymes and pathways are eagerly sought. Biodegradation functions have been improved by various protein engineering methods for cytochrome P450. A highly toxic and hazardous explosive Hexa-hydro-trinitro-triazine was recently studied for degradation with a P450 degrading system.

2.12 SUMMARY

The use of microorganisms or plants as accumulators, detoxifiers, degraders, or solvents is an amazing technology. Even when pollutants are present in low concentrations, bioremediation can transform and/or remove them. Research is still being conducted to understand how microbes and plants remove or transform pollutants. This chapter explored different perspectives on bioremediation by considering recent advances and new ideas for enhanced process scalability concerning larger-scale applications as well as types of bioremediations. The need to clean up environmental pollution led to the search for environmentally friendly, low-cost, and more efficient techniques to do so. Among the most promising alternatives, today is bioremediation, a type of environmental biotechnology.

The bioremediation process utilizes plants and their associated microbial flora to clean up the environment on a low-tech, cheap scale. Heavy metal polluted soil can be remedied using this biological method, as it utilizes the soil's indigenous microorganisms to break down the heavy metals into harmless components. Microbes have revolutionized environmental pollution degradation and remediation in the last few years thanks to their incredible metabolic potential. They aid in the reclamation of contaminated habitats by cleaning up trash in an environmentally friendly way in addition to producing safe plants and chemicals.

2.13 EXERCISES

Objective Response Questions

1. The bioaugmentation process involves ______________.

(a) eliminating sludge
(b) Using plants for bioremediation
(c) Addition of microbes in a cleanup site
(d) bioventing

2. The bioremediation process involves removing soil or groundwater from its natural environment.

(a) Bioaugmentation
(b) *in situ* bioremediation
(c) ex-situ bioremediation
(d) Phytoremediation

3. Methanogenic bacteria dominate this stage of wastewater treatment ________________.

(a) Sludge digestion
(b) Primary treatment
(c) Secondary treatment
(d) Biological oxidation

4. In this method of bioremediation, the contaminated water and soil are mixed with fertilizers and carbon dioxide, and then a bioreactor is used to promote biodegradation.

(a) Composting
(b) Slurry-phase bioremediation
(c) *In situ* hybridization
(d) Biopile treatment

5. Bioremediation ________________.

(a) use of microorganisms to create new organisms
(b) use of microorganisms to create new antibiotics
(c) use of microorganisms for destroying environmental pollutants
(d) use of aerobic microorganisms to create new vaccines

6. It is usually anaerobic bacteria that are involved in bioremediation. The electron acceptors used by anaerobic microbes during biodegradation are not among the following?

a) CO_2
b) NO_3^-
c) Fe (III)
d) H_2O

7. Which of the following bacteria can tolerate radiation doses higher than those human cells can?

a) *Staphylococcus aureus*
b) *Deinococcus radiodurans*
c) *Escherichia coli*
d) *Saccharomyces cerevisiae*

8. In-situ bioremediation involves introducing _______________ to contaminated areas.

a) Oxygen and nutrients
b) Carbon dioxide and methane
c) Nitrogen and CO^2
d) CO and methane

9. Which of the following method not included in In-situ bioremediation?

a) Biosparging,
b) Bioventing
c) Injection recovery
d) Bio-luminescence

10. What is the major advantage of in-situ treatment?

a) Low intrusion
b) High intrusion
c) High contamination risk
d) High cost

Practice Work

Qu 1. Define bioaugmentation.

Qu 2. Define bioremediation. Discuss the different process strategies for bioremediation through microbes.

Qu 3. What are the essential characteristics of microbes used for Bioremediation?

Qu 4. What is a land farming technique? Explain the advantages and limitations of it

Qu 5. Write a short note on bioventing.

Qu 6. Describe the technique of biosorption and how it will be helpful to remove contaminants?

Qu 7. Explain the concept of microbial transformation.

Qu 8. Explain metabolic processes involved in bioremediation?

Qu 9. What type of adaptions are seen in microorganisms in adverse conditions?

Qu 10. What are constructed wet lands and bio composting? Explain their role in bioremediation.

REFERENCES

Adams, G. O., Fufeyin, P. T., Okoro, S. E., & Ehinomen, I., (2015). Bioremediation, biostimulation, and bioaugmention: A review. *International Journal of Environmental Bioremediation & Biodegradation, 3,* 28–39. https://goo.gl/9XY7ni (accessed on 01 April 2023).

Azubuike, C. C., Chikere, C. B., & Okpokwasili, G. C., (2016). Bioremediation techniques-classification based on site of application: Principles, advantages, limitations, and prospects. *World Journal of Microbiology & Biotechnology, 32*(11), 180. https://doi.org/10.1007/s11274-016-2137-x.

Barr, D., (2002). *Biological Methods for Assessment and Remediation of Contaminated Land: Case Studies.* London: Construction Industry Research and Information Association.

Brim, H., Venkateswaran, A., Kostandarithes, H., Fredrickson, J., & Daly, M., (2003). Engineering *Deinococcus geothermalis* for bioremediation of high-temperature radioactive waste environments. *Applied and Environmental Microbiology, 69,* 4575–4582. doi: https://doi.org/10.1128/AEM.69.8.4575-4582.2003.

Chemlal, R., Abdi, N., Lounici, H., Drouiche, N., Pauss, A., & Mameri, N., (2013). Modelling and qualitative study of diesel biodegradation using biopile process in sandy soil. *Int. Biodeterior Biodegradation, 78,* 43–48. doi: 10.1016/j.ibiod.2012.12.014.

Frascari, D., Zanaroli, G., & Danko, A. S., (2015). *In situ* aerobic cometabolism of chlorinated solvents: A review. *J. Hazard. Mater., 283,* 382–399. doi: 10.1016/j.jhazmat.2014.09.041.

Höhener, P., & Ponsin, V., (2014). *In situ* vadose zone bioremediation. *Curr. Opin. Biotechnol., 27,* 1–7. doi: 10.1016/j.copbio.2013.08.018.

Khan, F. I., Husain, T., & Hejazi, R., (2004). An overview and analysis of site remediation technologies. *J. Environ. Manag., 71,* 95–122. doi: 10.1016/j.jenvman.2004.02.003.

Lee, J. H., (2013). An overview of phytoremediation as a potentially promising technology for environmental pollution control. *Biotechnol. Bioprocess Eng., 18,* 431–439. doi: 10.1007/s12257-013-0193-8.

Madhavi, G. N., & Mohini, D. D., (2012). Review paper on – parameters affecting bioremediation. *International Journal of Life Science and Pharma Research, 2,* 77–80. https://www.cabdirect.org/cabdirect/abstract/20123395716 (accessed on 01 April 2023).

Mrozik, A., & Piotrowska, S. Z., (2010). Bioaugmentation as a strategy for cleaning up of soils contaminated with aromatic compounds. *Microbiol. Res., 165*(5), 363–375.

Philp, J. C., & Atlas, R. M., (2005). Bioremediation of contaminated soils and aquifers. In: Atlas, R. M., & Philp, J. C., (eds.), *Bioremediation: Applied Microbial Solutions for Real-World Environmental Clean-Up* (pp. 139–236). Washington: American Society for Microbiology (ASM) Press.

Sasikumar, C. S., & Papinazath, T., (2003). Environmental management: Bioremediation of polluted environment. In: Bunch, M. J., Suresh, V. M., & Kumaran, T. V., (eds.), *Proceedings of the Third International Conference on Environment and Health* (pp. 465–469). Department of Geography, University of Madras and Faculty of Environmental Studies, York University, Chennai.

Sharma, S., (2012). Bioremediation: Features, strategies and applications. *Asian Journal of Pharmacy and Life Science, 2,* 202–213. https://goo.gl/q1JpAh (accessed on 01 April 2023).

Vidali, M., (2001). Bioremediation an overview. *Pure Appl. Chem, 73,* 1163–1172.

Zangi, K. M., Ben, D. E., Tiehm, A., & Kushmaro, A., (2015). Microbial community structure and dynamics in a membrane bioreactor supplemented with the flame retardant dibromoneopentyl glycol. *Environ. Sci. Pollut. Res. Int., 22,* 17615–17624. doi: 10.1007/s11356-015-4975-8.

CHAPTER 3

Immobilization for Bioremediation

3.1 IMMOBILIZATION

In history, the twentieth century is remembered as a period of vigorous civilizational and industrial development. Due to industrialization, war, and extensive use of heavy metals and synthetic xenobiotics, many environmental problems have been caused.

A variety of immobilization and support techniques are available based on the application. Our discussion in this chapter has focused on the possibility of degrading pollutants by immobilized microbial cells. In comparison to free-living bacteria, immobilized cells in experiments demonstrated improved degradation for the treatment of oil-contaminated areas and with other harmful contaminants. Cells that have been immobilized are more efficient, better, and more durable.

Presently, bioremediation is the least harmful and cheapest way to remove xenobiotics from the environment. It is more successful, less expensive to operate, and employs more biocatalysts when a bioremediation process immobilizes bacteria that degrade specific contaminants. As a result of its simplicity and non-toxicity, the adsorption method is mostly applied in bioremediation.

Choosing the right carrier for bioremediation is crucial to success. Aside from how the process was performed (*in situ* or *ex situ*), pollution type and immobilized microorganism properties must be considered. To illustrate how they might be used, the article summarizes new scientific reports on their efficiency, impacts on the environment, and types of research they have done.

The process of bioremediation increasingly uses immobilization methods. A method of immobilization is referred to as retarding the mobility of microbial cells or their enzymes while simultaneously preserving their viability and catalytic qualities. To perform this process, it is possible to use the naturally occurring capacity of microbes to associate and form biofilms

Bioremediation and Phytoremediation: Technologies for Toxic Pollution
Ashwini A. Waoo, PhD (Author)

on different surfaces. Bioremediation processes that involve immobilization are significantly cheaper and more effective than conventional ones.

Using immobilized cells, Brodelius and Vandamme reviewed the formation of amino acids, organic acids, antibiotics, steroids, and enzymes as well as their production using bioreactors. The commercial success of these processes demonstrates the value of using immobilization technology under controlled conditions.

It is a general term that refers to many different types of cell attachment or entrapment. Different types of cell attachment are flocculation, adsorption on surfaces, side-chain covalent bonding, cross-linking of cells, encapsulation in a polymer gel, and entrapment in a matrix. In the laboratory, one method that has proven successful in commercial applications uses a polymer gel matrix to encapsulate cells. The results of bioreactor studies have demonstrated that cells enclosed in a capsule have numerous advantages over those free.

3.2 MICROBIAL ENCAPSULATION

Different types of microorganisms have been encapsulated in a variety of matrices for a variety of applications. The encapsulation of fungi has been used for biocontrol and biodegradation. Several studies have investigated the use of encapsulated microalgae for metal sequestration and wastewater nutrient removal. Encapsulated bacteria are used for heavy metal accumulation and degradation of toxic compounds has been explored. Potential commercial applications have also been explored for encapsulated thermophilic, methanogenic, and other anaerobic bacteria. This technology has successfully encapsulated a range of microorganisms and has been explored across several applications, demonstrating its utility and versatility and its use in the environment. Microorganisms encapsulated in a gel matrix as a carrier for environmental applications, such as the in-situ bioremediation of soil or groundwater, are still in the early stages of development. An emphasis is placed in this chapter on applying this technology to *in situ* bioremediation of chemically contaminated soils, and the immobilization of microorganisms for pollutant biodegradation.

3.3 APPLICATIONS OF IMMOBILIZED MICROORGANISMS

Many industrial and scientific applications have been achieved using immobilized microbial enzymes, organelles, and cells. Several industrial applications have been investigated because of the economic value of immobilization.

Bioreactor fermentations employ immobilization technology extensively. Numerous reviews have examined how these technologies can be used to benefit patients and to increase enzyme production.

Researchers have studied microbial inoculants for soil applications including increasing the symbiotic or associative fixation of N_2, controlling soil-borne plant pathogens, and degrading hazardous chemicals. The soil has been inoculated with microorganisms as liquid suspensions or as particles adsorbing onto organic carriers, such as peat, charcoal, and vermiculite. The formulation of microorganisms can enhance their efficacy, stability, safety, and application ease. The liquid formulations of cell suspensions could be more susceptible to contamination, could be more difficult to contain in case of an accidental spill, as well as have a greater potential for off-site dispersion during application than the dry formulations.

A recent study has suggested the use of microorganism carriers as a method for improving the growth and survival of microorganisms in the soil after its inoculation. There has been a review of the advantages and disadvantages of using solid carriers in soil. The carrier must be effective and safe. An ideal carrier should not be toxic, should hold its purity, should be consistent in quality, should have a long shelf life, should permit adequate cell activity and density, and should allow for precise bacterial release at the target site. Several features of encapsulation make it a promising candidate to be used as a medium for the introduction and establishment of microbes. Increasing microbial survival via such a method may help cells survive soil's many changing conditions.

Buekwe and Al-Muttawa isolated an Arthrobacter sp. and bacillus from oil lakes in Kuwait, and Following incubation with foam, sawdust, or wheat bran, the bacteria grew resulting in steady exopolysaccharide mediated immobilized cultures. After six weeks at 45°C, the authors tested the ability of the bacteria to survive and degrade hydrocarbons. Freshly immobilized cells degrade crude oil less than free cells.

In another study, Quek et al., examined the immobilization and potential of *Rhodococcus* sp. Polyurethane foam (PUF) is used to bioremediate petroleum hydrocarbons. They degraded a variety of petroleum products, including Al-Shaheen crude, Arabian light crude, diesel, and oil slops.

According to Radwan et al., oil-consuming bacteria were immobilized in biofilms coating macroalgae. The natural immobilization of bacteria can protect them from being removed and diluted, while also providing them with oxygen, nutrient compounds, and vitamins for oil-consuming bacteria.

To immobilize *Rhodococcus corynebacterioides* QBTo, Gentili used chitin flakes and chitosan flakes.

Supports obtained from shrimps and crabs are natural, non-toxic, non-polluting, and biodegradable.

TPH degradation has been hampered by nutrient addition during phytoremediation. According to Hutchinson et al., (2001) grasses with N/P amendments degrade TPH better than grasses without inorganic amendments. Microbial bioremediation of TPH contaminants with nutrient addition also produced widely varying results. As a result, Graham et al., (1999) studied the reaction between Nitrogen and Phosphorus and hexadecane biodegradation and found that additives over stoichiometric requirements diminished degradation rates. Increased soil salinity may occur because excessive nitrogen can result in increased osmotic stress and lower hydrocarbon-degrading organism activity. Waldworth et al., (2003) demonstrated that soils with low levels of N14 or P demonstrate a more rapid decline when N or p is added. The activity of some PAH-degrading organisms may decrease with soil amendments because they are adapted to low nutrient conditions.

3.4 IMMOBILIZATION METHODS

In recent years, bioremediation processes have increasingly used immobilization methods. The term immobilization refers to limiting the mobility of microorganisms or their enzymes during the maintenance of their viability and catalytic functions. In this process, the microbes may utilize their ability to form biofilms on various surfaces, like what you see in nature. Increasing efficiency and reducing costs of bioremediation processes are significantly improved by immobilization. This approach has many benefits – higher efficiency in the degradation of pollutants, the use of more biocatalysts, reduced costs, a stable microenvironment for DNA and enzymes, reduced risk of mutations, no shear forces affected, etc., accelerated biocatalyst recovery from hazardous storage and increased endurance to high levels of pollutants.

Immobilization techniques include adsorption, electro statical or covalent bonding to the surface, flocculation (natural or artificial), encapsulation, and entrapment, as flocculation requires no carriers, it will not be discussed.

- ***Adsorption***

The adsorption of microbial cells and enzymes occurs as a result of their interaction with the surface of water-insoluble carriers. In bioremediation processes, this method is simple, quick, eco-friendly, and cost-effective. Weak bonds are formed on the surface of carriers to achieve adsorption. As a result, this method is not used for cell immobilization due to a high probability of cells escaping the carrier.

> **Methods of Immobilization**
>
> - **Adsorption**
> - **Binding on a Surface**
> - **Entrapment in Porous Metrix**
> - **Encapsulation**

- ***Binding on a Surface***

Physical adsorption is less likely to leak microorganisms than electrostatic adsorption. Using this method, the carrier is soaked in a buffer solution so that it becomes hydrophilic, which attracts cells or enzymes negatively charged. The process of immobilization in covalent binding is different since binding agents are needed. Chemically activated carriers containing amide, carbamate, and ether bonds are the only suitable vehicles for immobilization. Because binders are sometimes toxic to cells, and the result is lower microbial viability and activity, this method is mainly used for enzyme immobilization. As a result of the covalent bonds, the molecules cannot leak into the environment.

- ***Entrapment in Porous Matrix***

Microorganisms can be trapped and entrapped in a variety of ways for bioremediation. In the entrapped state, microorganisms are confined to a carrier. Cells cannot leak into the environment as a result but nutrients and metabolites can be limited. Microorganisms entrapped within the heterogeneous carrier have diverse physiological requirements. In contrast to the cells that are starved in the interior of the carrier, those with high metabolic activity are located near the surface. The method of entrapping is quick, non-toxic, inexpensive, and versatile. Encapsulated microorganisms are protected from environmental factors. Pore size ratio to cell size is the most important parameter in entrapping microorganisms. Immobilized cells can leak into the environment when pores are wider than immobilized cells.

- ***Encapsulation***

Like entrapment, encapsulation involves separating immobilized particles from the surroundings with a semi-permeable membrane. This method provides significant protection for biological material from adverse environmental conditions. Encapsulation is rarely used in ex-situ bioremediation due to the limited permeability of the membrane and the possibility of damage caused by growing cells.

Advantages of Encapsulation

- During storage, transport, and application there is less chance of inoculum contamination.
- Reduction of off-site drift during application.
- Beads are biodegradable, non-polluting, and non-toxic.
- Suitable for large-scale production, storage as dried beads for extended periods, and mechanical application.
- Ensures microbial survival in the face of environmental stresses biotic and abiotic.
- Enhanced metabolic activity of encapsulated cells.
- By reducing soil movement due to water flow, cells release more slowly.
- Plasmid stability increased.

Limitations of Encapsulation

- Gas and solute diffusion may be limited.
- Encapsulated cells may consume less oxygen.
- Modifications of the morphology or metabolism of cells can be harmful.
- Changing water activity can limit the effectiveness of the treatment.

Colonies may not form outside of beads. Beads may need to be applied repeatedly.

3.5 CARRIER MATERIALS

Both natural and synthetic polymers have been used to encapsulate microbes. The natural polymers include algal polysaccharides such as agar, agarose, alginate, and carrageenan, while the synthetic polymers include polyacrylamide, polystyrene, and polyurethane.

In groundwater remediation, polyurethane microbeads have been investigated to encapsulate cells. In controlled bioreactors, there are different requirements for gel beads that are applied directly to the soil. Even though synthetic polymers offer long-term stability and small pores, they are not ideal for use in environmental applications. The toxicity of polyacrylamide prepolymers and their instability in the soil can make the management of polyacrylamide difficult. Polymerization of synthetic matrices can also compromise cell integrity and function.

The use of natural polymers in the soil is recommended in general. By using biodegradable gel matrices like alginate, it is possible to introduce bacteria into the soil in a very safe and efficient manner. In addition to minimizing contamination and ensuring cell safety, encapsulation does not stress cells. The carriers used are biodegradable and non-toxic. Meanwhile, they protect cells until beads are degraded, releasing microorganisms over time. During the process of immobilization, some carrier material is used such as amendment in formulation, skim milk, alginate, and carrageenan carrier system.

- **Amendments to Formulations:** As carriers, both alginate and carrageenan can be used pure or with additional amendments which enhance their effectiveness. There have been amendments such as locust bean gum or carob bean gum that have improved the strength of the beads in bioreactor systems. Several amendments have been investigated as a protective and/or nutrient source to provide soil stability, but this is not necessary for soil applications.
- **Clay:** By adding clay to carrier formulations, cells can survive longer and be bulked up, providing additional mechanical strength. In laboratory studies, Bacteria survive in soil better with certain clays based on laboratory studies. Clay has been incorporated into beads in studies that have shown that the survival rates differ according to the type of clay and the microorganism interacting with that clay.
- **An Amendment Involving Skim Milk:** Skim milk can contribute to increased activity, growth, and survival in bead formulations. Adding skim milk and encapsulating bacteria in alginate enhanced bacterial persistence. When cells are encapsulated in skim milk, they may sustain longer during the drying and wetting cycles of the soil.
- **Carrier Systems Based on Alginates and Carrageenan:** The mild encapsulation process (e.g., ambient temperature) has led to increased use of alginate. In bioreactors, alginate is particularly useful. Cells encapsulated in alginate showed greater survival in soil than free cells.

There are many benefits of using encapsulated bacterial cells in the soil during bioremediation. As an alternative to free cells or other carriers such as vermiculite, peat, or charcoal, using encapsulated cells for soil applications provides numerous advantages. To ensure viability and activity, formulations must be carefully designed so they can be stored and distributed for longer periods.

When immobilized under aseptic conditions, only the desired cells can be entrapped. Incubation of cells on activated carbon and maintaining them at 4°C preserved their degradation capacity for 12 months or more. As a result of alginate carrier formulations, storage life was extended and mechanical abrasion was prevented during application.

3.6 IMMOBILIZED CELL BIOREACTORS

Cell immobilization and immobilized cell bioreactors find applications in a variety of fields, including environmental, industrial, and medical applications. The use of immobilized cells for the treatment of conventional wastewater is one of the earliest applications of environmental engineering and bioremediation (Figure 3.1). Various waste streams, or gaseous, that contain metals, sulfur dioxide, sulfates, and organic compounds, such as aromatic, heavy hydrocarbons, and polycyclic aromatic, have been treated using immobilized cell bioreactors, especially the tailings generated during the processing of oil sands (Nedovic and Willaert, 2005).

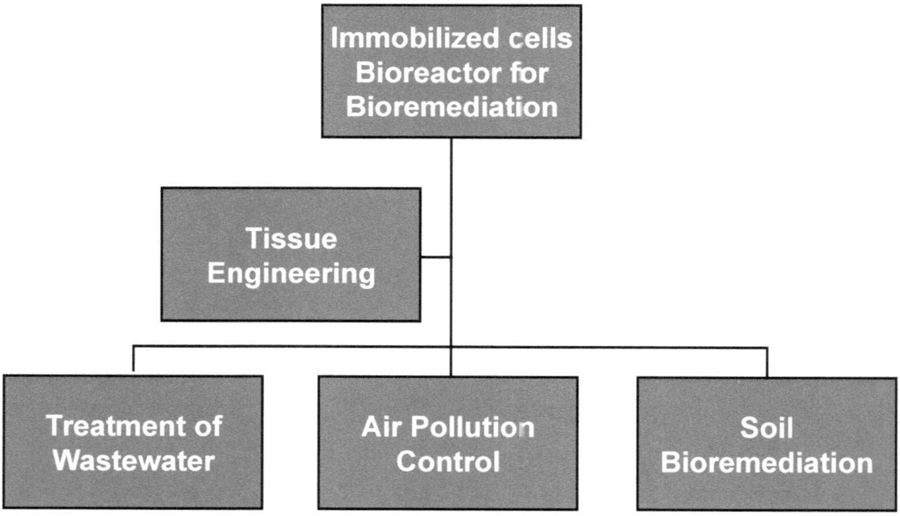

FIGURE 3.1 Applications of immobilized cell bioreactors.

3.6.1 CONFIGURATION AND DESIGN CHARACTERISTICS OF IMMOBILIZED CELL BIOREACTORS

Immobilized cells can be maintained in the bioreactor as a separate phase with independent hydrodynamic behavior, which combined with the flexibility of their shapes, sizes, and arrangements allow the creation of a variety of bioreactor configurations. A bioreactor with immobilized cells can also be operated in different modes, such as batch, semi-continuous, and continuous, as well as with various flow regimes, including well-mixed, dispersed plug, and ideal plug flows. Several factors influence the choice of the bioreactor, such as the immobilization method and characteristics of the particles (e.g., size, shape, mechanical strength, density, cost, reaction kinetics (Shuler and Kargi, 2002).

A properly designed immobilized cell bioreactor should meet the following criteria:

(i) Current shear forces should be as low as possible to preserve particle integrity;
(ii) The bioreactor design should allow maximum particle hold up while maintaining efficient mass and heat transfer; and
(iii) The bioreactor design should cater to the reaction kinetics.

Considering the inhibitory effect of a substance or product is important where it is significant. When the reaction is inhibited by the substrate, mixed flow is appropriate, as the incoming substrate is constantly diluted with the contents of the bioreactor. Plug flow regimes are suitable for reaction conditions where the inhibitory effects of the product are observed, as high concentrations of the product are observed only toward the end part of the bioreactor. However, there are many different types of bioreactor configurations that use immobilized cells, and they generally fall into the categories of stirred tank bioreactors (mechanical or gas agitating), fixed-bed bioreactors, fluidized-bed bioreactors, and membrane bioreactors. Each group is described in detail in the following sections. The following sections describe each group in detail.

- ***Stirred Tank Bioreactors***

In stirred tank bioreactors, inhibition of cell activity is mitigated when substrate concentrations are high, however, they are not appropriate for situations in which the product inhibits the reaction. Stirred tank bioreactors have one of the biggest problems, those using propellers and flat blade turbines is the strong shear forces created on the particles (Figure 3.2). Particularly gel particles could be severely damaged by these forces. It may also be possible to reduce this damage using spiral ribbon screws or anchor impellers since they mix more gently than flat blade turbines or propellers (Obradovic and Nedovic, 2004).

Another method for achieving effectual mixing as well as mass transfer in a bioreactor is to inject and circulate air or a gaseous stream in a mild shear regime. Rotating biological contactors (RBC), which are often used to treat wastewaters and mines drainage. In the case of RBC, there is an efficient amount of mass transfer without impairing the cells. In addition, by exposing cells to air often, oxygen requirements are typically met, thereby eliminating the additional cost associated with aeration system in a stirred tank bioreactor. RBCs, due to their open nature, are best suited for processes in which aseptic conditions are not necessary. It should be noted that RBC may also be used under anaerobic conditions, including denitrification and sulfate reduction.

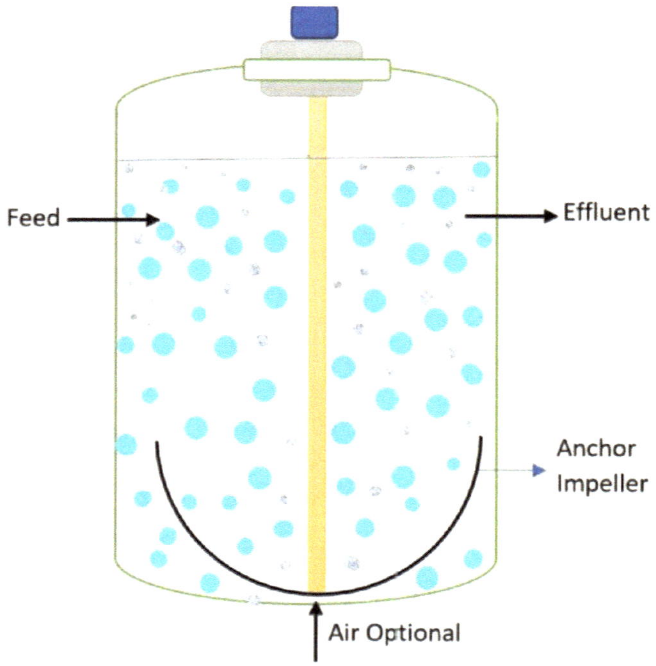

FIGURE 3.2 Stirred tank bioreactors.

- ***Fixed-Bed Bioreactors***

When the reaction product imposes a strong inhibitory effect, packed-bed bioreactors operated under the plug flow regime (once-through basis) are advantageous. A high level of reaction rate is possible with them.

Due to low fluid velocity and an absence of mixing, packed-bed bioreactors have poor heat and mass transfer. The slow release of gaseous metabolites such as CO_2 or NH_3 leads to stagnant slugs and the channeling of flow in it. Partially recycling liquid effluent could reduce gas accumulation and enhance heat and mass transfer. The system would also perform better if the cells' activity is decreased by substrate inhibition. Immobilized cells of a uniform size increase bed voidage and decrease pressure drop across the bed, as illustrated in Figure 3.3.

A trickle-bed bioreactor is generally used for three-phase operations, with the liquid phase flowing from top to bottom, while the gas phase flows counter-currently downward (Blanch and Clark, 1997). The partial recycling of effluent liquid improves the performance of systems by uniformly distributing the liquid. The trickle bed bioreactor's main feature is its effective mass transfer between gas and liquid.

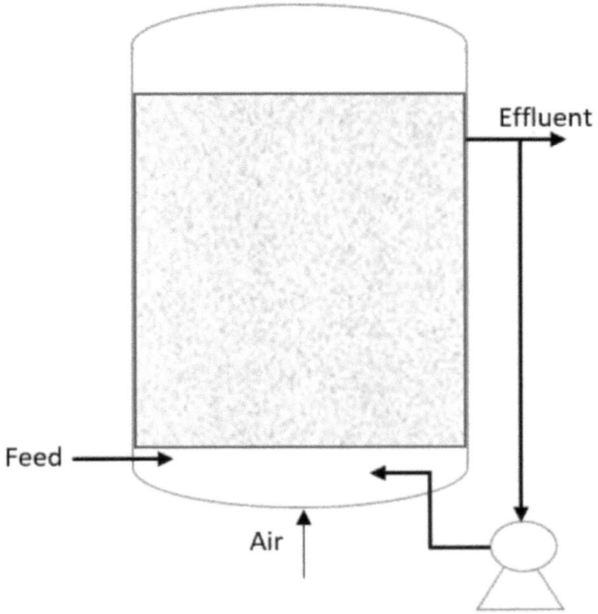

FIGURE 3.3 Fixed-bed bioreactors.

- ***Fluidized-Bed Bioreactors***

Fluidized-bed bioreactors are suitable for two- and three-phase processes (solid–liquid and solid–liquid–gas, respectively), where liquid or a mixture of gas and liquid is introduced into the bottom. The upward flow of the injected fluid causes the immobilized cell particles to be suspended and the bed to expand. In most systems, the fluid velocity that results in the optimum operation falls between two limiting levels: the minimum fluidization velocity which is required to keep suspended particles suspended, and the maximum allowable velocity, above which suspended particles are entrained by the flowing fluid and leave the system. In most cases, the proper control and provision of adequate velocity are achieved by partial recycling of the liquid phase (Baron et al., 1996). In this way, the liquid is externally aerated and the two-phase regime is simplified. A recycling process can also be used to regulate pH, and remove excess and detached biomass. The structure of the fluidized-bed bioreactor is schematically represented in Figure 3.4.

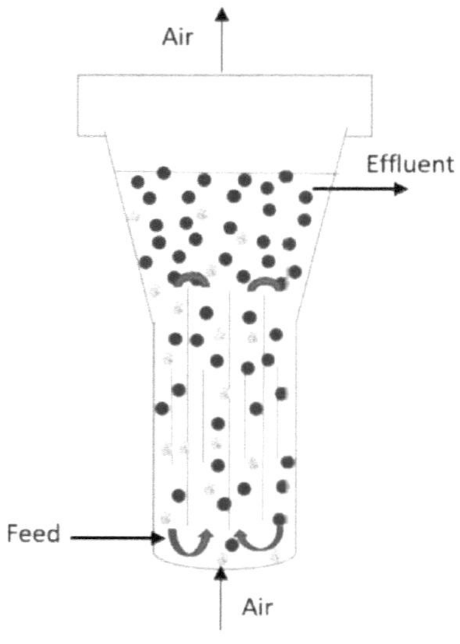

FIGURE 3.4 Fluidized-bed bioreactors.

- ***Membrane Bioreactors***

These membrane bioreactors, configured as either flat sheets or hollow fiber models, are very complex in structure and design, as well as more expensive due to the high costs of the membrane when compared with conventional bioreactors. In general, they provide bioconversion as well as separation of the product. Further justification of using membrane bioreactors is the elimination of costly purification steps, especially in the case of high-value biomolecules. When treating specific waste streams that contain toxic compounds or heavy metals, membrane bioreactors may be preferred because they prevent direct exposure of cells to these harmful compounds.

A typical example of this is the bioremediation of acid mine drainage (AMD). Membrane bioreactors can use biofilms to immobilize cells or bioreactors may separate cells from the bioreactor medium by the membrane and maintain them in a separate compartment. However, mammalian and plant cells should be protected from existing shear forces and bubble bursts regardless of the immobilization mode. Membrane bioreactors can use a wide variety of commercial membranes.

Choosing the right membrane for an application depends on factors, like and construction of the membrane and pore size. Membranes are classified as isotropic based on their structural properties and anisotropic depending upon composition. As shown schematically in Figure 3.5, One of the most common types of membrane bioreactor is the hollow fiber membrane bioreactor.

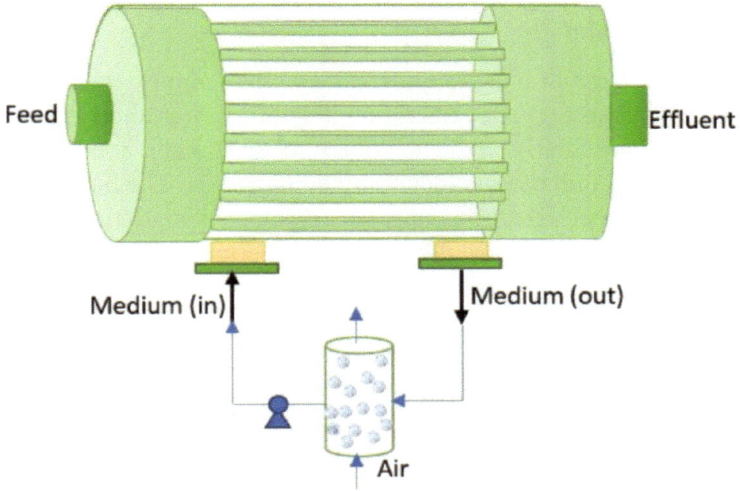

FIGURE 3.5 Membrane bioreactor.

3.6.2 MERITS OF IMMOBILIZED CELL BIOREACTORS

Merits of Immobilized Cell Bioreactors

- **Biological Stability**
- **Improved Biomass**
- **Support Matrix**
- **Cell Characteristics**
- **Environmental Conditions**
- **Enhanced Mass Transfer in Liquid**
- **Improved Yield**
- **Partitioning Effect**
- **Downstream Processing**
- **Cell Proximity and Reaction Selectivity**

There has been extensive research on immobilized cells and their several practical applications utilizing immobilized cell bioreactors. Another advantage of immobilized cell bioreactors is that they can be set up to create favorable microenvironment conditions, protect cells from damage caused by existing shear forces, etc., and improvements in biological stability. Following are some brief introductions to these positive attributes and the possible mechanisms that may explain them.

- ***Biological Stability***

Cells immobilized in bioreactors and those that employ them are generally considered to have greater operational stability than those that are freely suspended. Particularly when it comes to genetically modified cells, the instability of culture can hamper commercialization. The biological stability of immobilized cells is attributed to several mechanisms, although the underlying mechanism may vary depending on the metabolic and physiological state of the cells. In the case of nonviable cells, the immobilization matrix protects from harsh environmental conditions (e. g heavy metals, oxygen in the strictly anaerobic cells, Organic solvents,), and shear forces are one of the main components of biological stability.

- ***Improved Biomass***

In all bioreactors, regardless of how many cells are present, volumetric reaction rate or volumetric production rate is affected by biomass hold-up, i.e., free vs. immobilized. Immobilized cell bioreactors have a higher biomass hold-up than those that use freely suspended cells. Depending on the number of support particles used and the average biomass hold-up per particle, and any free cells that may be present in the system, the biomass hold-up in a bioreactor is calculated. Cell immobilization technology and hydrodynamic conditions of the bioreactor are the two main factors governing biomass hold-up per particle. The following attributes affect these criteria.

- ***Support Matrix***

Depending on the amount of support material occupied within the particle, biomass will be contained there. Electrical charge is a factor in determining cell attachment to support particles. Finally, the biological interaction between the cells and the support material is crucial in composting and converting waste cellulosic materials such as manure, wood chips, and compost.

- ***Cell Characteristics***

These factors impact the biomass hold-up and degree of immobilization of the cells according to their biological characteristics, such as history, type, size, shape, surface charge, cell wall composition, reproduction method, inoculation size, and oxygen requirements.

- ***Environmental Conditions***

In addition to pH and temperature, the concentration of dissolved oxygen and ionic strength also plays an important role.

- ***Enhanced Mass Transfer in Bulk Liquid***

Mass transfer in the bulk liquid is of great importance in a bioreactor. Transferring oxygen from the gaseous phase to the liquid phase, for example, is an essential step in any aerobic process. A similar approach is required for the bioremediation of waste streams contaminated with organic compounds such as heavy petroleum hydrocarbons, polycyclic aromatic hydrocarbons, and naphthenic acids.

- ***Improved Yield***

In addition to directing the flow of material through a particular pathway, cell immobilization has been shown to increase the yield of certain metabolites through changes in metabolic activity (Żur et al., 2016). The flow of nutrients from cellular synthesis to metabolite production facilitates the production of a desirable metabolite while also preventing excessive biomass growth.

- ***Partitioning Effect***

As a result of the carrier matrix's chemical composition, cells may be partitioned into a favorable microenvironment. The substrate can be concentrated near the cell surface by selecting the proper material for the matrix.

- ***Downstream Processing***

By using an immobilized cell bioreactor, it is possible to eliminate the need to separate the cells from the product-containing stream. By maintaining cells as discrete phases, the bulk liquid viscosity does not increase, thereby making the liquid easier to transport and process to recover the product.

- ***Cell Proximity and Reaction Selectivity***

It is advantageous for cells to have the ability to communicate if they are close together in an immobilized cell system. By placing the cells so close, plasmid DNA is also transferred, providing another method of conjugation. Co-immobilization of multiple species eliminates the need for multiple bioreactors when multiple reactions are taking place. Further, using carriers that are permeabilized in the desired way (such as semi-permeable membranes) could allow differing molecular sizes of reactants to be exploited to improve the selectivity of a specific reaction.

Cell immobilization is a cross-disciplinary subject that links the traditional and applied sciences. Several applications of cell immobilization technology have been found in environmental bioremediation, chemical and pharmaceutical synthesis, bioprocessing from biosensor development, as well as biomedical engineering and medicine. Various bioreactor configurations have been designed, developed, and utilized to suit these diverse applications. Immobilized cell bioreactors offer prolonged stability, increased biomass hold-up, increased reaction selectivity, and simplified downstream processing.

3.7 SUMMARY

Various industrial and scientific endeavors have used immobilized microbial cells extensively. Immobilized cells, however, have not been widely used in environmental applications. There is a special focus in this chapter on cells enclosed in biopolymer gels, which involve several scientific and technical considerations.

Immobilized cells are used in bioreactor studies to study some advantages and limitations.

The immobilization of viable cells has been studied extensively for the past four decades as an alternative to enzyme immobilization. Several applications for the basic and applied aspects of cell immobilization technology have been developed thanks to multidisciplinary research in areas such as environmental remediation and pollution control.

3.8 EXERCISES

Objective Response Questions

1. In which of the following does encapsulation of cells occur?

a) Immobilization
b) Enzyme reaction
c) Biofilm development
d) Pasteurization

2. Which of the following cannot be degraded by algae?

a) Alcohol
b) Organic pollutants
c) Carbonic acid
d) Water

3. Which of the following factor doesn't affect cell's Ecotoxicological evaluations?

a) Stability
b) Size
c) Color
d) Texture

4. What factor is not used to evaluate the response of the aquatic environment and sediment biota on immobilized cells?

a) Altered growth rate
b) Physiological activity at the organism level
c) Fermentation
d) Physiological activity at a cellular level

5. What factors affect the degradation rate of cells?

a) Residence time
b) Fermentation time
c) Instability
d) Growth

6. Which of the following does not responsible for the biodegradability of immobilized cells?

a) Easily degradable
b) Easily nondegradable
c) Potentially degradable
d) Fast degradable

7. Which of the following does not a polymer compounds that degrade very easily and are used for immobilization?

a) Sugar
b) Amino acids
c) Organic acids
d) PCB

8. In the process of immobilizing cells, which of the following methods is not applied for the degradation of pollutants?

a) Adaptation of bacteria
b) Outside and Inside enzyme reaction
c) Biofilm Production
d) Pasteurization

9. Immobilized cells require synthetic inoculation for degradation.

a) True
b) False

10. In addition to artificial inoculation, there is an extra substrate obligation.

a) True
b) False

Practice Work

Qu 1. What is immobilization? How immobilized cells are used for bioremediation?

Qu 2. Explain the various methods of immobilization.

Qu 3. Enlist the applications of immobilization.

Qu 4. What is an encapsulation of microbes? Explain their advantages and limitations.

Qu 5. What type of carrier material is used for immobilization?

Qu 6. Write short notes on immobilized cell bioreactors.

Qu 7. Explain the instrumentation of fluidized bed bioreactors.

Qu 8. What are membrane bioreactors?

Qu 9. Explain the advantages of membrane bioreactors.

Qu 10. Explain fixed-bed bioreactors in detail.

REFERENCES

Blanch, H. W., & Clark, D. S., (1997). *Biochemical Engineering*. New York: Marcel Dekker, Inc.

Brodelius, P., & Vandamme, J., (1987). Immobilized cell systems. In: Kennedy, J. F., (ed.), *Biotechnology: A Comprehensive Treatise in Eight Volumes* (Vol. 7a). VCH Verlagsgesellschaft mbH, Germany.

England, L. S., Lee, H., & Trevors, J. T., (1993). Bacterial survival in soil: effect of clays and protozoa. *Soil Biol. Biochem., 25*, 525–531.

Gentili, A. R., Cubitto, M. A., & Ferrero, M., (2006). Bioremediation of crude oil polluted seawater by a hydrocarbon-degrading bacterial strain immobilized on chitin and chitosan flakes. *Int. Biodeterior. Biodegradation, 57*, 222–228. doi: 10.1016/j.ibiod.2006.02.009.

Nedovic, V., & Willaert, R., (2005). *Focus on Biotechnology Applications of Cell Immobilization Biotechnology* (Vol. 8B). Dordrecht: Springer.

Obradovic, B., & Nedovic, V. A., (2004). Immobilized cell bioreactors. In: Nedović, V., & Willaert, R., (eds.), *Focus on Biotechnology: Fundamentals of Cell Immobilization Biotechnology* (Vol. 8A, pp. 411–431). Dordrecht: Kluwer Academic Publishers.

Obuekwe, C. O., & Muttawa, M., (2001). Self-immobilized bacterial cultures with potential for application as ready-to-use seeds for petroleum bioremediation. *Biotechnol. Lett., 23*, 1025–1032.

Quek, E., Ting, Y. P., & Tan, H. M., (2006). *Rhodococcus* sp. F92 immobilized on polyurethane foam shows ability to degrade various petroleum products. *Bioresour. Technol., 97*(1), 32–38.

Radwan, S. S., Hasan, R. H., & Salamah, S., (2002). Bioremediation of oily sea water by bacteria immobilized in biofilms coating microalgae. *Int. Biodet. Biodeg.*, *50*, 55–59. doi: 10.1016/S0964-8305(02)00067-7.

Shuler, M. L., & Kargi, F., (2002). *Bioprocess Engineering Basic Concepts* (2nd edn.). Upper Saddle River, NJ: Prentice-Hall, PTR.

Willaert, R., Baron, G. V., & Backer, L. De., (1996). *Immobilized Living Cell Systems: Modelling and Experimental Methods*. Chichester: Wiley.

Żur, J., Wojcieszyńska, D., & Guzik, U. (2016). Metabolic Responses of Bacterial Cells to Immobilization. *Molecules*. *21*(7), 958. doi: 10.3390/molecules21070958. PMID: 27455220; PMCID: PMC6273605.

CHAPTER 4

Bioremediation of Dyes and Plastics

4.1 TOXIC DYES

Rapid industrial growth and civilization result in the release of many chemicals, including dyes in the environment. The commercial dye industry produces around 100,000 dyes, including several types of chemical dyes such as acidic dyes, basic dyes, and reactive dyes including alkyds, azo, diazo, and anthraquinone-based dyes. A large proportion of the dyes are included in industrial effluents. Food, cosmetics, plastics, and other industries use them extensively. Living organisms may be adversely affected by some of these chemicals for a variety of reasons, including toxicity and carcinogenicity.

There are many environmental challenges today, including the generation of enormous volumes of wastewater due to human activities. The textile industries have extensively used man-made dyes because they are inexpensive and provide bright colors. Natural dyes, however, are not cost-effective and their preparation is time-consuming. As a result, synthetic colors are widely used and textile industries are generating large volumes of wastewater, causing aquatic pollution worldwide. Artificial dyes or imitation dyes are widely used in plastic toys, textile staining, the printing of papers, food, plastic products, pharmaceuticals, dye cinematography, cosmetics, and other important industries. Dyeing fabrics generate a great deal of colored wastewater as polluted effluent.

Water polluted with dyes can cause disastrous effects, viz. According to Suryavathi et al., (2005), bladder tumors are particularly common in dye industry workers. In the past, textiles have been dyed with natural pigments that fade with exposure to heat, light, and water. Fast and synthetic colors have been developed to replace the use of natural pigments.

A variety of combined biological, physical, and chemical decolorization processes are used to remove the color from colored effluents It is estimated

Bioremediation and Phytoremediation: Technologies for Toxic Pollution
Ashwini A. Waoo, PhD (Author)

that approximately 15–30% of dyes unused after dyeing and subsequent washing end up in wastewater (Rajamohan and Karthikeyan, 2006). Due to exorbitant costs and disposal issues, chemical and physical methods for dye wastewater treatment are not widely used in textile industries. Adsorption of dyestuffs on bacteria and fungi are environmentally friendly solutions to this problem or low-cost non-conventional adsorbents (Yang et al., 2009).

In recent years, dye removal has become a research topic of increasing interest, due to more stringent government regulations regarding the disposal of contaminated effluent. Dye removal from industrial effluents can be accomplished in various ways, including chemical coagulation with lime, alum, ferric sulfate, ferric chloride, and electrocoagulation, but these methods are extremely time-taken and expensive with low efficiency. Because fungal bioremediation is significantly more biomass-intensive than bacteria, research interest in this method for degradation of man-made or synthetic dyes is increasing significantly (Shahid et al., 2013).

Numerous studies have shown that microorganisms have the ability of biodegrading lethal components (Aksu, 2005). As an on-site method, bioremediation offers several benefits, including a cost-effective approach and minimal problems, as well as the capability of combining it with other approaches.

4.2 TYPES OF TEXTILE DYES

Colors are classified according to the Color Index (CI) system, devised by Zollinger (1987).

Reactive Dyes: A reactive dye consists of azo groups, such as triphen-dioxazine, metalized azo, formazan, phthalocyanine, and anthraquinone. Molecularly, these dyes differ and are structurally simpler than direct dyes. In addition, they produce brighter, more cheerful shades than direct dyes. Cotton fibers are traditionally dyed and printed with reactive colorants.

Direct Dyes: In the presence of electrolytes, these dyes are anionic and water-soluble. They bind well to cellulose fibers. The compounds in this group mostly consist of polyazo compounds, along with phthalocyanines and oxazines. They are also used with formaldehyde or a cationic dye–complexing resin.

Disperse Dyes: Dyes in this group are water-impermeable and mostly used to color polyester, polyamide, polyacrylonitrile, and polypropylene fibers.

Such dyes are not used for the dyeing of nylon, cellulose acetate, or acrylic fibers. Chemically, these dyes are chiefly prepared from azo and anthraquinonoid groups, with groups that support the formation of stable aqueous solutions.

Sulfur Dyes: Cotton cloth can be dyed with these dyes by reducing sodium salts using sodium salts that are insoluble in water. The dyes have a low cost, which makes them popular with industries.

Vat Dyes: The dyes in vats are not soluble in water. They are applied mainly to cellulose fiber by altering them to their leuco compounds. By reducing and solubilizing $Na_2S_2O_4$ with NaOH solution, the leuco compound is formed, this process is called a "vatting process." Anthraquinones are the chief chemical/structural group of vat dyes.

Cationic (Basic) Dyes: This group of dyes is cationic and water-soluble. For dyeing, they use paper, polyacrylonitrile, nylon, and polyester containing modified polymers. These can also be applied to silk fiber, wool, and tannin-mordant cotton. It is more important that they are light and washable.

Acid Dyes: Dyes in this group are water-soluble anionic dyes and are used to color wool, nylon, silk, and acrylics. These molecules are also used for inkjet printing as well as dyeing paper, leather, food, grease paint, and cosmetics.

Solvent Dyes: These dyes are not soluble in water but do dissolve in certain specific solvents. Their molecules are generally nonpolar or sometimes polar; moreover, they do not undergo a process of ionization. They are employed for coloring gasoline, plastics, oils, waxes, etc.

Mordant Dyes: Covalent and coordinate bonds enable these dyes to form strong bonds with metal residuals. There is also a chelate compound involved. As fixatives of metallic salts, Al, Cr, Cu, Co, Ni, Fe, and Sn have been used.

4.3 BIOREMEDIATION OF DYES

Dye is a chemical substance that has the property of coloring a substrate when applied to it. Textiles, plastics, therapeutics, food, cosmetics, graphic arts, and paper and pulp industries use them primarily. Adhesion to compatible surfaces can be achieved through the solution, by salts or metals forming covalent bonds, or by physical adsorption.

Fungal decolorization, bacterial degradation, and dye adsorption by (dead and living) microbiological biomass and detoxification systems constitute biological degradation. The method can be successfully used to degrade dyes present in textile effluent, since a variety of microbes, including yeasts, bacteria, fungi, and algae, can accumulate in textile effluent. This treatment, however, has certain limitations due to its technical limitations. Microorganisms like fungi and bacteria are difficult to handle due to contamination concerns (Barr and Aust, 1994), although their cultivation is easy (Banat et al., 1997). It is an economic method of degradation and Recent years have seen much attention given to this topic.

Anaerobic and aerobic degradation of effluents is possible in biological treatment. Anaerobic treatment of textile effluent uses anaerobic bacteria that produce powerful enzymes such as azoreductase. Anaerobic–aerobic treatment of textile effluent has been studied by Van Der Zee (2005) in a both sequential and integrated manner. Studies have examined the use of bacteria in decolorizing dyes and degradation of dyes by Santos et al., (2007).

According to several studies Microbes can biodegrade lethal components from the biosphere (Aksu, 2005). Bioremediation has several advantages, including the fact that it can be carried out on-site, it is cost-effective and causes few problems, and it can be combined with physical and chemical approaches.

It is believed that fungi are best adapted to remove dye from industrial effluent. It is beneficial for fungi to have mycelium instead of unicellular organisms. It solubilizes insoluble substrates using extracellular enzymes. Fungi have a greater cell to surface ratio; hence, they show better physical and enzymatic interactions with the environment. They also have a greater ability to degrade high concentrations of toxins. Biological treatment has limitations, as it needs a huge area and is affected by day-to-day variations (Bhattacharya et al., 2003). With the present conventional biodegradation processes, biological treatment for color removal sometimes becomes ineffective (Robinson et al., 2001). The complexity of several biological constituents also leads to the degradation of many of them and the recalcitrance of others (Ravi Kumar et al., 1998). Specifically, azo dyes are not completely degraded by biological methods, because of their xenobiotic nature. Textile effluent is difficult to treat; hence, combinations of biological treatment and physicochemical techniques have been used to decolorize textile effluent (Kannan and Sundaram, 2001; Rai et al., 2005; Solmaz et al., 2009; Wojnarovits and Takacs, 2008; Ince and Tezcanh, 1999; Chaudhari et al., 2011).

4.4 PLASTIC BIOREMEDIATION TECHNOLOGY

By 1940, plastic was discovered by an accident due to an instrument malfunction, and no one had imagined it would be an important material in packaging, construction, and medical applications (Thompson et al., 2009). The industrialization and urbanization of the world have brought many discoveries using plastic material act as a base to ease human lives. Since plastics have a long carbon backbone, a hydrophobic property, and high molecular weight, they cannot biodegrade. The annual global production of plastic waste reaches several million tons, evidently due to the non-degradability and the unlimited use of plastic products. It was discovered that polyvinyl chloride survived for a long time in the marine environment, but that thermal and physical damage resulted in morphological damage and the leakage of micro-sized particles (Tang et al., 2018). Plastic waste and its impact on the environment are persistently under-aware among government and public sectors. Furthermore, plastic is damaging to the environment, releasing toxins and volatile chemicals, and affecting animals by ingestion of plastic particles and entanglement in different ecosystems as a result of its presence (Diepens and Koelmans, 2018). Conventional methods have been developed for reducing plastic waste, such as landfilling and incineration; however, these methods have also led to the release of various toxic end products and to rising global temperatures that threaten our environment.

4.5 ENVIRONMENTAL EFFECTS OF PLASTIC POLLUTION

Plastics cause adverse effects on the environment due to their toxic content. The decomposition of plastics takes a very long time, say thousands of years, causing serious environmental damage.

Plastic pollution can cause many health problems. Here are some key points about plastic pollution. Plastic waste poses several serious problems in the landfills and garbage dumps that release pollutants into the soil, affecting both flora and fauna as well as groundwater. Fishing nets used for commercial purposes are typically made of plastic. In the case of long-term exposure of these nets to water for fishing, it releases toxins that affect the surrounding waters and fish. Whenever organisms in the water become poisoned by the plastic they swallow, it affects the larger organisms in the food chain, thereby affecting the entire food web. The plastics in landfills react with water, creating hazardous toxins that seep into the groundwater and affect its quality.

Throughout the surrounding area, animals might be suffocated by plastic. The burning of plastic in the open-air releases toxic substances that pollute the environment. Polluted air causes health hazards and respiratory problems to humans and animals. Plastics disposed of in the ocean decompose more quickly due to sunlight, rain, and other environmental conditions, by releasing toxic chemicals like bisphenol A (Teuten et al., 2009).

The plastic pieces in the ocean are taken up by organisms that filter feed on them, which in turn lead to bioaccumulation of these toxins in the food web and food chain. Polyethylene films have the most adverse environmental effects when they are swallowed by wild animals.

The major sources of plastic pollution in the oceans are plastic pellets, polystyrene, and plastic bags. Chemically treated plastic emits harmful chemicals that seep into groundwater and create serious problems for aquatic species that consume that water.

A breakthrough in biotechnology enables the rational engineering of microbial genomes and metabolic pathways for the specific degradation of pollutants. Most of the studies do not adequately describe their field application and lack fundamental knowledge. A microbial genetic engineering tool for bioremediation technology would increase its efficacy. Initial excitement created in this area is expected to encourage the development of eco-friendly microbes, enzymes, or exact compositions of genes to produce novel remedies with maximum efficacy (Kulshreshtha, 2013).

Microorganisms degrade plastics by producing extracellular enzymes which break down long-chain polymers into oligomers and monomers that are utilized by microbial cells. Microbes attach to polymers and then colonize the surfaces of these polymers during the biodegradation of plastic. A polymeric substrate is hydrolyzed by enzymes attached to its surface. The aerobic and anaerobic metabolic processes further break down oligomers and monomers. Aerobic degradation results in carbon dioxide and water, while anaerobic metabolism results in carbon dioxide, methane, and water (Ahmed et al., 2018).

4.6 FACTORS AFFECTING PLASTIC BIODEGRADATION

An amalgamation of microbiology and genetic engineering can be used to develop new bioremediation techniques that encompass the expression and DNA profiling studies of microorganisms. Currently, systemic biology approaches allow researchers to broaden bioremediation agents and exploit cellular catabolic mechanisms.

The biodegradation process is influenced by several factors, such as polymeric properties, environmental parameters, enzyme properties, O_2, sunlight, moisture content, pH and temperature, pH and temperature affect not only the growth of microorganisms but also the activity of enzymes. Degradation rates are affected by polymer characteristics such as molecular weight, shape, and size. A plastic's hydrophobic nature prevents microorganisms from attaching and forming biofilms. The presence of biosurfactants and additives also affects the activity of microbes (John and Rukhiya Salim, 2020).

Thus, systemic biology technologies provide a chance to resuscitate bioremediation using anthropogenically enhanced bacterial degraders, as well as translational implications (Ahmad and Ahmad, 2014). These systems use databases to select chemically and energetically viable traits in desired polymeric substrates.

A multi-omics analysis provides valuable information that can be used to select a host that will be suitable for a particular substrate or a broader range of substrates. Genetic engineering enables addressing specific genes, reducing metabolic burden, and targeting specific metabolic pathways, at the genome-scale and kinetic scale, as well as at the pathway level.

4.7 PROCESS OF POLYMER BIODEGRADATION

Biodegradation is the process by which microorganisms change a chemical in the environment. Plastics are non-biodegradable, though there are a few microbes that can degrade and assimilate them. This led to the solution of reclaiming plastics more sustainably. In the laboratory, it is impossible to emulate an exact natural environment, especially when one considers the efficiency of diverse microbial communities and their ability to use and transform a variety of plastics. Developing proficient field applications requires an in-depth understanding of microbe-material interaction and microbial cellular mechanisms. There are four major phases of microbial degradation of polymeric materials, as shown below, i.e., microbial adherence, mineralization, and fragmentation (Sen and Raut, 2015).

4.7.1 *MICROBIAL ADHERENCE*

As abiotic and biotic factors influence microbes' interaction with plastic surfaces (Shah et al., 2008), many abiotic factors also apply to plastic surfaces. Many factors contribute to this phenomenon, including surface roughness, hydrophobicity, topography, salinity, pH, temperature, and oxygen content. Polymers undergo morphological changes due to abiotic factors, which facilitate the colonization of microbes. Biological factors may be exogenous or endogenous. Based on the nutrient source of the cell, these factors affect adherence and film formation, starting with the bacterial physical interaction with the substrate and ending with the irreversible settling of a cluster of cells onto the surface. A biofilm arises from three distinct stages: the accumulation of cells on the surface, the formation of attachment forces between them, and the development of properties.

1. The deposition of cells on the surface;
2. The attachment to the substrate via polymeric bridges; and
3. The colonization by bacteria followed by cellular growth and division (Garrett et al., 2008) resulted in biofilm formation.

The firmness of adherence reflects the induction of appropriate bridging between the polymer surface and microorganisms by time as well as the physical properties of the surface.

Temperature is also a key factor in determining how bacteria adhere to surfaces. In the process of biodegradation, surface energy, charge, and roughness have a vital impact on the adhesion of cells to surfaces. As part of these events, pioneer cells may form an initial conditioning film that helps promote growth as well as the Formation of the microbial community in a biofilm. A change in any of the factors will affect the degradation and rate, which will result from either acceleration or inhibition of cellular key steps.

4.7.2 *FRAGMENTATION*

Large polymeric structures are primarily broken down by dynamic and chemical forces. By enhancing the hydrophilicity of plastic surfaces through pre-treatments or abiotic processes, microbes are more likely to infect (da Luz et al., 2014). Depolymerizing enzymes found outside of the cell are responsible for the breakup of terminally available branches into smaller molecular fragments that can be ingested by the cell. Polymer chains are cut and oxidized primarily at the edges by extracellular enzymes resulting in a reduction in molecular weight and a change in carbonyl bond index. The enzymes involved in polymer fragmentation are mostly oxidoreductases and hydrolases (Mueller, 2006).

4.7.3 ASSIMILATION

Low molecular weight fractions formed on plastic surfaces, the microbial process fragments the surface, allowing small oligomers to pass through cell membranes and enter the main metabolic pathways (Sen and Raut, 2015). As a result of chain fragmentation, the process of biodegradation of plastic is primarily an electron transfer process.

Most monomers fragmented by degradation are analogs of natural substrates, like organic acids, alcohols, and amide compounds. Several microbe-produced enzymes, including laccase, lipases, cutinases, and esterases, have been found to hydrolyze several synthetic plastics, even though their original substrates are phenolic, lipid, and ester compounds (Kawai, 2010). Cellular metabolism depends solely on enzymes catabolizing molecules.

A plastic biodegrades primarily through the exertion of extracellular enzymes that act on its surface. This is usually a surface erosion process. Polymers in polymer chains deteriorate when prooxidants catalyze the formation of free radicals, which then react with oxygen.

4.7.4 MINERALIZATION

Biomass is converted into gases, salts, minerals, water, and remaining biomass during mineralization. The mineralization process is complete when all the carbon-based solids are converted to CO_2 and water, and methane by either erobic or anaerobic degradation (Sen and Raut, 2015). Model studies and field experiments must improve our understanding of microbial community structure and its impact on carbon and nutrient cycling from plastics.

Dussud and Ghiglione illustrate biodegradation in four steps (Figure 4.1).

Bio-deterioration: In response to the release of acids and the formation of biofilms, microbial metabolic activity will cause cracks in the plastic or alter its physical properties.

Bio-fragmentation of the long polymer chain: Microorganisms produce enzymes that cause oligomerization.

Degradation of oligomers to monomers: Upon entering the cells, oligomers are assimilated by secondary degradation enzymes, increasing the biomass of microbial cells.

Assimilation: The absorption of individual oligomers and expulsion of the oxidized metabolites as H_2O, CO_2, N_2, and CH_4.

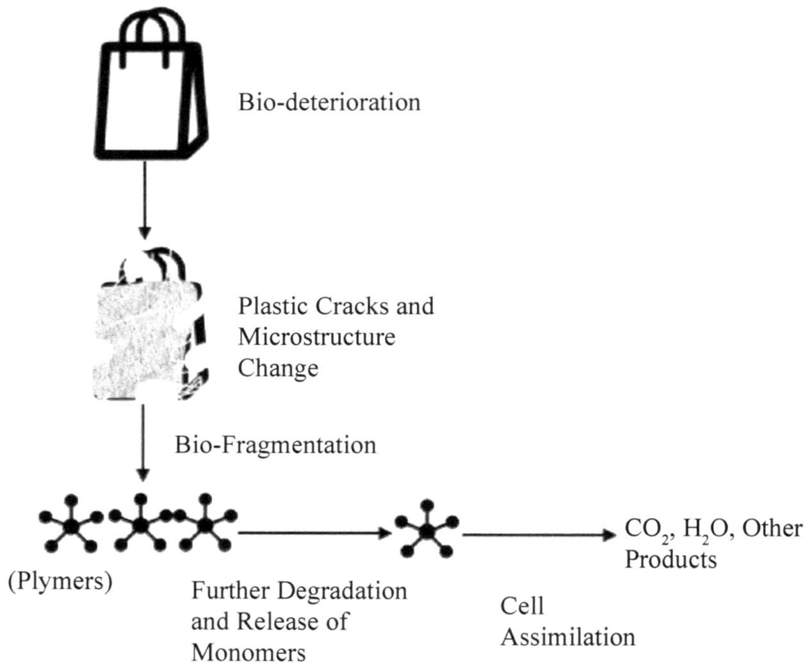

FIGURE 4.1 Biodegradation of plastic.

4.8 BIODEGRADATION OF SYNTHETIC PLASTICS

In the era of modern biotechnology, microorganisms of all kinds contribute to the prevention of bioaccumulation of several inorganic and organic contaminants in the environment, so it is important to understand how they break down plastics to make new materials. Synthetic plastics are degraded by biotic processes that are primarily driven by microbes; they have been documented as potential degraders of chemical xenobiotics based on their ability to adapt and consume these chemicals as food sources. Organisms in this group use the enzymes in their enzymatic systems to degrade polymers into intermediates that can then be assimilated and metabolized to provide them with energy.

In the last decades, several studies have been conducted to investigate how microorganisms can degrade various plastics. It appears that synthetic polymers degrade relatively slowly by a variety of microbes, so this process cannot be used in real-time in industry. Furthermore, this has led to the development of various biodegradable additives, as well as the rapid development of more biodegradable bioplastics (Thiruchelvi et al., 2020).

- ***Actinomycetes***

In the soil, actinomycetes are filamentous bacteria that can be found in diverse forms, marine environments, and plant tissues. They have been well-studied for their metabolic adaptability and various biotechnology applications, including bioremediation, medicine, and food processing. *Streptomyces* groups, *Rhodococcus ruber, Actinomadura* species, and *Thermoactinomyces* have been isolated from diverse ecosystems and shown to degrade plastics (Jabloune et al., 2020). They produce a variety of different hydrolytic enzymes.

Their hydrolytic enzymes are important factors that help them grow on a variety of plastic polymers by degrading high molecular weight compounds to simpler ones. Additionally, the cells produce extracellular polymers such as glycogen, dextran, levan, and slime polysaccharides, which may facilitate their attachment to plastic surfaces. (Pujic et al., 2015). In a microbial consortium with a substantial fraction of actinomycetes, actinomycetal plastic degradation has also been shown to be effective.

- ***Algae***

The significant role of algal photosynthetic and heterotrophic forms in bioremediation, amongst their many industrial applications, has been well studied. By accumulating, adsorbing, or metabolizing contaminants into relatively safer levels, they can degrade both inorganic and organic pollutants from diverse environments (Hwang et al., 2020). Comparatively, only a few studies have been conducted on the potential of algae to remove synthetic plastic polymers. Many of the studies have focused on their use in the production of green plastics. Even though reports have shown that marine algae can ingest plastic materials.

Algae such as *Anabaena, Spirogyra, Chlorella, Oscillatoria, Nostoc,* and *Spirulina* have been found to colonize different types of plastic on the ground, though no evidence exists that they consume the plastic. (Sarmah and Rout, 2018).

Kumar et al., (2017) showed that *Anabaena spiroides, Scenedesmus dimorphus,* and *Navicula pupula* could degrade both high density and low-density PE, while A *Spirulina* species also degraded PET and PP after 112 days, but at a significantly slower rate than bacteria and fungi.

- ***Bacteria***

In the environment, bacteria are responsible for the transformation and cycling of nutrients as they are the frontline for nutrient cycling. Through their role in decomposition, they release carbon and nutrients from a variety of complex polymers. Their significance in bioremediation has been studied, and they can destroy a wide variety of materials, including antibiotics, metal compounds, petroleum, plastics, and other compounds that have become more prominent in our Anthropocene age. Different bacterial species from *Pseudomonas*, *Escherichia*, and *Bacillus* genera have been revealed to be capable of degrading plastic using metagenome analysis, cloning, pure culture, and bioinformatics methods (Gan and Zhang, 2019). It is also interesting to note that these bacteria have been isolated from different ecological niches, including urban dumpsites.

There is evidence that the formation of biofilms plays an important role in the decomposition of plastics by bacteria, enhancing their adhesion to the plastic surface and their persistence. As a result of their homopolymeric nature, thermoplastics are resistant to microbial degradation. Polystyrene and polycarbonate are also common thermoplastics that degrade, which are degradable by *Pseudomonas aeruginosa*, *Rhodococcus ruber*, *Bacillus megaterium, Serratia marcescens*, *Streptococcus pyogenes Staphylococcus aureus*, and other strains (Arefian et al., 2020).

- ***Fungi***

The biogeochemical cycles and the uptake of nutrients on earth are maintained in large part by fungi and bacteria. Different fungal species are capable of degrading various plastic polymers considering their ability to utilize these synthetic polymers as their sole source of carbon or energy. Numerous strains of fungi that come in a variety of classes, etiologies, and forms have been found to degrade plastic. The biodegradation of synthetic plastics has been studied. Recently, the *Aspergillus* genus has emerged as the most prominent fungus. (Gajendiran et al., 2016).

It is also crucial to their initial colonization that fungal hyphae can diffuse and penetrate They also produce hydrophobins for hyphal attachment to hydrophobic surfaces. Various factors have been demonstrated to enhance the fungal biodegradation of plastics, such as photo-treatment and temperature, acid pretreatment), and various additives (Sánchez, 2020, Andrew, 2012).

- ***Extremophilic Bacteria***

In terrestrial and marine environments, millions of tons of plastic accumulate annually, which represents 20–30% (by volume) of municipal solid waste. The environment is seriously endangered by these pollutants. Plastic accumulation is delayed by environmental degradation. There are several mechanical and chemical methods of treatment, which depend on several factors. Mechanical aging is affected by temperature, moisture, and sunlight, and can lead to changes in a plastic structure like cracking, discoloring, changing shape or optical characteristics, and flaking.

4.9 METHODS FOR DETECTING PLASTIC BIODEGRADATION

Despite biodegradation's reputation as an innovative solution for plastic disposal, Biodegradation takes a long time. Scientists have searched for materials without harmful environmental effects since the 1980s, so they can design them. Furthermore, trying to biodegrade the already accumulated plastics and their traces in the environment.

Microbial activity can be detected using the following methods:

1. Examination of visible changes in plastics, such as the presence of holes, a change in surface appearance, or the presence of biofilms, is illustrated in Figure 4.2.
2. For more sophisticated observations, scanning electron microscopy (SEM) and atomic force microscopy (AFM) is recommended.
3. Monitoring changes in properties of a polymer, such as tensile strength and mass loss, and chemical properties, such as molecular weight.
4. Measuring the rate at which oxygen and carbon dioxide are utilized.
5. An assessment of growth is performed by: measuring the biomass accumulated, usually in minimal media and using polymers as carbon sources; observing the formation of a clear halo around the colonies, which depolymerizes the polymer.

6. Determination and characterization of depolymerization products by enzyme assay.

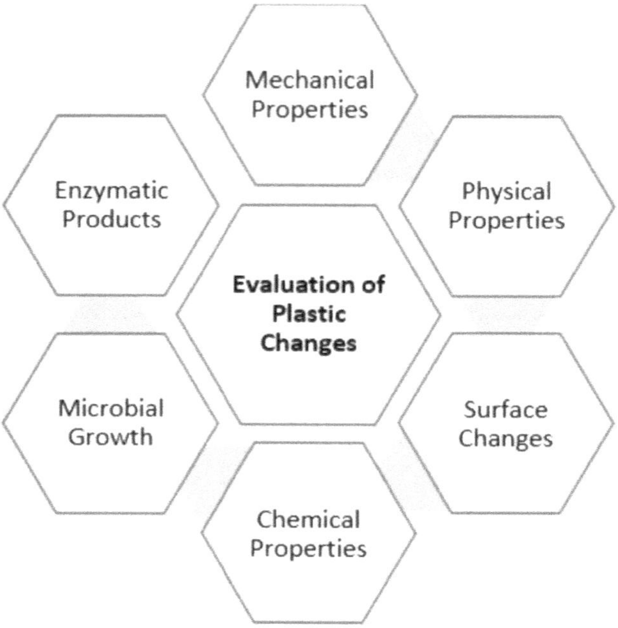

FIGURE 4.2 Evaluation of plastic changes.

4.10 ENZYMES IN PLASTIC BIODEGRADATION

Plastic-degrading enzymes have the same classification as other enzymes responsible for the degradation of polymers, i.e., extracellular and intracellular enzymes. One of the groups of enzymes studied the most in the extracellular ones, which are incredibly reactive, from hydrolytic oxidative to functionality. They break down long carbon chains of plastic polymers into a mixture of oligomers, dimers, and monomers. There is growing biotechnological evidence to suggest that these enzymes act the same as microbial peroxidases, laccases, lipases, cutinases, esterases, etc. Additionally, extracellular enzymes may involve in heterogeneous reactions at the surface of solid/liquid interfaces, since they interact with macromolecules on the surface of solid plastic while in the liquid state (Chinaglia et al., 2018).

Plastic is degraded by a variety of microbial enzymes. Alkane hydroxylase from *Pseudomonas aeruginosa* is used to biodegrade polyethylene, laccase

(*Rhodococcus* ruber), lipases, lignin- and manganese dependent peroxidase (LiP and MnP), protease, serine hydrolases, polyurethanases, heme peroxidase, and esterases (lignin peroxidase) are some important enzymes among them. Oxygenase increases the solubility of plastic polymer, making it easier for bacteria to digest. Endopeptidases with carboxylic groups and amine groups are attacked by bacterial lipases and esterases.

4.11 SUMMARY

Artificial plastics can be considered xenobiotics. Our routine lives are flooded with these materials, and they are ubiquitous. Since they clog the various sites of the environment, they have become a matter of ponderance. As they are so recalcitrant and non-degradable, the question of how to eliminate them has become a matter of urgency.

As a result of this, plastic and polythenes have accumulated in the natural environment and are causing a great deal of environmental damage. Researchers have investigated different ways to degrade synthetic polymers, including chemicals, physical methods, and microbes. Microbe participation has resulted in marvelous results for the biodegradation of plastic and polymer.

Researchers have studied the mechanism of microbial enzymes degrading synthetic products, and this area of research is gaining more attention.

A lot of investigations are still underway to explore these phenomena, such as bio-fragmentation, bio-mineralization, and bio-assimilation. To eliminate plastic pollution from the environment, further future studies are needed to identify effective microbial strains that produce an array of hydrolytic enzymes. As technology advances, only biodegradable products will likely be available in the future, but we should plan to get rid of the stack of polymers and plastics previously designed and manufactured by learning new enzymes and microbial characteristics.

4.12 EXERCISES

Objective Response Questions

1. **In the ocean, microplastics can be caused by which of the following?**

   a) Some types of toothpaste
   b) Plastic bottles

c) Plastic cutlery
d) All of these

2. It's better to recycle plastic instead of refusing it in the first place. Is it true or false?

a) True
b) False

3. The term which refers to the breakup within a compound due to microbial activity is?

a) Biodegradation
b) Agro-degradation
c) Phyto-degradation
d) Decomposition

4. Benzene, a _____ substance, is an industrial solvent utilized in pharmaceuticals, plastics, and dyes production?

a) Odorless
b) Non-flammable
c) Biodegradable
d) Carcinogenic

5. _______ generate a great deal of colored wastewater as polluted effluent.

a) Dyeing fabrics
b) Glass wastes
c) Plastic Waste
d) Pharmaceutical Waste

6. Which of the following is not a type of textile dye?

a) Reactive Dyes
b) Disperse Dyes
c) Sulfur Dyes
d) Toxic Dyes

7. The decomposition of plastics takes _______.

a) Very long time

b) Very short time
c) Few seconds
d) Few minutes

8. Large polymeric structures are primarily broken down by dynamic and chemical forces, this process is called as _______.

a) Fragmentation
b) Assimilation
c) Polymerization
d) None of the above

9. What is the full form of AFM?

a) Atomic force microscopy
b) Atomic fluorescence microscopy
c) Atomic flow microscopy
d) None of the above

10. Determination and characterization of depolymerization products carried out by _______.

a) Enzyme assay
b) Molecular assay
c) Electrical assay
d) Titration assay

Practice Work

Qu 1. What is a bioreactor? Explain the use of bioreactors for bioremediation purposes.

Qu 2. Which harmful contaminants are present in textile dyes? Explain the technique of bioremediation of these dyes.

Qu 3. Explain in detail toxic dyes and their adverse effects on the environment.

Qu 4. What are the classes of dyes?

Qu 5. Explain the bioremediation process of textile dyes from industries.

Qu 6. Describe plastic pollution and its consequences.

Qu 7. Explain the factors affecting plastic biodegradation.

Qu 8. Explain the process of polymer biodegradation.

Qu 9. Describe the role of enzymes in plastic biodegradation.

Qu 10. Explain the process of detecting plastic degradation

REFERENCES

Ahmad, M., & Ahmad, I., (2014). Recent advances in the field of bioremediation. In Book: *Biodegradation and Bioremediation, Chapter: Recent Advances in the Field of Bioremediation* (Vol. 11, pp. 1–42). Publisher: Studium Press LLC, U.S.A.

Ahmed, T., Shahid, M., Azeem, F., Rasul, I., Shah, A. A., Noman, M., & Muhammad, S., (2018). *Environ. Sci. Pollut. Res., 25*(8), 7287–7298.

Aksu, Z., (2005). Application of biosorption for the removal of organic pollutants: a review. *Process Biochem., 40*, 997–1026.

Arefian, M., Tahmourespour, A., & Zia, M., (2020). Polycarbonate biodegradation by newly isolated *Bacillus* strains. *Arch. Environ. Prot., 46*(1), 14–20. doi: 10.24425/aep.2020.132521.

Chinaglia, S., Tosin, M., & Degli-Innocenti, F., (2018). Biodegradation rate of biodegradable plastics at molecular level. *Polym. Degrad. Stab., 147*, 237–244. https://doi.org/10.1016/j.polymdegradstab.2017.12.011.

Da Luz, J. M. R., Paes, S. A., Bazzolli, D. M. S., To´tola, M. R., Demuner, A. J., & Kasuya, M. C. M., (2014). Abiotic and biotic degradation of oxo-biodegradable plastic bags by *Pleurotus ostreatus*. *PLoS One, 9*(11). https://doi.org/10.1371/journal.pone.0107438.

Diepens, N. J., & Koelmans, A. A., (2018). Accumulation of plastic debris and associated contaminants in aquatic food webs. *Environ. Sci. Technol., 52*, 8510–8520. https://doi.org/10.1021/acs.est.8b02515.

Gajendiran, A., Krishnamoorthy, S., & Abraham, J., (2016). Microbial degradation of low-density polyethylene (LDPE) by *Aspergillus clavatus* strain JASK1 isolated from landfill soil. *3 Biotech, 6*, 52. https://doi.org/10.1007/s13205-016-0394-x.

Gan, Z., & Zhang, H., (2019). PMBD: A comprehensive plastics microbial biodegradation database. *Database, 2019*, baz119. https://doi.org/10.1093/database/baz119.

Garrett, T. R., Bhakoo, M., & Zhang, Z., (2008). Bacterial adhesion and biofilms on surfaces. *Prog. Nat. Sci., 18*(9), 1049–1056. https://doi.org/10.1016/j.pnsc.2008.04.001.

Hwang, J. H., Sadmani, A., Lee, S. J., Kim, K. T., & Lee, W. H., (2020). Microalgae: An eco-friendly tool for the treatment of wastewaters for environmental safety. In: Bharagava, R. N., & Saxena, G., (eds.), *Bioremediation of Industrial Waste for Environmental Safety* (pp. 283–304). Springer, Singapore. https://doi.org/10.1007/978-981-13-3426-9_12.

Jabloune, R., Khalil, M., Moussa, I. E. B., Simao-Beaunoir, A. M., Lerat, S., Brzezinski, R., & Beaulieu, C., (2020). Enzymatic degradation of p-Nitrophenyl esters, polyethylene terephthalate, cutin, and suberin by Sub1, a suberinase encoded by the plant pathogen Streptomyces scabies. *Microbes Environ., 35*, ME19086. https://doi:10.1264/jsme2.ME19086.

Kawai, F., (2010). The biochemistry and molecular biology of xenobiotic polymer degradation by microorganisms. *Biosci. Biotechnol. Biochem., 74*(9), 1743–1759. https://doi.org/10.1271/bbb.100394.

Kulshreshtha, S., (2013). Genetically engineered microorganisms: a problem-solving approach for bioremediation. *J. Bioremed. Biodeg., 4*, e133. https://doi.org/10.4172/2155-6199.1000e133.

Kumar, R., Kanna, G., & Elumalai, S., (2017). Biodegradation of polyethylene by green photosynthetic microalgae. *J. Bioremediat. Biodegrad., 8*, 2. doi: 10.4172/2155-6199.1000381.

Mahalakshmi, V., & Andrew, S. N., (2012). Assessment of physicochemically treated plastic by fungi. *Ann. Biol. Res., 3*, 4374–4381.

Mueller, R. J., (2006). Biological degradation of synthetic polyesters—Enzymes as potential catalysts for polyester recycling. *Process. Biochem., 41*, 2124–2128. https://doi.org/10.1016/j.procbio.2006.05.018.

Nayomi, J., & Rukhiya, S., (2020). "Biodegradation of plastics by microbes: A review." *AIP Conference Proceedings, 2263*, 020003. https://doi.org/10.1063/5.0016828.

Pujic, P., Beaman, B. L., Ravalison, M., Boiron, P., & Rodríguez-Nava, V., (2015). Nocardia and actinomyces. In: Tang, Y., & Sails, A., (eds.), *Molecular Medical Microbiology* (pp. 731–752). Elsevier, Amsterdam. https://doi.org/10.1016/B978-0-12-397169-2.00040-8.

Rajamohan, N., & Karthikeyan, C., (2006). *Kinetic Studies of Dye Effluent Degradation by Pseudomonas stutzeri.* ECO Services International. https://eco-web.com/edi/061103.html (accessed on 01 April 2023).

Sánchez, C., (2020). Fungal potential for the degradation of petroleum-based polymers: An overview of macro-and microplastics biodegradation. *Biotechnol. Adv., 40*, 107501. https://doi.org/10.1016/j.biotechadv.2019.107501.

Santos, A. B., Cervantes, F. J., & Lier, J. B., (2007). Review paper on current technologies for decolourisation of textile waste waters: perspectives for anaerobic biotechnology. *Bioresour. Technol., 98*, 2369–2385.

Sarmah, P., & Rout, J., (2018). Algal colonization on polythene carry bags in a domestic solid waste dumping site of Silchar town in Assam. *Phykos, 48*, 67–77.

Sen, S. K., & Raut, S., (2015). Microbial degradation of low-density polyethylene (LDPE): A review. *J. Environ. Chem. Eng., 3*(1), 462473. https://doi.org/10.1016/j.jece.2015.01.003.

Shah, A. A., Hasan, F., Hameed, A., & Ahmed, S., (2008). Biological degradation of plastics: a comprehensive review. *Biotechnol. Adv., 26*(3), 246–265. https://doi.org/10.1016/j.biotechadv.2007.12.005.

Shahid, A., Singh, J., Bisht, S., Teotia, P., & Kumar, V., (2013). Biodegradation of textile dyes by fungi isolated from North Indian field soil. *Env. Asia, 6*(2), 51–57.

Suryavathi, V., Sharma, S., Sharma, S., Saxena, P., Pandey, S., Grover, R., Kumar, S., & Sharma, K. P., (2005). Acute toxicity of textile dye wastewaters (untreated and treated) of Sanganer on male reproductive systems of albino rats and mice. *Reproductive Technol., 19*, 547–556.

Tang, C. C., Chen, H. I., Brimblecombe, P., & Lee, C. L., (2018). Textural, surface, and chemical properties of polyvinyl chloride particles degraded in a simulated environment. *Mar. Pollut. Bull., 133*, 392–401. https://doi.org/10.1016/j. marpolbul.2018.05.062.

Teuten, E. L., Saquing, J. M., Knappe, D. R. U., Barlaz, M. A., Jonsson, S., & Björn, A., (2009). Transport and release of chemicals from plastics to the environment and to wildlife. *Philos. Trans. R Soc. Lond. Ser. B Biol. Sci., 364*(1526), 2027–2045.

Thiruchelvi, R., Das, A., & Sikdar, E., (2021). Bioplastics as better alternative to petro plastic. To be published in mater. *Today: Proc. (Preprint), 37*(2), 1634–1639. https://doi.org/10.1016/j.matpr.2020.07.176.

Thompson, R. C., Swan, S. H., Moore, C. J., & Vom, S. F. S., (2009). Our plastic age. *Philos. Trans. R. Soc. Lond. B Biol. Sci., 364*, 1973–1976. https://doi.org/10.1098/rstb.2009.0054.

Van, D. Z. F. P., & Villaverde, S., (2005). Combined anaerobic-aerobic treatment of azodyes—a short review of bioreactor studies. *Water Res., 39*, 1425–1440.

Yang, X. Q., Zhao, X. X., Liu, C. Y., Zheng, Y., & Qian, S. J., (2009). Decolorization of azo, triphenylmethane, and anthraquinone dyes by a newly isolated *Trametes* sp. SQ01 and its laccase. *Process Biochem., 4*, 1185–1189. http://biotechnology.sxu.edu.cn/docs/20100521094306428505.pdf (accessed Aug 01, 2021).

Zollinger, H., (1987). Synthesis, syntheses, properties and applications of organic dyes and pigments. In: *Color Chemistry* (2nd edn., pp. 2–100). Wiley VCH.

CHAPTER 5

Biochar for Remediation of Contaminated Soils

5.1 BIOCHAR

In soil, thousands of living organisms exist with dynamic interplay. The standardization of ecosystem functioning is maintained by them, yet anthropogenic disturbances, such as industrialization, urbanization, and improper agricultural practices, lead to soil pollution, finally resulting in contaminants contaminating the environment. Even though there is no single factor to blame for creating these challenges, there are several other factors that contribute directly and indirectly to environmental pollution. Therefore, a newly developed sustainable, cost-effective, and carbon-rich by-product called "biochar." It can be used in the bioremediation of soils contaminated by petroleum refinery wastes containing aliphatic, aromatic, and other complex hydrocarbons as well as heavy metals due to their long-term resistance to microbial degradation.

The use of biochar for the remediation of organic and inorganic pollutants is potentially cheaper because the waste source materials are relatively inexpensive and the production of biochar is relatively low cost (Hale et al., 2011). The addition of biochar to contaminated soil, however, has also been shown to increase the likelihood of pollutant breakdown by their microbial community, due to its large surface area and microspores (Rhodes et al., 2008). Several recent and past studies have examined the feasibility of using biochar in petroleum hydrocarbon-contaminated soil remediation (Beesley et al., 2011). This chapter describes how biochar application on contaminated soils containing both organic and inorganic pollutants could be a sustainable way to reduce the concentration of soil pollutants.

Bioremediation and Phytoremediation: Technologies for Toxic Pollution
Ashwini A. Waoo, PhD (Author)

In this context, biochar is defined as a carbon-rich byproduct of different biomasses, generated by thermal degradation of biomass at higher temperatures (450–600°F) and partial oxygen, including rice straw, grass, wood chips, crop residue, and sugarcane bagasse, which have remarkable nutritional benefits, as well as the ability to significantly remediate soil pollutant levels.

Adding biochar to soil can significantly improve soil properties by reducing methane emissions, enhancing soil pH and organic carbon, enhancing soil nutrients, and increasing methanotroph populations. (Milla et al., 2013). A high temperature is necessary to produce high-quality biochar, which differs from charcoal because it can be used in soil amendments for various purposes, such as fertility improvement, crop growth, pollution control, etc. (Gaunt and Johannes, 2008; Lehmann and Joseph, 2009).

Biochar is a porous carbonaceous material produced through thermochemical processes (e.g., pyrolysis, gasification) and hydrothermal processes (Aziz et al., 2020; Sajjad et al., 2020). As a soil amendment, biochar is increasingly being considered as it not only mitigates climate change by sequestering carbon from the atmosphere into the soil but also enhances soil properties by improving microbial activity, moisture retention, and nutrient retention, thereby increasing crop production.

The main mechanisms for removing organic and inorganic pollutants from the soil are sorption by the biochar and interactions between its active functional groups. Biochar can retain moisture in its pores and increase soil water holding capacity, thereby improving soil fertility (Kamali et al., 2020; Mohamed et al., 2016). Porous biochar is usually characterized by abundant surface functional groups. As a result of these properties, biochar is very effective at absorbing both organic and inorganic contaminants (Mirhosseinian et al., 2020).

Its highly porous structure lends a high surface area to biochar, which explains many of its benefits. A gram of biochar can have a surface area of over 2,000 sq ft. For comparison, the part of a tennis court that is used for singles matches measures 2,106 square feet.

Biochar is a carbon-rich scaffold that boosts soil vitality by:

- Enhancing water infiltration and retention.
- Preventing soil nutrient leaching.
- Increasing soil biodiversity by diversifying and cultivating helpful microorganisms.
- Modifying the acidity of the soil – for example, to increase the pH of highly acidic soils by removing unwanted aluminum ions.
- Adsorption of ammonium ions and reduction of fertilizer runoff.

5.1.1 BIOCHAR PRODUCTION

Enhancing soil fertility can also be achieved by modifying microbial life and nutrient content in the soil. One of the most promising technologies for generating biochar is pyrolysis, a method of heating biogenic waste at temperatures between 250 and 850°C in the absence of oxygen or with a limited amount of oxygen (Zhang et al., 2020; Ahmed et al., 2020). Pyrolysis also produces syngas and hydrocarbon-rich bio-oil. A pyrolysis reaction can be divided into three main categories: slow pyrolysis, fast pyrolysis, and flash pyrolysis. Among the main factors affecting pyrolysis classification and pyrolysis products distribution, biomass particle size, retention time, reaction temperature, and heating rate are critical (Salehi et al., 2020).

5.1.2 FACTORS AFFECTING BIOCHAR QUALITY

The pyrolysis temperature and plant material both affect the properties of biochar. Increasing the final temperature resulted in improved surface areas but reduced yields of biochar. Interestingly, the largest surface area at biochar was formed from the wood mixture and the smallest one was formed from hay from grass.

Biochars or their feedstocks can be modified to alter their physicochemical properties; the pyrolysis process can also change biochar structure and composition, thereby altering its adsorption capacity. Pyrolysis temperature affects the adsorption properties of biochar by modifying its surface area, pore distribution, and mineral concentration. Optimal pyrolysis temperatures are required to maximize inorganic and organic contaminant adsorption in biochar. The effect of pyrolysis temperature on the properties and adsorption capacity of biochar was studied. Their findings indicate that high-temperature BC has a larger surface area, and previously blocked pores are no longer blocked. The biochar surface consists of more aromatic groups and fewer oxygen-containing functional groups.

5.2 BIOCHAR FOR AGRO-ENVIRONMENTAL DEVELOPMENT

Due to its long-term persistence in soil, biochar can play an important role in agricultural and environmental sustainability as shown in Figure 5.1. There was a high amount of stable organic matter, with carbon contents of about 70–80% (Singh et al., 2017), and mineral matter, including nutrients, in larger quantities. Agricultural waste in form of pyrolyzed biomass has a high surface area, porosity, variable charge, and functional groups which increase soil water-holding capacity, pH, cation exchange capacity (CEC), base saturation, bioremediation of hydrocarbons and heavy metals, and crop resistance to disease.

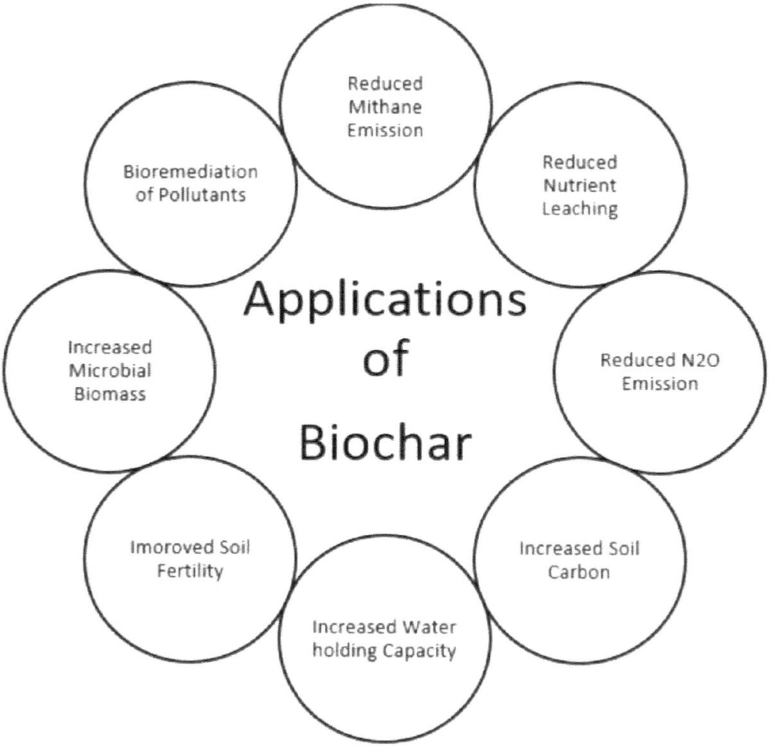

FIGURE 5.1 Biochar applications in agriculture.

Biochar offers several benefits, including carbon sequestration, soil fertility improvement, pollutant immobilization, and pollutant remediation. Further, growing biochar on soil can increase bacterial abundance that contributes to soil bioremediation, and decrease soil bulk density (Gundale and DeLuca, 2006).

There is a significant reduction of salt's effect on plants in agricultural, urban, and contaminated soils when biochar is applied (Thomas et al., 2013). A soil amended with biochar has increased its accessibility to P and Zn, as well as its total N content. Crop yields can be increased by incorporating biochar into the soil without using synthetic fertilizers (Blackwell et al., 2010). Additionally, biochar can minimize the effect of soil erosion and nutrient leaching due to the strongly binding pattern to soil particles.

5.3 BIOCHAR FOR CONTAMINATED SOILS

The use of biochar may be beneficial for both commercial and small-scale farming due to its ability to improve soil fertility and to adsorb toxic pollutants. The combined contamination of metalloids, organic pollutants, and metals is difficult to treat since each contamination involves different treatment methods. In contrast, biochar can remediate combined forms of pollutants (organic and inorganic); both contain numerous types of pollutants, with electrostatic interaction precipitating heavy metals, and adsorption, partition, and sequestration of organic compounds.

Compared to other composts, biochar was more effective at reducing heavy metals and highly toxic PAHs. As a carrier of immobilized microorganisms, biochar could enhance the bioremediation of PAHs in contaminated soil. In any case, selecting the appropriate category of biochar to enhance biodegradation, bioremediation, and organic degradation is imperative. In amended soil, biochar can reduce the concentration and mobility of arsenic (As) compared to unamended soil. A previous study showed that biochar derived from orchard residues significantly increased as concentrations in pore water (500–2000 g L^{-1}) (Beesley et al., 2013).

The land indeed has most ecosystems on this planet, but anthropogenic activities are affecting those ecosystems in various ways; therefore, environmental protection policies are needed, such as the proper treatment of wastewater from industries and the proper management of wastewater from households. In this regard, some biotechnological techniques can be used to manage environmental pollution, such as bioremediation and biodegradation, with the use of biochar as a carrier. Degraded and contaminated land requires cooperation, integration, and assimilation of different biotechnological advances (Mani and Kumar, 2014)

It was reported by Belyaeva and Haynes (2012) that biochar had little to no effect on the composition of soil microbial communities, soil microbial biomass, N fertility, or plant growth during revegetation. In industrial wastewater treatment, biochar could adsorb soil Cd and Cu but is not able to promote the growth of a wetland plant species, possibly because waterlogged plants do not interact well with biochar. A deeper understanding of these mechanisms is therefore needed.

5.4 BIOCHAR FOR HEAVY METAL REMOVAL FROM WASTEWATER

Many techniques for removing these heavy metals from wastewater exist, including adsorption, ion exchange, chemical precipitation, membranes, reverse osmosis, and electrochemistry (Demirbas, 2008). In the context of these technologies, adsorption is the most convenient and fast way to treat heavy metal-contaminated solutions.

In this treatment technology, there is little residue generation and the adsorbent can be recovered and recycled. Furthermore, biosorbents are not only eco-friendly but are readily available in sufficient quantities. and Biochar is a well-known biosorbent. Biochar is an organic material derived from the thermal degradation of organic compounds under oxygen-limited conditions. Biochar is also effective in absorbing all kinds of organic and inorganic pollutants from aqueous solutions. Studies have shown that biochar can be used as a biosorbent for the removal of heavy metals from contaminated water as well as immobilizing metal ions in metal-contaminated soil (Xu et al., 2013). Researchers have focused on biochar modification and characteristics in response to the growing need for clean water.

Biochars can be used to capture metals from aqueous solutions and have gained attention during the past decade. Despite this, studies are generally carried out at a lab scale, focusing on the sorption of single metals from spiked solutions. Several heavy metals coexist in natural waters; therefore, metals and other ions or organic pollutants compete for sorption sites on biochar surfaces. Biochar has not been extensively studied for its competitive sorption of metals. Recently, there is no report of the use of biochar in field-based wastewater treatment.

Contaminated water is more complex than the simulated water used in current studies. To test the suitability of biochar wastewater treatment, physicochemical conditions must be simulated or actual contaminated water should be used for studies. Furthermore, to support field application, future studies should consider factors related to metal removal efficiencies, such as pH, dose and recovery methods, and regeneration and disposal of metal-sorbed biochars.

5.5 SUMMARY

In highly challenging fields such as bioremediation, biochar can be used as a sorbent with high strength, reducing the bioavailability of organic and inorganic pollutants. Several studies have shown that the application of biochar to contaminated soils reduces the bioavailability of pollutants but may also increase the rate of soil diffusion.

Biochar created from different feedstocks could be used to remediate agricultural contaminated soils. It is still necessary to develop new emerging approaches in the field of bioremediation to prevent environmental risks since every year pollution levels are rising due to indiscriminate human activities such as industrialization, urbanization, improper agricultural waste management practices direct discharging of households, are some serious concerns to all living beings.

5.6 EXERCISES

Objective Response Questions

11. The process of biochar production is called as __________.

c) Pyrolysis
d) Paralysis
e) Pyrogenification
f) Biochar Synthesis

12. PAHs stands for ___________.

a) Polycyclic Aliphatic Hydrocarbons
b) Polycyclic Aromatic Hydrocarbons
c) Poly Aromatic Hydrocarbons
d) Polyhdroxides

13. During Biochar production biogenic waste at heated at temperatures ____________.

a) between 250 and 850°C
b) between 255 and 850°C
c) between 250 and 858°C
d) between 250 and 850°C

14. Biomass is complex and usually composed of natural biomacromolecules, which is not part of biochar _________.

a) Cellulose
b) Hemicellulose
c) Lignin
d) Biotin

15. How biochar reduces fertilizer runoff?

a) Adsorbing beneficial ammonium ions
b) Adsorbing beneficial aluminum ions
c) By absorbing Water
d) By degrading pollutant

16. Adding biochar to soil can significantly improve soil properties by _________.

a) Reducing methane emissions
b) Enhancing soil pH
c) Enhancing organic carbon
d) All the above

17. A pyrolysis reaction can be divided into three main categories except one _________.

a) Slow pyrolysis
b) Fast pyrolysis
c) Flash pyrolysis
d) Flow pyrolysis

18. Biochar offers several benefits, excluding this _________.

a) Carbon sequestration
b) Soil fertility improvement
c) Pollutant immobilization, and remediation
d) Increasing heavy metal concentration

19. This the techniques used for removing heavy metals from wastewater _________.

a) Reverse Osmosis
b) NMR

c) TLC
d) Microscopy

20. The biochar surface consists of __________.

a) More aromatic groups
b) Less aromatic groups
c) More aliphatic groups
d) None of the above

Practice Work

Qu 1. How does charcoal differ from Biochar?

Qu 2. What are the properties of Biochar?

Qu 3. Is biochar considered organic or inorganic as a soil amendment?

Qu 4. Explain applications of biochar in environmental bioremediation

Qu 5. Describe the process of biochar production.

Qu 6. Biochar can be made from what types of biomass?

Qu 7. Why is biochar beneficial to the soil?

Qu 8. What are the environmental benefits of biochar?

Qu 9. What is the role of biochar in heavy metal remediation?

Qu 10. What are the advantages and disadvantages of biochar?

REFERENCES

Ahmed, M. J., & Hameed, B. H., (2020). Insight into the co-pyrolysis of different blended feedstocks to biochar for the adsorption of organic and inorganic pollutants: A review. *J. Clean. Prod., 265,* 121762.

Aziz, S., Ali, M. I., Farooq, U., Jamal, A., Liu, F. J., He, H., Guo, H., et al., (2020). Enhanced bioremediation of diesel range hydrocarbons in soil using biochar made from organic wastes. *Environ. Monit. Assess., 192,* 1–14.

Beesley, L., Marmiroli, M., Pagano, L., Pigoni, V., Fellet, G., Fresno, T., Vamerali, T., et al., (2013). Biochar addition to an arsenic-contaminated soil increases arsenic concentrations in the pore water but reduces uptake to tomato plants (*Solanum lycopersicum* L.). *Sci. Total Environ., 455,* 598–603.

Beesley, L., Moreno-Jiménez, E., Gomez-Eyles, J. L., Harris, E., Robinson, B., & Sizmur, T. (2011). A review of biochars' potential role in the remediation, revegetation, and restoration of contaminated soils. *Environ. Pollut., 159,* 474–480.

Belyaeva, O. N., & Haynes, R. J., (2012). Comparison of the effects of conventional organic amendments and biochar on the chemical, physical and microbial properties of coal fly ash as a plant growth medium. *Environ. Earth Sci., 66*, 1987–1997.

Blackwell, P., Krull, E., Butler, G., Herbert, A., & Solaiman, Z., (2010). Effect of banded biochar on dryland wheat production and fertilizer use in south-western Australia: an agronomic and economic perspective. *Aust. J. Soil Res., 48*, 531–545.

Demirbas, A., (2008). Heavy metal adsorption onto agro-based waste materials: A review. *Journal of Hazardous Materials, 157*(2, 3), 220–229. doi: 10.1016/j.jhazmat.2008.01.024.

Gaunt, J. L., & Johannes, L., (2008). Energy balance and emissions associated with biochar sequestration and pyrolysis bioenergy production. *Environ. Sci. Technol., 42*(11), 4152–4158.

Gundale, M. J., & DeLuca, T. H., (2006). Temperature and source material influence ecological attributes of ponderosa pine and Douglas-fir charcoal. *Forest Ecol. Manag., 231*, 86–93.

Hale, S., Hanley, K., Lehmann, J., Zimmerman, A., & Cornelissen, G., (2011). Effects of chemical, biological, and physical aging as well as soil addition on the sorption of pyrene to activated carbon and biochar. *Environ. Sci. Technol, 45*, 10445–10453.

Kamali, N., Mehrabadi, A. R., Mirabi, M., & Zahed, M. A., (2020). Synthesis of vinasse-dolomite nanocomposite biochar via a novel developed functionalization method to recover phosphate as a potential fertilizer substitute. *Front. Environ. Sci. Eng., 14*, 70. https://doi.org/10.1007/s11783-020-1249-6.

Lehmann, J., & Joseph, S., (2009a). Biochar for environmental management: an introduction. In: *Book*: *Biochar for Environmental Management: Science and Technology*. ES-BEM-16, 1–9. https://www.css.cornell.edu/faculty/lehmann/publ/First%20proof%2013-01-09.pdf (accessed on 12 May 2023).

Lehmann, J., & Joseph, S., (2009b). Biochar systems, biochar for environmental management. In *Book: Biochar for Environmental Management: Science and Technology*. Earthscan, London. ES_BEM_13-1, 147–164. https://www.css.cornell.edu/faculty/lehmann/publ/First%20proof%2013-01-09.pdf (accessed on 12 May 2023).

Mani, D., & Kumar, C., (2014). Biotechnological advances in bioremediation of heavy metals contaminated ecosystems: an overview with special reference to phytoremediation. *Int. J. Environ. Sci. Technol., 11*, 843–872.

Milla, O. V., Rivera, E. B., Huang, W. J., Chien, C. C., & Wang, Y. M., (2013). Agronomic properties and characterization of rice husk and wood biochars and their effect on the growth of water spinach in a field test. *J. Soil Sci. Plant Nutr., 13*, 251–266.

Mirhosseinian, N. S., Anbia, M., & Salehi, S., (2020). Preparation and characterization of superhydrophobic melamine and melamine-derived carbon sponges modified with reduced graphene oxide–TiO_2 nanocomposite as oil absorbent materials. *J. Mater. Sci., 55*, 1536–1552.

Mohamed, B. A., Ellis, N., Kim, C. S., Bi, X., & Emam, A. E., (2016). Engineered biochar from microwave-assisted catalytic pyrolysis of switchgrass for increasing water-holding capacity and fertility of sandy soil. *Sci. Total Environ., 566*, 387–397.

Rhodes, A. H., Carlin, A., & Semple, K. T., (2008). Impact of black carbon in the extraction and mineralization of phenanthrene in soil. *Environ. Sci. Technol., 42*, 740–745.

Sajjad, A., Jabeen, F., Farid, M., Fatima, Q., Akbar, A., Ali, Q., Hussain, I., et al., (2020). Biochar: a sustainable product for remediation of contaminated soils. In: *Plant Ecophysiology and Adaptation Under Climate Change: Mechanisms and Perspectives II* (pp. 787–799). Springer. https://doi.org/10.1007/978-981-15-2172-0_30.

Salehi, S., Mandegarzad, S., & Anbia, M., (2020). Preparation and characterization of metal-organic framework-derived nanoporous carbons for highly efficient removal of vanadium from aqueous solution. *J. Alloys Compd., 812*, 15205.1. doi: 10.1016/j.jallcom.2019.152051.
Singh, C., Tiwari, S., & Singh, J. S., (2017). Impact of rice husk biochar on nitrogen mineralization and methanotrophs community dynamics in paddy soil. *Int. J. Pure App. Biosci., 5*, 428–435.
Sizmur, T., Fresno, T., Akgül, G., Frost, H., & Moreno, J. E., (2017). Biochar modification to enhance sorption of inorganics from water. *Bioresour. Technol., 246*, 34–47.
Thomas, S. C., Frye, S., Gale, N., Garmon, M., Launchbury, R., Machado, N., Melamed, S., et al., (2013). Biochar mitigates negative effects of salt additions on two herbaceous plant species. *J. Environ. Manag., 129*, 62–68.
Xu, X. Y., Cao, X. D., & Zhao, L., (2013). Removal of Cu, Zn, and Cd from aqueous solutions by the dairy manure-derived biochar. *Environmental Science and Pollution Research, 20*(1), 358–368. doi: 10.1007/s11356-012-0873-5.
Zhang, J., Jin, J., Wang, M., Naidu, R., Liu, Y., Man, Y. B., Liang, X., et al., (2020). Co-pyrolysis of sewage sludge and rice husk/bamboo sawdust for biochar with high aromaticity and low metal mobility. *Environ. Res., 191*, 110034.

CHAPTER 6

Cyanobacterial Bioremediation

6.1 CYANOBACTERIA: INTRODUCTION

Development countries face major challenges due to increased population, industrialization, and excessive wastewater generation. Pollutants/polluted water is discharged without any pre-treatment into freshwater resources, resulting in the degradation of natural water bodies. Due to increasing population, industrialization, and agricultural use, along with varying water availability in developing countries, the water demand is increasing continuously. Therefore, eco-friendly, and economic technologies for wastewater treatment are urgently needed, which require simple infrastructure and fewer inputs, while also being commercially feasible.

In phytoremediation, plants, or microorganisms (microalgae/cyanobacteria/bacteria/fungi) are used to remove contaminants (nutrients, organic compounds, heavy metals) from wastewater. It is promising to use photosynthetic microorganisms and microalgae as they play a significant role in the decontamination of natural waters. The microalgae use solar energy to provide oxygen for aerobic degradation, recycling nutrients responsible for eutrophication into biomass that may be useful for human consumption. Microorganisms like microalgae and anaerobic bacteria play an important role in the degradation of various halogenated xenobiotics in various environments to overcome the high persistence of these compounds.

The cyanobacteria are widely distributed and can survive in adverse conditions. The unique characteristics they possess further demonstrate their potential.

- They have simple growth requirements, so N and P rich wastewaters are good for use as cultivation media.

Bioremediation and Phytoremediation: Technologies for Toxic Pollution
Ashwini A. Waoo, PhD (Author)

- They are photoautotrophic organisms that fix atmospheric CO_2, for which they are responsible for mitigating greenhouse gas emissions.
- They serve dual roles of bioremediation of wastewater and generating biomass for various applications viz. feed, pharmaceutical, nutraceutical, industries biofuel, and biofertilizers.

Among the key actors in bioremediation, cyanobacteria have the following advantages over other organisms. Some adapted species are resistant to desiccation, can fix nitrogen easily, and can switch quickly to mixotrophic nutrition, among other features.

The cyanobacteria, which are mixotrophs, can metabolize a wide range of organic compounds, including persistent organic pollutants, breaking them down into less toxic or non-toxic compounds and using them as nutrients. The ability of cyanobacteria to use ammonium ions as a nitrogen source is vital for agricultural and domestic wastewater treatment. In bioremediation, the ability of cyanobacteria to adsorb microelements by activating their membrane transport system is essential.

Metals are removed from polluted environments via these mechanisms, resulting in active bioaccumulation. Metal ions are transformed into less toxic forms through several mechanisms, such as nanoparticle formation, chelation, and inclusion in organic compounds.

Furthermore, the accumulation of heavy metals by cyanobacteria from polluted water, in addition to having positive environmental impacts, could provide an opportunity to obtain metal nanoparticles both on the cell surface and inside the cell. Moreover, it is possible to obtain biomass with metals (Fe, Zn, Cu, Cr, etc.) included in some fractions –suitable raw material to produce some preparations for metal deficiency treatment.

Metal ions are transformed into less toxic forms through several mechanisms, such as nanoparticle formation, chelation, and inclusion in organic. A potential role for cyanobacteria may be in the reclamation of salt-affected (generally known as Usar in some parts of India), arid or semi-arid soils. It is not cost-effective or environmentally friendly to use chemical methods of using gypsum, sulfur, or excessive irrigation for the amelioration of salt-affected lands. Due to excessive salts in the upper layers, salt-affected soils (alfisol/sodic/alkaline/saline) are rigid, less productive soils. The salt content determines whether they are alkaline or saline. In alkaline soils, pH is high, exchangeable Na is high, carbonates are measurable, and clay is extensively dispersed (due to a high zeta potential of active Na^+).

Cyanobacteria possess certain physiological advantages that can help them survive such stresses:

a) limiting the inflow of Na^{+}.
b) uptake of inorganic (K^{+} ions) or organic osmoregulatory compounds (sugars, quaternary amines, etc.

6.2 REMOVAL OF ORGANIC POLLUTANTS BY CYANOBACTERIA

Globally, wastewater contributes to environmental pollution. Through the sanitary sewer system, domestic wastewater, agricultural wastes, and industrial effluents are spilled into open water. Domestic wastewater is often contaminated with paper, urine, feces, and synthetic detergents. According to the type of industry, there are different inorganic and organic pollutants in industrial effluents. The effluents of agricultural operations include animal waste, fertilizer, and various pesticides. Natural waters can be polluted by untreated sewage on purpose or by accident. As a result, domestic wastes can pollute different ecosystems through sewage leaks and blowouts, while oil is spilled mainly due to wreckage. Natural basins are often contaminated by pesticides due to the washing out of soil layers on top or precipitation resulting from spraying.

Wastewater that contains organic suspensions or dissolved organic substances negatively impacts the water environment. Particles suspended in water prevent light from penetrating deep into the water and inhibit photosynthesis. Sediments form on the bottom of the ocean, inhibiting microbiomes and enabling water to self-clean. Organic pollutants can form a favorable environment for pathogenic organisms when they decompose in water ecosystems. Organisms such as fungi, bacteria, yeast, and algae metabolize organic compounds using catabolic processes. These pathways help decompose organic compounds and bioorganic compounds. So, the anabolic process of algae doesn't help in the decomposition of bioorganic compounds. Decomposition of organic sediments can lead to compounds that are harmful or poisonous, and the amount of dissolved oxygen in water drops drastically. Fats, oils, and lubricants form tension-active reagent films on water surfaces, which inhibit gas exchange between water and atmosphere and reduce the oxygen content in water.

Organic wastewater treatment is a necessary process, which includes various biological approaches. Aerobic and anaerobic bacteria play a major role in this process. Also, cyanobacteria have a positive and negative influence on water quality. In some cases, cyanobacteria clean the water by synthesizing surfactants, substances that precipitate suspended particles, including organic ones. Some species produce antibiotics that have herbicidal properties, inhibiting the development of other species and preventing algae blooms. In addition, cyanobacteria absorb biogenic elements from effluents and inhibit bacterial growth, including pathogenic bacteria.

6.3 PURIFICATION OF WASTEWATER WITH CYANOBACTERIA

In the last few years, cyanobacterial cultures have been studied extensively as a means of obtaining biologically active substances (Gluck, 2009). Several biogenic elements are present in wastewaters that may lead to an increase in cyanobacteria growth. Therefore, they can be used to cultivate cyanobacteria (Larsdotter, 2006). Among the algae in the biological river basins of the purification systems, firstly predominate the cyanophytes, which makes these waters suitable for the cultivation of microorganisms of this type (Gronlund, 2004).

Under laboratory conditions, cyanobacteria cultivation experiments were conducted to determine their role in the biological purification process, the experiments were carried out for 20 days in a temperature range between 25 and 33°C with daylight and daily stirring. Throughout 5, 10, 15, and 20 days, observations were made. Control samples for studying the peculiarities of species growth are the mineral media specific to each species: *Synechocystis salina* species were cultured on Bristol medium, while *Phormidium foveolarum* species were cultured on Gromov 6 medium.

The effect of *Synechocystis salina* and *Phormidium foveolarum* on wastewater purification is explained by their overall reduction in ammonium, nitrate, nitrite, and phosphate ions in the first 10–15 days from the inoculation on the medium. These results concluded that the studied ions were accumulated by cyanobacteria because the amount of algal biomass had increased in this period since they were not detected in sediments.

Several organic pollutants are potential targets for degradation by cyanobacteria. Chemicals that are used in industry and agriculture include phenols, pesticides, polycyclic aromatic hydrocarbons (PAHs), and polychlorinated biphenyls (PCBs) (Suresh et al., 2013). Color does not affect microalgal degradation rates; microbial density, diversity, and association are much more important for degradation rate by algae.

The most extensively studied from this point of view are the cyanobacterial genera Anabaena and Nostoc, which include soil species that grow well in the aquatic environment. From this perspective, *Anabaena* and *Nostoc* have been the most extensively studied cyanobacterial genera, which include soil species that thrive in the aquatic environment.

An important group of environmental pollutants, nitrophenols, are widely distributed in the environment. These compounds are widely used in insecticides, pesticides, dyes, plastics, and explosives. Anabaena variabilis removes 100% of a-nitrophenols and p-nitrophenols (PNP), 51% of 4,6-trinitrophenol, and 23% of bisphenols. *Cyanobacteria Anabaena* sp. can degrade highly chlorinated aliphatic pesticides, lindanes (g-hexachlorocyclohexane). Pesticides that contain organophosphorus compounds are toxic and applied to crops for crop protection. In consequence, they have been detected in the water and air.

The researchers are constantly working on the development of methods and techniques of biodegradation of these compounds. Several studies have demonstrated that *Anabaena* sp. can grow under aerobic, photosynthetic conditions. transforms methyl parathion first to o,o-Dimethyl o-p-Nitrosophenyl thiophosphate and then to o,o- Dimethyl o-p-aminophenyl thiophosphate by reducing the nitro group. The cyanobacteria are toxic to methyl parathion in the dark, however, it does not affect their viability to light (Barton et al., 2004).

6.4 DETOXIFICATION BY CYANOBACTERIA

Due to their ubiquitous nature, existing cost-effective technologies for their cultivation, and high absorption capacity, cyanobacterial biomass is one of the most efficient bio accumulators (Opeolu et al., 2010).

Cyanobacteria act as biosorbents and are considered complex biological objects. Depending on the type of microorganism and the conditions of growth, this ability is affected by the composition of their cell walls, which act as a structure to immobilize metals. In addition to their omnipresence, cyanobacteria exhibit a highly flexible metabolism that makes them ideal biosorbents. Although cyanobacteria are highly diverse, only a few species are used in biotechnology. Following is a discussion of two genera – *Arthrospira* and *Nostoc*. *Arthrospira*, also known as *Spirulina*, is a well-studied organism.

There are already several technologies for its cultivation and extraction of various preparations. The bioaccumulation ability of *spirulina* helps it become a source of biomass that contains metals. The leftovers of various types of biomass processing can also act as biosorbents. *Nostoc* genus representatives are among the least demanding of all biological objects since they fix nitrogen as well as provide photosynthesis.

6.4.1 ARTHROSPIRA (SPIRULINA) AS DETOXIFYING AGENT

The biosorption capacity of *Spirulina platensis* largely depends on the technology used. Fixed biomass technologies are the most effective. Cells were immobilized by layering polyelectrolyte (PEM) on a silica and sio2 surface. According to the study of cyanobacteria biofilms, adsorption of cadmium and mercury induces cell damage at a concentration of 10–6 M (Tekaya et al., 2013).

The study focused on the biosorption of copper and nickel by *Arthrospira platensis* biomass with different biochemical compositions. Both metals were found to biosorb very rapidly. The biomass filled most of its metal sorption capacity within 15–30 minutes, and equilibrium was reached within 30–60 minutes.

Copper and nickel biosorption kinetics were not affected by cultivation conditions. The capacity for copper ions biosorption was significantly positively influenced by the accumulation of carbohydrates in the dry biomass. Living biomass has a higher biosorption capacity than dry biomass in both metals. According to Markou et al., (2015), copper and nickel biosorption on *Arthrospira platensis* biomass primarily occurs as a result of ion exchange and complexation rather than physical adsorption.

As a result, *Arthrospira (Spirulina)* biomass can absorb metals from dilute solutions and different wastewaters. Biosorption can be applied to fresh, dry, or lyophilized biomass, as well as to biomass residues. It is crucial to note that *Arthrospira (Spirulina)* biomass is extremely valuable. After processing, the remaining *Arthrospira (Spirulina)* biomass is more valuable, especially because of its high biosorption capacity.

6.4.2 NOSTOC AS DETOXIFYING AGENT

Cyanobacteria fixed on some carriers considerably facilitate the bioremediation and serve a recyclable sorbent. As a result, several types of sorbents for chromatography columns were prepared based on Nostoc: Cyanobacteria can act as a sorbent and a catalyst in the bioremediation process. Biofilms of cyanobacteria (Methylformamide Nostoc slurry) on polymer-modified silica gel, biofilms of cyanobacteria on bare silica gel, and onto polymer-modified silica gel. The multi-metal probe detected 100% affinity for Cd^{2+} ions across all sorbents.

An isolate from a mining site was used to remove Cu (II), Zn (II), Pb (II), and Cd (II) from aqueous solutions containing a mixture of these metals. Maximum removal of 96% Pb (II) and 97% Cu (II) was observed, followed by 80% Cd (II) and 71% Zn (II). At the end of the 60-hour cultivation period, the data were collected. By passive biosorption and metal accumulation in the biomass during Nostoc culture, these metals have been removed. The accumulation of zinc and lead inhibits copper and cadmium oxidation (Roy et al., 2015).

Nutrient supplementation technology was used to remove arsenic from the aquatic environment. This enhances the ability of *Nostoc* sp. to accumulate this metal. So, when arsenic-contaminated waters are supplemented with nutritive elements (P, Co, Fe, Mo, and N) at optimized concentrations, the amount of free arsenic decreases significantly (Maeda et al., 1987).

Only a few studies have been conducted in the area of cyanobacteria *Nostoc sp*'s application to removing heavy metals from the environment. However, the obtained results confirm the efficiency of studying strains for reducing heavy metals content in aquatic and terrestrial environments.

6.5 METAL BIOSORPTION BY CYANOBACTERIA

The process of biosorption is dependent on factors such as the concentration of metal ions in the solution, the concentration of biomass, the biosorption capacity, the temperature, the incubation time, and the pH of the solution. The parameters vary from one metal to another and from one cyanobacteria strain to another. Biosorption of heavy metal ions by cyanobacteria is a two-step process.

During the first 'rapid' phase, metal ions are mainly adsorbing onto cyanobacteria's surface by functional amino, carboxylic, sulfhydryl, phosphate, and thiol groups. In the second "slow" phase, metal ions are reduced to metallic forms by biomolecules on the microbial surface or are partially transported through the microbial membrane (Chojnacka et al., 2005).

A review of different cyanobacterial strains' ability to biosorb various metal ions (iron, zinc, selenium, chromium, and copper) is presented here.

- ***Biosorption of Iron***

Few studies have examined the removal of iron from wastewater. Zinicovscaia et al., (2015) conducted experiments on a pure culture of *Spirulina platensis* to examine metal removal from wastewater from a machinery company. Results from neutron activation analysis identified zinc as well as iron as the main adsorbents by *Spirulina platensis* biomass. Within the first five minutes following contact, iron was absorbed in its highest concentration in the biomass, after which it gradually decreased. Within 30 minutes, *Spirulina platensis* accumulated approximately 60 times more iron than native biomass. During a bioaccumulation experiment in which the biomass was in contact with wastewater for 3 days, the biomass accumulated 3850 times more iron than its concentration in the control biomass. In large-scale applications, bioaccumulation is very costly because it requires the maintenance of specific living conditions for microorganisms.

Nostoc linckia was tested for metal uptake by chemically complex wastewater effluents and found that after 30 min of wastewater contact with biomass, the iron content of the biomass increased 60-fold.

- ***Biosorption of Zinc***

There are several studies on the use of spirulina biomass for zinc removal from model solutions and wastewater. One such study, by Ferreira et al., (2011), focused on the sorption capacity of dried Spirulina platensis in the removal of zinc ions from single metal aqueous solutions at increasing concentrations (0.5–3.0 mM) at the contact time of 120 min. Obtained results showed that zinc ion sorption on dried *Spirulina platensis* biomass was influenced by the metal concentrations in the solution, and at increasing concentration values, the removal efficiency decreased.

In addition to working with model solutions, Zinicovscaia and co-workers (2015) explored the removal of zinc from industrial wastewater by *Spirulina platensis*. *Spirulina platensis* adsorbs approximately 30% of zinc from wastewater in 1 hour. The lower percentage of zinc removed in comparison with the study on zinc removal from model solutions can be explained by the complex chemical composition of wastewater and the competition between metals for binding sites.

- ***Biosorption of Copper***

Researchers have conducted several studies using spirulina strains to remove metals from solutions. The adsorptions of Cu (II) on native, non-living, and chemically modified cells of the cyanobacterium *Spirulina platensis* were studied by Fang et al., (2011). A native cell's binding capacity was greater than that of a non-living cell. Due to the formation of new carbonyl ligands, the sorption capacity of chemically modified biomass was slightly higher than that of non-living biomass. *Spirulina platensis* has carboxyl groups on its cell wall that are the main site for copper ion binding.

The biosorption of copper and chromium by four types of *Spirulina* spp. Chojnacka et al., (2005) researched this topic. Different initial concentrations of the sorbent and constant sorbent concentrations were used in the biosorption experiments. After approximately 10 minutes, equilibrium was reached for all types. Copper and zinc accumulation in *Spirulina platensis* biomass was studied as a function of pH and metal concentration in solution, both under free (living cells, lyophilized cells, and oven-dried cells) and immobilized conditions. For both free and immobilized biomass, maximum metal biosorption was observed at a pH of 6.0.

Across all investigated cases, metal biosorption was dependent on metal concentration and increased with increasing metal concentrations. The maximum metal uptake was observed in living cells, followed by lyophilized and oven-dried biomass of *Spirulina* (Vannela and Verma, 2006)

6.6 ENGINEERING CYANOBACTERIA FOR BIOREMEDIATION

In the case of heavy metal contaminated tailings and soils, bioremediation is still the most cost-effective treatment, even though some heavy metals cannot be bioaccumulated by microorganisms. The toxicity of these compounds may have prevented microorganisms from evolving appropriate bioaccumulation pathways; populations of microorganisms responsible for this bioaccumulation are too small or inactive to completely remove compounds, or a mixture of pollutants is too complex for existing pathways to handle.

The addition of exogenous microorganisms to indigenous populations may improve their ability to remove pollutants. The process, commonly known as bioaugmentation, involves adding microorganisms that naturally contain catabolic genes or those that are genetically modified (GMOs). Furthermore, this strategy can result in the transfer of plasmids containing the necessary genetic material between the different populations.

Microorganisms have been subjected to molecular biology advances to create new strains with desirable bioremediation properties. A few of these include the construction or adaptation of catabolic pathways; redirecting carbon flux to prevent harmful intermediates from being produced; improving the genetic stability of catabolic activities; increasing the bioavailability of pollutants; and improving monitoring, yield, control, and efficiency of processes.

GMOs have many advantages when it comes to bioremediation, but their use is still limited because of the instability of the inserted genetic material.

There are two main reasons for this: first, GMOs only function well if they can stably carry the genetic material; and second, genetic transfer to native species is often perceived as a negative attribute, even though it happens often among native species. Due to these factors, there has been increasing interest in examining how GMOs survive in the environment, compete, and persist and the potential risks involved in their use.

There is a potential for *Cyanobacteria* to produce genes that can express high amounts of stress-resistant proteins. Abiotic stress resistance crops can be produced by genetically engineering the target plants to carry stress resistance genes (Jyotsna et al., 2014).

As a model organism for photosynthesis and circadian rhythm, *cyanobacteria* have long been studied, however, the engineering of cyanobacteria for applied purposes remains underdeveloped. Engineered cyanobacteria were previously considered for applications related to their roles as aquatic, photosynthetic microorganisms that naturally thrive in a wide range of environments, including use as an enhanced food source in the aquaculture industry. Recently, there has been renewed interest in the use of engineered cyanobacteria.

To manage pollution caused by humans, such as organic compounds and heavy metals, bioremediation tools are needed. A cyanobacterium is an excellent candidate for bioremediation of lakes, rivers, oceans, and wastewater due to its ability to grow in aquatic environments. Particularly, filamentous diazotrophs *Anabaena* sp. PCC 7120, which has been engineered to degrade organic pollutants lindane and 4-chlorobenzoate as well as remove heavy metals such as cadmium and zinc, is an ideal host for this process. The use of cadmium-inducible promoters and expression of mutant heavy metal-binding proteins with enhanced binding capacity has been used in improving the removal of cadmium from wastewater. Cyanobacteria are ideal hosts for biosensing and bioremediation in aquatic environments (Guo et al., 1999; Shao et al., 2002).

Due to the accumulation of toxic metals in the food chain and their persistence in the environment, environmental contamination by toxic metals is a problem worldwide. For removing heavy metals at low concentrations (below 50 mg/l), conventional technologies, like ion-exchange or lime precipitation, are often inefficient and expensive. Most of these techniques rely on a physical displacement or chemical replacement, which produces toxic sludge, whose disposal further burdens the technology's economic viability. Thus, new technologies are needed to meet the environmental standards at an affordable cost.

6.7 SUMMARY

Cyanobacteria are autotrophic, requiring only low levels of nutrients, and they produce large amounts of biomass in comparison to other microbial. A high sorption capacity has also been shown with these algal strains for removing heavy metals. Adsorption of heavy metals or integration into cells is used to remediate heavy metal-polluted effluent with algae biomass. The use of cyanobacteria and algae in phytoremediation provides either removal of contaminants or degradation of the pollutants that they contain. Selecting appropriate target species is therefore the first critical step in implementing an algal bioremediation program. If algal bioremediation is to be used at multiple locations, then the second critical step is to consider whether a single strain of the target species is used across all locations or local strains are isolated and used at each location.

In recent years, industrial effluents have released toxic heavy metals into the environment and public health, which has made the field of biotechnology in the environment possible. By adsorption onto the surface of dead biomass, pollutants can be concentrated or disintegrated into less harmful compounds. It should be recognized that microalgae, rightly referred to as a "wonder organism," are capable of bioremediation. Algal blooms can grow in polluted water and absorb a variety of pollutants. Following harvesting and lipid/protein extraction, algal biomass could be used as an efficient biosorbent. An integrated approach based on bioremediation parameters such as growth conditions, cellular structure, pretreatment of biomass, etc. can result in the development of low-cost microalgal-based biomass with high bioremediation potential to neutralize and adsorb heavy metal ions. Thus, this chapter discusses the various avenues for bioremediation using cyanobacteria or algae.

6.8 EXERCISES

Objective Response Questions

1. **The anabolic processes of algae contribute to the decomposition of bioorganic compounds.**

   a) True
   b) False

2. Microalgae can overcome the high persistence of halogenated xenobiotics.

a) True
b) False

3. What factor does not affect the rate of degradation by microalgae?

a) Association
b) Microbial diversity
c) Microbial density
d) Color

4. Which of the following statements is true regarding the regulation of metabolic pathways?

a) Most of the metabolic pathways are regulated
b) Most of the metabolic pathways are not regulated
c) Regulation of metabolic pathways always involves changing the number of enzymes
d) Metabolic regulation always depends on control by hormones

5. In this bioremediation technique, contaminated water and soil, fertilizers, and carbon dioxide are mixed in a bioreactor to stimulate biological degradation

a) *In situ* hybridization
b) Biopile treatment
c) Composting
d) Slurry-phase bioremediation

6. A pollutant that degrades readily can be degraded by algae through direct metabolism.

a) True
b) False

7. Which of the following organisms cannot be used to degrade very easily degradable pollutants during algal biomass harvesting and drying?

a) Bacteria
b) Fungi

c) Algae
d) Dioxins

8. Which of the following is the best method to degrade xenobiotic compounds?

a) Microbial degradation
b) Chemical degradation
c) Water degradation
d) Physical agents

9. Which of the following microbes are known to degrade xenobiotic compounds?

a) Planktonic algae
b) Cyanobacteria
c) Fungi
d) Goldfish

10. Tolerant cyanobacteria strain has higher esterase activity.

a) True
b) False

Practice Work

Qu 1. Explain various distinguishing characteristics of cyanobacteria.

Qu 2. Describe the mechanism of removal of organic pollutants by cyanobacteria.

Qu 3. Describe the valuable role of cyanobacteria in the remediation of heavy metals.

Qu 4. How will you purify waste water using cyanobacteria?

Qu 5. Explain detoxification of xenobiotics by cyanobacterial strains.

Qu 6. Describe the process of biosorption of metals by cyanobacteria.

Qu 7. What type of engineering can be done to enhance cyanobacterial bioremediation?

Qu 8. Why cyanobacteria are called "Wonder Organisms"?

Qu 9. What are the physiological advantages associated with cyanobacteria which enable them to withstand these stresses?

Qu 10. Explain the applications of genetically engineered cyanobacteria.

REFERENCES

Barton, J. W., Kuritz, T., O'Connor, L. E., Ma, C. Y., Maskarinec, M. P., & Davison, B. H., (2004). Reductive transformation of methyl parathion by the cyanobacterium *Anabaena* sp. strain PCC7120. *Appl. Microbiol. Biotechnol., 65*, 330–335.

Chojnacka, K., Chojnacka, A., & Gorecka, H., (2005). Biosorption of Cr^{3+}, Cd^{2+} and Cu^{2+} ions by blue-green algae *Spirulina* sp.: Kinetics, equilibrium and the mechanism of the process. *Chemosphere, 59*(1), 78–84.

Fang, L., Zhou, C., Cai, P., Chen, W., Rong, X., Dai, K., Liang, W., Gu, J. D., & Huang, Q., (2011). Binding characteristics of copper and cadmium by cyanobacterium *Spirulina platensis*. *J. Hazard Mater., 190*, 810–815.

Gluck, S., (2009). Producing fuels and chemicals from algae and the intersection of the water industry. In: *Proceedings of the Water Environment Federation, WEFTEC 2009* (Vol. 1, No. 7, pp. 122, 7916–7923).

Guo, X. X., Shi, D. J., Xu, X. D., Ouyang, Y. X., & Ru, B. G., (1999). Metal-induced expressing of mammal metallothionein-1 gene in cyanobacteria to promote cadmium-binding preferences. *Appl. Microbiol. Biotechnol., 52*, 806–810. doi: 10.1007/s002530051596.

Jyotsna, C. H., Prakash, J. S. S., Attipalli, R., & Kirti, P. B., (2014). Functional analysis of cyanobacterial groEL2 in tobacco plants for abiotic stress tolerance. *2nd International Conference on Agricultural & Horticultural Sciences Radisson Blu Plaza Hotel.* Hyderabad. India. https://www.walshmedicalmedia.com/proceedings/functional-analysis-of-cyanobacterial-groel2-in-tobacco-plants-for-abiotic-stress-tolerance-15026.html.

Larsdotter, K., (2006). Wastewater treatment with microalgae: A literature review. *Vattan, 62*, 31–38.

Markou, G., Mitrogiannis, D., Çelekli, A., Bozkurt, H., Georgakakis, D., & Chrysikopoulos, C. V., (2015). Biosorption of Cu^{2+} and Ni^{2+} by *Arthrospira* platensis with different biochemical compositions. *Chem. Eng., 259*, 806–813. doi: 10.1016/j.cej.2014.08.037.

Opeolu, B., Bamgbose, O., Arowolo, A. T., & Adetunji, M. T., (2010). Utilization of biomaterials as adsorbents for heavy metals removal from aqueous matrices. *Sci. Res. Essays, 5*, 1780–1787.

Roy, A. S., Hazarika, J., Manikandan, N. A., Pakshirajan, K., & Syiem, M. B., (2015). Heavy metal removal from multicomponent system by the cyanobacterium *Nostoc muscorum*: kinetics and interaction study. *Applied Biochemistry and Biotechnology, 175*(8), 3863–3874. doi: 10.1007/s12010-015-1553-y. PMID: 25725800.

Shao, Q., Shi, D. J., Hao, F. Y., Ma, L. N., Chen, Z. J., & Yu, M. M., (2002). Cloning and expression of metallothionein mutant alpha-KKS-alpha in *Anabaena* sp. PCC 7120. Mar. *Pollut. Bull., 45*, 163–167, PMID:12398381, doi: 10.1016/S0025-326X(02)00105-4.

Shigeru, M., Kosuke, K., Mitsuru, M., Shiro, H., & Toshio, T., (1987). Bioaccumulation of arsenic by freshwater algae (*Nostoc* sp.) and the application to the removal of inorganic arsenic from an aqueous phase. *Appl. Organomet. Chem., 1*, 363–370. https://doi.org/10.1002/aoc.590010411.

Suresh, R. S., Ramakrishnan, B., Megharaj, M., Venkateswarlu, K., & Naidu, R., (2013). Mixotrophic cyanobacteria and microalgae as distinctive biological agents for organic pollutant degradation. *Environ. Int.*, *51*, 59–72.
Tekaya, N., Gammoudi, I., Braiekm, M., Tarbague, H., Moroté, F., Raimbault, V., Sakly, N., et al., (2013). Acoustic, electrochemical and microscopic characterization of interaction of *Arthrospira* platensis biofilm and heavy metal ions. *J. Environ. Chem. Eng.*, *1*, 609–619.
Vannela, R., & Verma, S. K., (2006). $Co^{2+,}$ Cu^{2+}, and Zn^{2+} accumulation by cyanobacterium *Spirulina platensis*. *Biotechnol. Prog.*, *22*, 1282–1293.
Zinicovscaia, I., Duca, G., Cepoi, L., Chiriac, T., Rudi, L., Mitina, T., Frontasyeva, M. V., et al., (2015). Biotechnology of metal removal from industrial wastewater: A zinc case study. *Clean Soil Air Water*, *42*, 1–6.

CHAPTER 7

Bioremediation of Mine Waste

7.1 MINE POLLUTION

Mining activities have been driven by urbanization and population growth, as well as an increase in energy and mineral demands. The mining industry aims to meet the needs for energy and mineral resources, improve infrastructural development, and improve the quality of life for the population. In addition to enhancing human civilization and national economies, mining and mineral extraction have caused serious environmental degradation. The various wastes, sludge, and water produced during mineral extraction have significant environmental implications Mining produces vast quantities of waste streams that pose severe pollution challenges to the environment. Most conventional treatments are inefficient and unsustainable. Using microorganisms for the treatment of mine wastes and recovery of heavy metals is an alternative approach. Mining waste is detoxified, extracted, or sequestered using microorganisms. Mine wastes are controlled and treated by sulfate-reducing microorganisms, which generate alkalinity and neutralize acidic wastes. Creating engineered sulfate-reducing bacteria (SRB) consortia is an effective way to process acid mine waste.

Mining and mineral processing produce large quantities of toxic, corrosive, materials. Whenever there is a mining site, nearby waste ponds and tailing ponds, along with chemical processing practices surround it. Mining waste has no economic value. Ash, gangue, industrial minerals, flue dust, loose sediment, metallurgical slags, ore, metals, particulate emissions, and roasted ore or rock all fall into this category. At least one ton of mining waste is generated for every ton of metal ore extracted. To reduce, recover, dispose, and protect the environment from these waste streams, informed planning and decision-making are essential (Hudson-Edwards and al., 2011). When

Bioremediation and Phytoremediation: Technologies for Toxic Pollution
Ashwini A. Waoo, PhD (Author)

released into the environment, these wastes can significantly impact surface water, underground water, air, and land resources.

Bioremediation uses plants and microorganisms to decontaminate polluted sites and restore them to their former appearance. Recent advances in microbiology and biotechnology have improved bioremediation techniques. Using biotechnological and microbiological techniques, this study reviews and discusses heavy metal bioremediation from mine sites contaminated with heavy metals. Future study strategies are also discussed.

7.2 BIOREMEDIATION FOR TREATMENT OF ACID MINE DRAINAGE WATER

Mine sites with sulfur-bearing rocks experience acid mine drainage (AMD). It can seriously damage plants, animals, and the flora in a region if it is left to accumulate in an area as acid mine drainage. Anthropogenic activities like mining, which expose sulfur-bearing rocks, are the primary cause of mine water drainage. As a result of this drainage, groundwater, surface water, and soil are all polluted. In the past, many remediation measures have been implemented, but none have been able to solve the issue completely. Phytoremediation and nanoremediation are the new and better methods being developed to treat acid mine drainage water. Additionally, phytoremediation is considered a better option than nanoremediation because it is more cost-effective and environmentally friendly.

7.3 ACID MINE DRAINAGE (AMD)

Acid mine drainage (AMD) is a phenomenon characterized by pollution of the environment especially of water bodies caused by water flowing from sites containing sulfur-bearing rocks. It is a major concern these days. AMD can occur naturally, but anthropogenic activities such as mining and processing metal ores and on a large scale, coal can contribute significantly to its generation. Sulfide minerals are exposed to the environment during the mining process, resulting in excessive amounts of acid being produced, which can have immediate and long-term effects on the environment. There has been a continued threat to the environment from acid mine drainage. The metal mining industry must deal with this issue as it has a major environmental concern.

AMD can negatively impact aquatic ecosystems and equipment, corrosion of mining equipment, degrade soil quality, and contaminate groundwater due to the leaching of heavy metals in acid mine water. The focus has been on the serious environmental and health risks posed by acid mine drainage water. There has also been an effort to highlight new emerging technologies, such as phytoremediation and nanoremediation, and their application for the proper treatment of said pollutants.

7.3.1 FORMATION OF AMD

As iron pyrite or Iron disulfide is exposed to air or water, they naturally oxidize, producing acid mines drainage. Besides natural processes, the earth's surface is exposed and acidic drainage is generated in mining. In the reaction between metal sulfides and oxygen and water, metal sulfates are formed, as well as sulfuric acid. Further oxidation of the metals can increase acidity. Mines receive fresh water from rainwater or water used for dust control, drilling, and other mining operations. Groundwater can seep into underground mines through cracks and fissures.

Aqueous environments can be used to transport oxidized sulfide mineral products into nearby rivers and other water bodies, from which they are carried away further. When water and oxygen react with pyritic sulfur, sulfuric acid and iron sulfate are formed. An acidic environment such as that created by coal mines promotes the growth and activity of acidophilic bacteria like *Thiobacillus ferroxidans*. Bacteria catalyze acid production, making it much faster than chemical oxidation, which is the cause of the acidic water (Singh and Bhatnagar, 1985). As a result of the sulfuric acid formation and the hydrolysis of pyrite products, mine drainage water is acidic. Alternatively, sulfur content can be used to estimate the amount of reactive pyrite present in a particular stream (Singh, 1987).

7.3.2 ADVERSE EFFECTS OF AMD ON ENVIRONMENT

Trace (heavy) metals may be present in acid mine drainage water. The drainage water can dissolve heavy metals found in the surrounding environment at low pH levels. A heavy metal's solubility is generally determined by its reaction with iron sulfate and sulfuric acid. The drainage water is acidic and contains toxic heavy metals, as a result, it is unfit for consumption. In addition to the high amount of hardness, sedimentation of ore particles and bacterial contamination make the water unfit for drinking. Water contaminated by this type of substance can also result in a wide range of diseases in both humans and animals. (Carlson et al., 2002).

If the drainage water enters main streams, such as rivers, it can contaminate not only the local area but even further away. Surface water is less affected by acid mine drainage than groundwater. Permeable rocks and deep soil allow acidic water to percolate to the groundwater. Through groundwater movement, the polluted water spreads to a wider area. Either from wells or bore wells, humans end up consuming the polluted groundwater. The acidic mine water also corrodes the equipment of mine plants and affects the mine surface texture. The high acidity of acid mine drainage water may also negatively affect fish growth and reproduction. If fish swim in acidic water, their blood is depleted of sodium ions. Furthermore, it adversely affects the function of the gills, which ultimately results in death. The ferric hydroxide will precipitate from the drainage water, and this layering will prevent benthic organisms from growing on streambeds (Hoehn and Sizemore, 1977).

7.4 PHYTOREMEDIATION FOR AMD

In both soils and waters, plant-based bioremediation can be used to remove acid mine drainage. In the treatment of environmental contaminants, green plants and their associated microbes and soil amendments are used in conjunction with agronomy and soil amendments.

Pollution may occur in nearby water bodies as a result of acid mine drainage impacted soil. It is, therefore, crucial to remediate both the soil and the water near mining sites. Out of all the phytoremediation techniques analyzed, only phytoextraction and phytostabilization are considered in this study. The plants store heavy metals in several different parts after extracting them from the soil, such as roots, stems, leaves, and fruits. The phytostabilization process works to make acid sulfate contaminated soils more resistant to erosion by providing a vegetative cover. Phytoremediation of mining sites generally involves plants tolerant to certain metals. Metals can also be accumulated 100 times in hyperaccumulator plants.

Plants that are considered hyperaccumulators have bioaccumulation factors greater than 1 which is more than a normal plant. High accumulation and high translocation factors (Baker et al., 1994) are characteristics of hyperaccumulator plants. The high accumulation factor is also known as the bioaccumulation factor.

Mine drainage is acidic water discharged from an active or abandoned mine. Mines can discharge highly acidic wastes if the ore contains sulfides or pyrites. The dominant metal ion in these cases could be zinc, copper, or nickel instead of iron. There are a variety of approaches available for preventing and cleaning acid mine drainage. To stop bacteria from catalyzing the necessary reactions, these strategies neutralize acid, prevent oxygen exposure, and restrict fluid flow. Here are a few of the more common methods, as illustrated in Figure 7.1. The use of hyperaccumulators is one of the promising and cost-effective approaches.

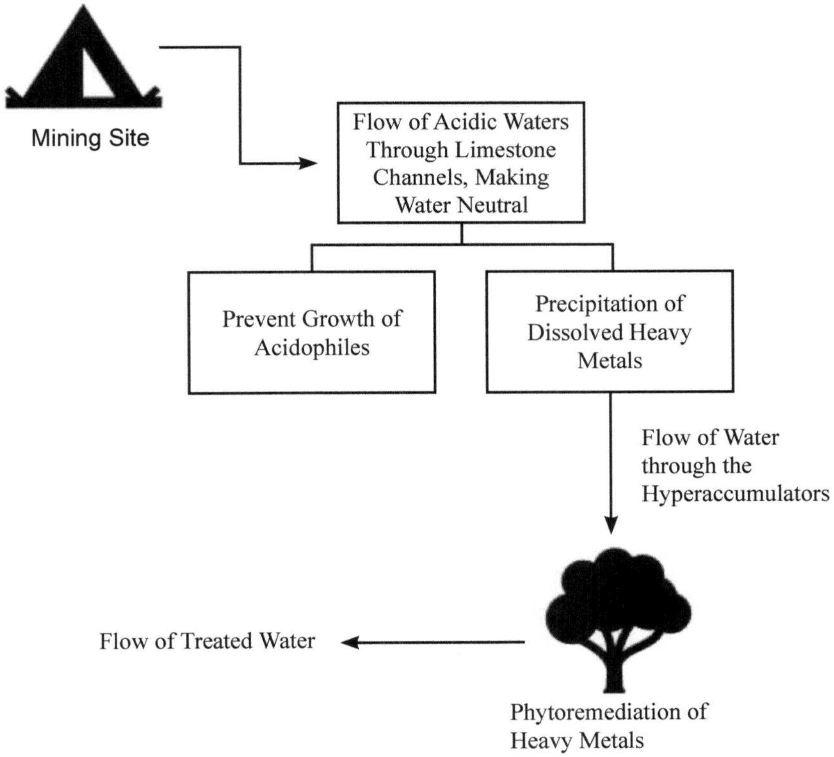

FIGURE 7.1 Treatment of mines drainage water.

Metal bioaccumulation factor is derived from the ratio between metal concentrations in plants roots and atmospheric pollution of metals. Translocation is defined as the ratio of metal pollutant concentrations in the shoots and the roots of a plant. Plants with a translocation factor greater than 1 are efficient in translocating metals from roots to shoots (Chowdhury et al., 2001). Following plant species exhibit the efficiency of remediating acid mines drainage.

1. *Cyperus alternifolius;*
2. *Chrysopogon zizanioides;*
3. *Chrysopogon aciculatus;*
4. *Sesbania rostrate;* and
5. *Cynodon dactylon.*

Using plants with rhizospheric microbes and soil amendments to treat acid mine drainage water still has some drawbacks, but they are considerably fewer than other methods. In this case, the first drawback is choosing the right plants. The acidic environment polluted by sewage water and another contaminant will not be conducive to growing every plant. Therefore, ideal plants for the purpose should be chosen that will grow with low nutrient concentrations, in various weather conditions, and be able to accumulate more contaminants than normal plants can. Because of these specific characteristics, it is difficult to select plants that are suitable for bioremediation of AMD. One drawback to the remediation process is that there is a possibility that animals will eat the contaminated vegetation.

7.5 NANOREMEDIATION

The process of nano-remediation involves introducing nano-sized particles (less than 100 nanometers in diameter) into polluted water or soil to neutralize contaminants. This technology is relatively new, and currently in its nascent form. In the field of nano-remediation, zero-valent iron is an emerging and important tool that can effectively treat acidic water polluted with several heavy metal pollutants. By immobilizing dissolved heavy metals, zero-valent iron effectively and rapidly neutralizes the acid in mine water. In studies, heavy metal removal is primarily caused by the adsorption of these metals on nanosized iron particles. A corrosion product of iron present on an unreacted metal surface also can act as an adsorption site.

An oil liquid membrane may be used to coat the surface of zero-valent iron to form an emulsified particle. Bimetallic nanoparticles are also used as nanoparticles. Nanoparticles of this type usually contain a combination of metals and metal catalysts. The bimetallic nanoparticles catalyze redox reactions by increasing the kinetics. The combination of iron and palladium is the most used nanoparticle of such a type. Nanomaterials, such as carbon nanotubes, have recently been introduced for the treatment of polluted water. A unique adsorption property and affinity for target make these nanotubes highly effective. In addition to being chemically and thermally stable, the carbon tubes act as an alternative to activated carbon. They are used mainly to remove metals such as lead, chromium, and zinc. Additionally, they help remove various biological impurities and a wide variety of organic and inorganic compounds (Rajan, 2011). Despite being successful in the remediation of mine water, nanoremediation still raises many questions that remain unsolved. There is a need to discover the toxicological effects of nanoparticles on other components of the environment (Köber et al., 2014). Nanoparticles have many drawbacks and risks associated with them.

7.5.1 *RISKS ASSOCIATED WITH THE USE OF NANOPARTICLES*

Many applications have been found for nanoparticles in recent years, offering many benefits. Research shows that it is very efficient and quick in treating contaminated soil and groundwater. Because of their smaller size and greater mobility, nanoparticles can easily disperse in the environment and therefore cause ecotoxicity, which can cause large distances to be affected. Furthermore, nanoparticles in nature are highly persistent, and they are capable of bioaccumulation within living organisms (Grieger et al., 2010).

Green nanoparticles can reduce ecological toxicity by reducing the release of toxic by-products into the environment. They are obtained from plants. The utilization of bio nanoparticles derived from microbes can be used to biodegrade heavy metals found in acid mine water quickly and efficiently. Nanofactories, also known as fungi, are excellent for synthesizing metal nanoparticles (Dhillon et al., 2012).

7.6 REMEDIATION BY SULFATE-REDUCING BACTERIA

The prevention of acid mines is typically achieved by introducing alkaline materials to sulfide-rich mine waste. It is a chemical treatment that promotes acid neutralization, metal solubility reduction, and the subsequent precipitation of metals in solution. By adding alkaline materials to AMTW, ferrous iron oxidation is accelerated. Moreover, this increases pH and precipitates metal hydroxides and carbonates. In terms of capital costs and operating costs, conventional treatment processes such as ion exchange, chemical precipitation, alkaline addition, and fluidized bed-ion require a significant amount of investment. Some of these techniques involve the use of chemicals that can pollute the environment and require more cleanup, while others are non-sustainable (Hilson and Murck, 2001). A low-cost treatment that removes contaminants with greater efficiency becomes an alternative. Plants and microorganisms are currently the most cost-effective and environmentally friendly option for treating AMTW (Acid Mine Tailing Waste), though it has not been successful in some cases due to geographical and/or technological limitations. The ability of biological reactors to function without a constant feed of biomass has been demonstrated in some cases.

7.7 INFLUENCE OF ACID MINE TAILING WASTE (AMTW) ON MICROORGANISMS

AMTW's toxicity is primarily caused by its acidic pH and heavy metal content. Acidic pH increases heavy metal bioavailability, so they are more toxic because of the synergistic effect. Ayangbenro and Babalola in 2017 illustrate the impact of heavy metal exposure as a wide range of disorders and diseases. As a result of acid mine tailings, ecosystems are destroyed, landscapes deteriorate, biodiversity decreases, and pollutants are deposited in the environment. Reduced habitat for microorganisms leads to nutritional, ecological, and evolutionary issues (Ayangbenro and Babalola, 2017), fragmentation of habitat, and invasion by wild organisms and weeds species. The consequences are dire for the environment as well (Jain et al., 2016).

Microbial cells are stressed when metals and metalloids, as well as low pH, are present in the AMTW (Chakravarty and Banerjee, 2008). Inhibition of cell growth occurs as a result of these changes. In addition, acid denatures microbial enzymes, thereby stopping metabolic activity. Loss of macromolecular building blocks results in growth being slowed down. Further, the hydronium ions in rainwater leach important nutrients and minerals like magnesium (Davies and Mundalamo, 2010) and therefore inhibit growth. In addition, organisms with low pH tolerances are destroyed, which results in a loss of biodiversity.

7.7.1 AMTW REMEDIATION BY SULFATE-REDUCING BACTERIA

Algal cells can provide a carbon source and oxygen source for sulfate-reducing bacteria (SRBs) to remediate acid mine tailings (Hallberg, 2010). Sulfate acts as a terminal electron acceptor, converting $SO4^{2-}$ into H_2S.

Developing innovative treatment methods has been made possible with the isolation of acid-tolerant SRB. These bacteria reduce sulfate to sulfide so that sulfuric acid is converted to hydrogen sulfide and generates alkalinity. As many toxic metals form highly insoluble sulfides, the reductive reaction is a rate-limiting procedure for selectively removing them from acid mine tailings. In the bioreactor, pH is controlled to regulate the concentration of the reactant sulfate, resulting in metal recovery. Mine tailings wastes are therefore effectively cleaned and converted into useful products by removing toxic metals. The benefit of bioremediation is that it has very low maintenance costs, as well as the fact that solid-phase organic matter is retained within the sediments of wetland areas (Johnson and Hallberg, 2005).

Typically, composted organic substrates such as manure, wood chips, sawdust, spent mushroom compost, sugarcane waste, and yeast extract are used in SRB remediation of AMTW sites. To achieve maximum metal and sulfate removal, these mixtures must be optimized (Kefeni et al., 2017). Organic matter and calcium carbonate help to reduce the initial pH of AMTW, which affects the growth of SRB. Sulfate and organic matter are converted into hydrogen sulfide and hydrogen carbonate, respectively, in remediation. After reacting with the hydrogen sulfide gas, the heavy metal ions form insoluble metal sulfides, which can be precipitated out by the sulfide reduction process (Kefeni et al., 2017).

The goal of genetic engineering is to improve or redesign microorganisms by improving their intrinsic properties, specificity, and resistance to environmental conditions. In prokaryotes and higher organisms, metals are responded to by cysteine-rich proteins such as glutathione, phytochelatin, and metallothionein, which can bind and seize metal ions. Metal accumulation will be increased by cells expressing metallothioneins over their usual levels, providing a promising strategy for developing biosorbents based on microbes for remediating metal contamination (Perpetuo et al., 2011).

A negative perception of environmental biotechnology may hinder the application of genetically modified microorganisms in the field (Davison, 2005), requiring well-grounded studies.

7.8 ALGAL BIOREMEDIATION OF ACID MINE DRAINAGE

There is a significant environmental concern worldwide regarding acid mine drainage (AMD). Heavy metals, which cause environmental damage, are present at high levels in this acidic waste stream. AMD may not respond to conventional treatment methods. Heavy metals must be removed from the environment using cost-effective technologies. This study examined and analyzed algae-based systems to identify their potentials and gaps for future research. Many algae strains have been demonstrated to be capable of eliminating heavy metals from AMD, such as *Spirulina sp., Chlorella, Scenedesmus, Oscillatoria, Anabaena,* and *P. tricornutum.* Algae form hyperaccumulators and hyperadsorbents with high selectivity for different elements, as well as high alkalinity, which is necessary for heavy metal precipitation during treatment. However, algae are not the perfect solution to the problem, and further research is required.

Algae are good decontamination agents due to several factors, including low costs, easy manipulation, non-polluting, relatively easy recycling of the metal contaminants, and they do not produce secondary waste (Kalin et al., 2006). Algae grow in contaminants, and then the biomass and water are separated and dried to recover the content of the concentrated metal by converting the metals to oxides or other recoverable salts. It is also possible to store the dried algal biomass for future use.

Municipal waste landfills can be used to dispose of algae biomass and enhance biofuel production (Edmundson and Wilkie, 2013).

Phycoremediation is a wastewater treatment method that uses macro- or microalgae. Compared to conventional methods, it has many advantages, including being less expensive, consuming less energy, and producing less sludge, so that it has been widely accepted worldwide (Ghosh and Singh, 2005; Abdel-Raouf et al., 2012). By using macro or microalgae, pollutants, including nutrients and xenobiotics, can be removed from wastewater and biotransformed (Ahmad et al., 2013). Studies investigating the use of microalgae for wastewater treatment have been conducted in the past few decades (Dubey et al., 2011; Sharma and Khan, 2013).

7.9 ALGAL INFLUENCE ON AMD MANAGEMENT

AMD can be influenced by algae in different ways (Figure 7.2). They influence the acidity of the water directly by assimilating nitrate. The algae can scavenge metals and potentially reduce the concentrations of metals.

They contribute directly to improving perceived water quality. They act as a carbon source for SRB, which in turn produce alkalinity. Diverse effects of algae often occur in combination, not separately.

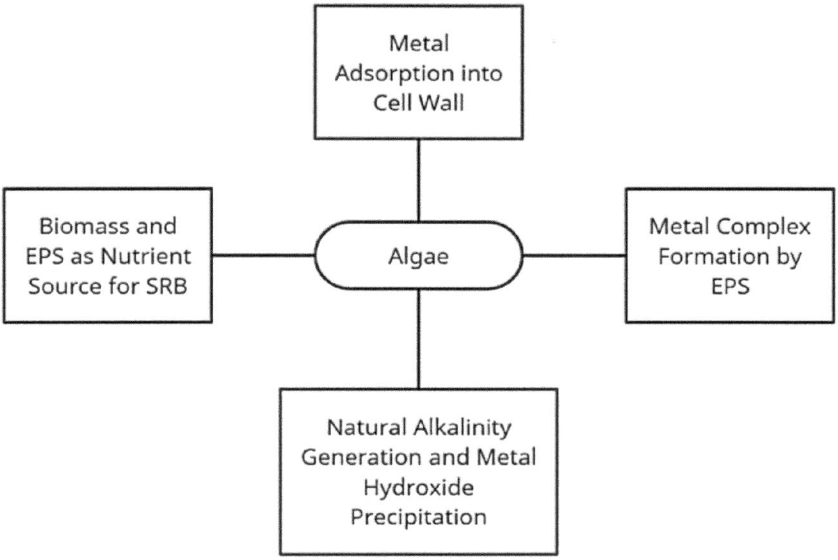

FIGURE 7.2 Algal Influence on acid mine drainage.

Approximately 30% of heavy metals in the environment are absorbed by algae, but how they absorb them depends on several factors (Novis and Harding, 2007). Seasonal differences also affect metal removal and productivity (Brake et al., 2004). Algae produce extracellular polysaccharides (EPSs) under nutrient stress conditions. polymeric EPSs are usually made of polysaccharides, protein, nucleic acid, and lipid, whereas algal EPSs are mainly made of glycoproteins (Karsten et al., 2007). The production of EPS by organisms is associated with the ability to uptake metal, according to several studies (FreireNordi et al., 2005). Different types of anionic functional groups (e.g., carboxyl, hydroxyl, pyruvate, succinate, uronate, phosphate, sulfate, glycerate, etc.) in the EPS interact with other cations as well as metals (Beech, 2004).

7.9.1 PRECIPITATION OF METAL HYDROXIDES AND ALKALINITY GENERATION

Algae produce alkalinity directly by assimilation of nitrate. Alkalinity gain is sustainable if the produced biomass is permanently removed from the system, e.g., by burying in the sediment.

During this process, nitrate dominates inorganic nitrogen, but ammonia synthesis produces acidity since ammonium is assimilated. An English Lake with weakly acidic pH was shown to be benefited from nitrate assimilation when producing primary nitrate through experimental phosphate fertilization (Davison et al., 1995).

7.9.2 METAL ADSORPTION AND ABSORPTION

Depending on the metal, the taxon, the age of the organism, and other factors, algae adsorb heavy metals to varying degrees (Novis and Harding, 2007). Seasonal variations also affect metal removal and productivity. Extracellular polysaccharides (EPS) are released by algae, especially in nutrient-stressed environments. A wide variety of macromolecules make up EPS, including polysaccharides, proteins, nucleic acids, and lipids; however, algal EPS consists predominantly of glycoproteins. The production of EPS by organisms is associated with the ability to uptake metal, according to several studies (Rose et al., 1998; Freire Nordi et al., 2005).

7.9.3 DIVERSITY AND PRESENCE OF ALGAE

It is also observed that bright-green algae such as Klebsormidium, Microspora, Mougeotia, Ulothrix, and Stigeoclonium often thrive (Stevens et al., 2001) in the AMD environment. Chara, Nitella, Mougeotia, Ulothrix, Chlamydomonas, and other algae are natural inhabitants of AMD (Orandi et al., 2007). AMD can be affected by algae in different ways. As they assimilate nitrate, they can directly affect the acidity of water. Metals are scavenged by algae, which decrease metal concentrations. Even though AMD contains a wide variety of algal species. The algae provide nutrients that support SRB, but there is no evidence that they cause disease.

Alkalinity is produced by microbial sulfate reduction, while metal sulfides are precipitated from the solution. Carbon is typically a limited resource in AMD environments (Koschorreck, 2008). Nutrients are essential for survival and growth, as well as sulfate reduction and metal precipitation. A cost and availability constraint has hindered the development of technologies to use SRBs for AMD remediation (Molwantwa et al., 2000). Kalin et al., (2004) suggest that SRB metabolism could be supported by photosynthetically produced dissolved organic carbon (PDOC). A cheap approach to driving bacterial sulfate reduction might be primary production by algae.

7.9.4 TOXICITY ASSESSMENT

AMD can be detected using algae as bioindicators. Regel et al., (2002) demonstrated that algal esterase activity could be used to rapidly evaluate the biological impacts of AMD. Observations of cyanobacteria (*Microcystis aeruginosa*), green alga (*Selenastrum capricornutum*), and their responses to AMD exposure time of 1 hour and 24 hours were conducted. The activity of algal esterase was influenced by increased conductivity as well as AMD-affected water. Algae bioassays were performed using the nonfluorescent substrate fluorescein diacetate (FDA), which was transformed into fluorescein via esterase production. At very low esterase activities, they successfully measured individual algal fluorescence by using flow cytometry.

7.10 SUMMARY

Studies of mining soil bioremediation should start with a screening of such highly resistant microorganisms, accounting for the natural intraspecific variation in bacteria's resistance to high metal concentrations in the soil. In mining areas, soil contamination is not caused by one chemical compound, but by a combination of different heavy metals, which could lead to the synergistic effects of bacterial mixtures on bioremediation. As a result of deforestation in degraded areas, the use of tree and bacteria combinations should be investigated further to fully restore soil and the environment. Considering the advancement of biotechnology in recent years, it is essential to test genetically modified bacteria for bioremediation enterprises in different soils and for the ability to absorb, precipitate, or accumulate various heavy metal contaminants.

Mining effluent has a profound impact on aquatic ecosystems and remediation attempts at monitoring data have been costly, at best temporary, at best. Acid mine drainage can be treated effectively with bioremediation, which utilizes naturally occurring environmental microbes. In AMD, the growth of algae is restricted by factors such as high metal concentrations, acidity, and low nutrient availability (mainly organic carbon and phosphate). Mutualistic relationships, however, increase their tolerance capacity. A combination of nitrate assimilation and inorganic carbon uptake is thought to generate alkalinity in the AMD environment. SRBs generate organic carbon from algal and fungal biomass; however, biomass utilization has not been as efficient as expected in technology. Fungi degrade complex organic carbon under AMD conditions. The application of knowledge about the roles of algae to technology development would require the selection of species, enzymatic processes, substrate design, and operational parameters such as pH, temperature, pressure, flow rate, and residence time.

7.11 EXERCISES

Objective Response Questions

1. **The organic compounds applied to the soil may be degraded by _______.**
   a) Leaching to Groundwater
   b) Adsorption to Organic Matter
   c) Chemical Decomposition
   d) Biological Degradation

2. **In terms of disposing of solid waste, what plan should we adopt?**
   a) Use of waste management plan
   b) Recycling of waste management plan
   c) Reducing of the waste management plan
   d) All the above

3. **Which one of the following methods does not generate mine waste?**
   a) Extraction of minerals
   b) Beneficiation of minerals
   c) Processing of minerals
   d) Remediation of mine waste

4. **_______ is the first phase that leads to ore removal.**
   a) Extraction
   b) Beneficiation
   c) Processing
   d) Remediation

5. **The tailings are the materials left after separating the valuable fraction from the ore.**
   a) True
   b) False

6. **By releasing toxic metals into the water, tailings can damage the environment.**
   a) True
   b) False

7. **Which of the following term is not used in place of the term "Tailings"?**
   a) Mine dumps
   b) Culm dumps
   c) Terra-coned
   d) Mine drainage

8. **Which of the following condition is occurred due to tailings?**
   a) Acid mine drainage
   b) Nutrient absorption
   c) Air contamination
   d) Afforestation

9. **By planting trees, the mining industry can try to stabilize environmental pollution.**
   a) True
   b) False

10. **Phytostabilization involves using plants to sequester pollutants near their roots to remove tailings from the environment.**
   a) True
   b) False

Practice Work

Qu 1. What is water pollution? Explain point and nonpoint sources of water pollution.

Qu 2. Describe a biotreatment technology for industrial wastewater.

Qu 3. Explain the bioremediation treatment of Acid Mine Drainage Water.

Qu 4. What are the adverse impacts of AMD on the environment?

Qu 5. Describe the phytoremediation technique for the treatment of AMD Water.

Qu 6. Describe the process of formation of Acid Mine Drainage.

Qu 7. Explain the nano-remediation of mine's waste.

Qu 8. Explain the action of sulfate-Reducing Bacteria on mines waste.

Qu 9. Explain different risks associated with nano-particles.

Qu 10. What are hyperaccumulator plants? Explain their role in mine waste remediation.

REFERENCES

Ayangbenro, A. S., & Babalola, O. O., (2017). A new strategy for heavy metal polluted environments: a review of microbial biosorbents. *Int. J. Environ. Res. Public Health, 14*, 94. doi: 10.3390/ijerph14010094.

Ayangbenro, A. S., Olanrewaju, O. S., & Babalola, O. O., (2018). Sulfate-reducing bacteria as an effective tool for sustainable acid mine bioremediation. *Front. Microbiol., 9*, 1986. doi: 10.3389/fmicb.2018.01986.

Baker, A. J. M., Reeves, R. D., & Hajara, A. S. M., (1994). Heavy metal accumulation and tolerance in British populations of the metallophyte Thlaspicaerulescens J. & C. Presl (*Brassicaceae*). *New Phytol., 127*, 61–68.

Baker, B. J., Lutz, M. A., Dawson, S. D., Bond, P. L., & Banfield, J. F., (2004). Metabolically active eukaryotic communities in extremely acidic mine drainage. *Applied and Environmental Microbiology, 70*, 6264–6271.

Beech, I. B., (2004). Corrosion of technical materials in the presence of biofilms current understanding and state-of-the-art methods of study. *International Biodeterioration & Biodegradation, 53*, 177–183.

Carlson, L., Bigham, J. M., Schwertmann, U., Kyek, A., & Wagner, F., (2002). Scavenging of as from acid mine drainage by Schwertmannite and ferrihydrite: A comparison with synthetic analogues. *Env. Sci. Tech., 36*, 1712–1719. doi: 10.1021/es0110271.

Chakravarty, R., & Banerjee, P. C., (2008). Morphological changes in an acidophilic bacterium induced by heavy metals. *Extremophiles, 12*(2), 279–284. Epub 2008 Jan 10. PMID: 18193380. doi: 10.1007/s00792-007-0128-4.

Chowdhury, A. R., Sarkar, D., & Datta, R., (2001). Remediation of acid mine drainage-impacted water. *Curr. Pollution Rep., 1*, 131–141. doi: 10.1007/s40726-015-0011-3.

Das, P. K., (2018). Phytoremediation and nanoremediation: Emerging techniques for treatment of acid mine drainage water. *Defense Life Science Journal, 3*(2), 190–196. doi: 10.14429/dlsj.3.11346.

Davies, T., & Mundalamo, H., (2010). Environmental health impacts of dispersed mineralization in South Africa. *J. Afr. Earth Sci., 58*, 652–666. doi: 10.1016/j. Jafrearsci.2010.08.009.

Davison, J., (2005). Risk mitigation of genetically modified bacteria and plants for bioremediation. *Journal of Industrial Microbiology and Biotechnology, 32*, 639–50.

Davison, W., George, D. G., & Edwards, N. J. A., (1995). Controlled reversal of lake acidification by treatment with phosphate fertilizer. *Nature, 377*, 504–507.
Dhillon, G. S., Brar, S. K., Kaur, S., & Verma, M., (2012). Green approach for nanoparticle biosynthesis by fungi: Current trends and applications. *Crit. Rev. Biotechnol., 32*, 49–73.
FreireNordi, C. S., Vieira, A. A. H., & Nascimento, O. R., (2005). The metal binding capacity of *Anabaena spiroides* extracellular polysaccharide: an EPR study. *Process Biochemistry, 40*, 2215–2224.
Grieger, K. D., Fjordbøge, A., Hartmann, N. B., Eriksson, E., Bjerg, P. L., & Baun, A., (2010). Environmental benefits and risks of zero-valent iron nanoparticles (nZVI) for *in situ* remediation: risk mitigation or trade-off? *J. Contam. Hydrol., 118*(3, 4), 165–183. doi: 10.1016/j.jconhyd.2010.07.011. Epub 2010 Aug 6. PMID: 20813426.
Hallberg, K., (2010). New perspectives in acid mine drainage microbiology. *Hydrometallurgy, 104,* 448–453. doi: 10.1016/j.hydromet.2009.12.013.
Hilson, G., & Murck, B., (2001). Progress toward pollution prevention and waste minimization in the North American gold mining industry. *J. Clean. Prod., 9*, 405–415. doi: 10.1016/S0959-6526(00)00083-4.
Hoehn, R. C., & Sizemore, D. R., (1977). Acid mine drainage (AMD) and its impact on a small Virginia stream. *Water Resour. Bulletin, 13*, 153–160.
Hudson-Edwards, K. A., Jamieson, H. E., & Lottermoser, B. G., (2011). Mine wastes: Past, present, future. *Elements, 7*, 375–380. doi: 10.2113/gselements.7.6.375.
Jain, R. K., Zengdi "Cindy" Cui, & Domen, J. K., (2016). Environmental impacts of mining. In; *Environmental Impact of Mining and Mineral Processing* (pp. 53–157). Boston: Butterworth-Heinemann. doi: 10.1016/B978-0-12-804040-9.00004-8. https://searchworks.stanford.edu/view/11640694 (accessed on 12 May 2023).
Johnson, D. B., & Hallberg, K. B., (2005). Acid mine drainage remediation options: A review. *Sci. Total Environ., 338*, 3–14. doi: 10.1016/j.scitotenv.2004. 09.002.
Karsten, U., Schumann, R., & Mostaert, A. S., (2007). Aeroterrestrial algae growing on man-made surfaces: what are the secrets of their ecological success? In: Seckbach, J., (ed.), *Algae and Cyanobacteria in Extreme Environment* (pp. 84–600). Springer, The Netherlands.
Kefeni, K. K., Msagati, T. A. M., & Mamba, B. B., (2017). Acid mine drainage: prevention, treatment options, and resource recovery: A review. *J. Clean. Prod., 151*, 475–493. doi: 10.1016/j.jclepro.2017.03.082.
Köber, R., Hollert, H., Hornbruch, G., Jekel, M., & Kamptner, A., (2014). Nanoscale zero-valent iron flakes for groundwater treatment. *Environ. Earth Sci., 72*(1). doi: 10.1007/s12665-014-3239-0.
Koschorreck, M., (2008). Microbial sulfate reduction at a low pH. *FEMS Microbiology Ecology, 64*, 329–342.
Molwantwa, J. B., Molipane, N. P., & Rose, P. D., (2000). Biological sulfate reduction utilizing algal extracellular products as a carbon source. In: *WISA 2000 Biennial Conference*. Sun City, South Africa.
Novis, P. M., & Harding, J. S., (2007). Extreme acidophiles: Freshwater algae associated with acid mine drainage. In: Seckbach, J., (ed.), *Algae and Cyanobacteria in Extreme Environment* (pp. 443–463). Springer, the Netherlands.
Orandi, S., Yaghubpur, A., & Sahraei, H., (2007). Influence of AMD on aquatic life at Sar Cheshmeh copper mine. Abstract. In: *Goldschmidt Conference*. Cologne, Germany.
Perpetuo, E. A., Souza, C. B., & Nascimento, C. A. O., (2011). Engineering bacteria for bioremediation. In: Carpi, A., (ed.), *Progress in Molecular and Environmental*

Bioengineering from Analysis and Modeling to Technology Applications (pp. 605–632). In Tech Open Access Publisher, Rijeka, Croatia.
Rajan, C. S., (2011). Nanotechnology in groundwater remediation. *Int. J. Environ. Sci. Develop., 2*(3), 182–187. doi: 10.7763/IJESD.2011.V2.121.
Regel, R. H., Ferris, J. M., Ganf, G. G., & Brookes, J. D., (2002). Algal esterase activity as a biomeasure of environmental degradation in a freshwater creek. *Aquatic Toxicology, 59*, 209–223.
Rose, P. D., Boshoff, G. A., Van, H. R. P., Wallace, L. C., Dunn, K. M., & Duncun, J. R., (1998). An integrated algal sulphate reducing high-rate ponding process for the treatment of acid mine drainage wastewater. *Biodegradation, 9*, 247–257.
Singh, G., (1987). Mine water quality deterioration due to acid mine drainage. *Int. J. Mine Water, 6*(1), 49–61. doi: 10.1007/BF02498139.
Singh, G., & Bhatnagar, M., (1985). Bacterial formation of acid mine drainage, causes, and control. *J. Sci. Ind. Res., 44*, 478–485.
Stevens, A. E., McCathy, B. C., & Vis, M. L., (2001). Metal content of Klebsormidium-dominated (*Chlorophyta*) algal mats from acid mine drainage waters in southeastern. *Ohio Journal of the Torrey Botanical Society, 128*, 226–233.

CHAPTER 8

Biosensors and Bioindicators in Bioremediation

8.1 BIOSENSORS

> **Biosensors**
> **A biosensor is tool that uses biological specificity to identify the target molecule.**

A major aim of environmental biotechnology is to work towards the restoration of the quality of our ecosystem by applying natural science and engineering. It has been almost a century since biotechnological processes have been used to protect the environment, long before the term "biotechnology" existed. Waste can be treated with biotechnological tools before or after it has been released into the environment. Biotechnology can also be applied to develop products and processes that produce less waste, use less non-renewable resources, and require a smaller amount of energy.

Biosensors detect target molecules based on their biological specificity. Molecular biosensors are reviewed in this chapter for use in bioremediation and their design and application. Biosensors should be applied to a wider range of scientific fields, according to the research. To integrate the biological recognition element with signal analysis systems, specialized research is needed in bringing it within proximity to the target molecules.

Biosensors are analytical devices combining a biological sensing element, e.g., enzymes or antibodies with a physical transducer, whereby the interaction between the target and the bio-recognition molecules is translated into a measurable electrical signal.

An alternative to conventional analytical methods could be optical biosensors using absorption, luminescence, fluorescence, reflectance,

Bioremediation and Phytoremediation: Technologies for Toxic Pollution
Ashwini A. Waoo, PhD (Author)

refraction, Raman scattering, and transmission. Biosensors that provide rapid, high-sensitivity, real-time, and high-frequency monitoring are free from the need for time-consuming sample pre-treatments and concentration steps. Optical biosensors can also be used in food and environmental monitoring, drug development, biomedical research, and diagnosis. Optical sensors are particularly useful for monitoring pollution and developing early warning systems. Thus, biosensors are classified according to their recognition elements as either optical, electrochemical, piezoelectric, or enzymatic, such as immunosensors, aptamers, nucleic acids, and enzymes, respectively.

Processes can be monitored using biosensor technology. Microelectronic elements pick up biosensor signals and process them based on the type of processor used. As biosensors, biomolecules interact with other molecules selectively, with reactions always following the same kinetics. The concept of molecular specificity is the basis for designing biosensors. Consequently, biosensors can be case-specific or can use genetic engineering and be more flexible. In addition to the signal generating surface, the analytical device also features the signal-processing unit for capturing the data. In the literature, there are different schematic diagrams of biosensors (Gronow et al., 1998; Cass, 1995). A biosensor is composed of different components and is designed in a manner to suit the shelf life and stability of the biological component. One can transform, hydrolyze the target molecule. A signal generated by a biological reaction may also include light, temperature, and a change in conductivity.

8.2 MOLECULAR BIOSENSORS

Molecular biosensors are constructed from recombinant plasmids. A promoter is used to generate the signal, whose expression is regulated by the target molecule and a reporter system. By acting on specific molecules, promoters control the signal generation process. Signal generation is directly correlated with the expression of the promoter. Genetically engineered recombinant plasmids can lead to biosensors that are case-specific. Sometimes the promoter is divided in a way that part of its activity is carried by the host chromosome.

8.2.1 REPORTER SYSTEMS

A reporter system produces a protein and it is part of a gene expression vector. This system produces a signal by itself or by catalyzing a biochemical reaction. A few common reporter systems are described here.

- ***The bacterial luciferase reporter system***

Luciferase, bioluminescence is caused by this enzyme. The same system can emit light at different wavelengths in different organisms based on the chemistry of the reaction (Hastings, 1996). In the lux operon, a series of reactions lead to the emission of light. *Vibrio fischeri*, *Photorhadbus luminescence*, and others were used as sources to clone the lux operon (luxCDABE) (Meighen, 1991).

- ***Green Fluorescent Protein***

The green fluorescent protein (GFP) in the *Aquorea victoria* jellyfish absorbs light with an extinction maximum of 395 nm and emits light with an emission maximum of 510 nm. The Ca^{2+}-activated photoprotein Aequorin uses this protein as an accessory in the bioluminescence pathway to cause jellyfish to glow (Hastings, 1996; Morise et al., 1974). GFP emits 510 nm of green light when excited with 395 nm of ultraviolet or blue light.

- ***Promoters as Biosensors***

An operon or gene promoter is a series of five flanking sequences that respond to cellular physiology and transcribe genetic information in messenger RNA form. Promoter molecules either take over a target molecule's response directly or by interacting with its receptor. In either case, the response can be modulated kinetically, which depends on whether mRNA or any other reporter molecule is expressed. The 5-region of the reporter system can be used to design a biological component for a biosensor, reflecting the target molecule whose presence will be monitored in the sample. The promoter sequence can either be a high or low activity sequence, depending on the target molecule. In this regard, biosensors' promoters are the sensing components at the molecular level. The main components of these biosensors, therefore, are three. The vehicle or vector molecule (plasmid) allows the biosensor to be maintained in the host and to survive. In addition, there are two other components, the promoter, which provides the genetic sensing system, and the reporter system, which is either the lux operon, GFP, or some other type of signal agent. Various genes have been utilized to produce promoters that have been used in environmental monitoring.

8.3 WHOLE-CELL BIOSENSORS FOR MONITORING BIOREMEDIATION

These sensors detect certain chemical and physical characteristics of various environmental samples using whole-cell microorganisms. Microbial detection produces a signal that can be decoded by the user in the form of a color change on a paper strip or as complex as a quantitative digital display (Figure 8.1). A multi-step assay can be performed separately or with embedded whole-cells. Genetically modified and non-genetically modified cells were used. Analyzing metabolism reporters detects toxicity that inhibits cell metabolism while analyzing catabolic reporters detects specific contaminants. Biosensors can collect data about monitoring bioremediation by determining the bioavailability of contaminants. The commercial availability of several sensor technologies has proven successful in measuring environmental contaminants after they were extensively studied and tested.

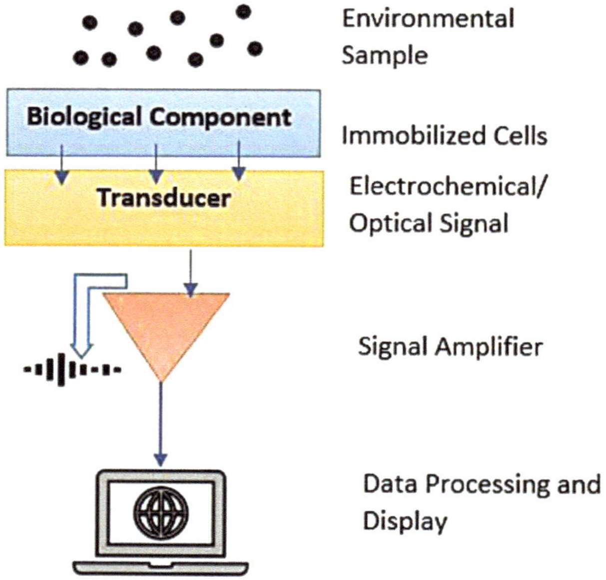

FIGURE 8.1 Environmental monitoring using whole-cell biosensors.

- ***Principle of Whole-Cell Biosensors***

The intact living cells offer more benefits than other bio components due to their affordability, accessibility, and potential for manipulation. They can provide information about processes like protein synthesis, apoptosis, and necrosis, and the bioavailability of a target analyte, i.e., how much is absorbed into a cell when it passes through the membrane. Whole-cell biosensors may contain either prokaryotic or eukaryotic cells as biocomponents. Microbial biosensors can be developed by adapting the microorganisms to a specific substrate (target) through selective cultivation (Garcia-Balboa C and Lopez Rodas, 2012) or by exploiting specific metabolic features (Zhang and Keasling, 2011)

- ***Whole-cell Immobilization strategies***

A vital step in the fabrication of microbial biosensors is embedding the cells onto the transducer to maximize viability, increase reproducibility, and prevent spreading genetically modified organisms into the environment for field use. Microorganisms can also be preserved by immobilization without the need for continual maintenance or cultivation. Whole cells are Immobilized by physical means (adsorption and encapsulation) or by chemical means (covalent bonds and cross-linking). The use of natural biofilms has gained some attention in recent years.

- ***Artificial films***

By encapsulation, adsorption, or covalent binding, artificial films can immobilize cell suspensions within a defined internal architecture. The concept of encapsulation refers to the concept of encasing living cells within the confines of a matrix which preserves their viability but stops their growth. The immobilization matrix must function at normal room temperature, retain all cells completely, be nontoxic, and be able to transmit nutrients and oxygen. A matrix should ensure that the cells will remain viable for a long time. Even though the organic polymer opacity alters optical signal detection, organic (e.g., hydrogels like agar) and inorganic polymers (e.g., sol-gel) have shown good performance in cell encapsulation.

8.4 CELL MICROARRAYS

Cell microarrays are miniaturized whole-cell biosensor systems that collect and analyze samples in parallel using platforms such as microchips, glass slides, etc. (Elad et al., 2010).

To overcome restrictions from cell preservation and reproducing the natural cell environment on solid support is critical to attaining an automated, fully automated environmental monitoring device. This support is often referred to as a "chip," and it can be made from diverse materials such as gold, silicon, glass, etc. As the cells are captured on specific spots on the chip, they interact with a variety of biocompatible materials, including oligonucleotides, antibodies, proteins, and lipids (Ben-Yoav et al., 2011).

Protein or oligopeptides immobilized by simple adsorption can be affected by several factors, such as their desorption or denaturation on the surface and their orientation. Therefore, covalent binding of the peptide sequences to the support is necessary.

There have been different strategies for immobilizing reporter strains in microarrays to monitor environmental toxicity for the past two decades. The biosensor strains of *E. coli* were captured on a chip or in a 384-well plate using nutrient agar for the detection and classification of toxicity (Lee et al., 2005).

8.5 SINGLE-CELL IMMOBILIZATION

Cellular biosensors produce a signal based on the average response of all the cells in a population. Since a mixed community will not be made up of identical cells, their responses will differ. In the same way, in clonal populations, even though the cells originate from a single parent, stochasticity in gene expression can introduce genetic variability that can alter their response as well. Thus, measurements at a single-cell level can provide additional information that whole-population analysis would not reveal (Cai et al., 2006). Single bacterial cell biosensors are promising for analyzing microenvironments because a chemical target produces measurable signals when it is exposed to them. Microphysiological analysis of bacterial cells can be useful in understanding the function of macroscale processes such as contaminant biodegradation, bacteria-eukaryotic beneficial interactions, and infections (Tecon & van der Meer, 2006).

8.6 COMMERCIAL BIOSENSORS

Recently, a few biosensors have been applied to real samples after most have only been tested on non-real samples (like distilled water or buffer solutions). They are also used for the determination of environmental quality parameters such as biological oxygen demand (BOD), toxicity, or endocrine effects using a variety of matrices. Biosensors must be applied to real samples before they can be commercialized, which, in general, is the goal of device development. A comparison of the results with those obtained with standard protocols is also necessary to get the acceptance of end-users.

Shortly, Micro-Electro-Mechanical Systems, or MEMS, may be able to meet these demands. A silicon substrate is used to integrate sensors, mechanical elements, actuators, and electronics. They provide the capability to simultaneously detect several toxic chemical and biological agents with these MEMS-based devices. This allows for the development of biochips and sensor arrays. This also introduces the concept of labs-on-chips. Immobilizing and stabilizing biomolecules on these nanodevices may be a greater challenge. There have already been some works in these areas.

To micro-immobilize enzymes on silicon wafers for biosensor applications, the molecular recognition ability of biomolecules like avidin-biotin or streptavidin-biotin is also studied in association with lithography (Gau et al., 2001).

8.7 ENVIRONMENTAL APPLICATIONS OF BIOSENSORS

Pollutants negatively impact human health as acknowledged by international organizations like WHO (Malcolm et al., 2015). The need to monitor dangerous substances in the environment grows in tandem with their concerns. Residues of pollutants can accumulate in food and drinking water. The purpose of food quality control is to prevent cases such as these. The ideal biosensor will be easy to use, cheap, stable when stored, and able to detect even tiny amounts of analytes, as an alternative to traditional chemical monitoring methods. Detection limits must be low and the response time must be short (Kumar et al., 2006). The transfer of a laboratory-tested biosensor to the field requires extensive optimization and testing. This chapter highlights the current trends in monitoring environmental microbes.

Biosensors offer interesting possibilities. With the current advances in biological sciences and other scientific and engineering disciplines, many analytical applications will inevitably be replaced by biosensors. By fusing biological sciences with other disciplines, Future developments will be able to maximize the potential of this technology.

8.8 BIOINDICATORS

The examination of living organisms provides information about the environment either qualitatively (bioindication) or quantitatively (biomonitoring). The field of bioindication/biomonitoring includes the observation and measurement of chemical compounds (both inorganic and organic) in well-defined bioindicator plants (including microbes) or animals (including humans). Biomonitoring and bioindicators may be applied alternatively in common usage, but in the scientific world, the concepts are more specific. Biological indicators qualitatively assess biotic responses to environmental stress Biomonitors determine the qualitative response to a given contaminant, while *Lecanora conizaeoides* indicate air quality problems, for example, a decrease in chlorophyll content of lichen or their diversity point to air pollution. As a result, "bioindicator" refers collectively to all terms relating to both the detection and the diagnosis of biotic responses to environmental stress. Bioindicators serve three main functions within this framework: 1. Environmental monitoring (physical or chemical changes), 2. Ecological processes monitoring, or 3. Biological monitoring.

A "bioindicator" is an organism that indicates the quality, or state, of the environment, or how it is changing over time. Bioindicators are often used to detect changes in the environment due to anthropogenic factors (e.g., pollution or land-use changes) or natural stress factors (e.g., drought or late spring freeze), although the research focuses primarily on anthropogenic stress factors. Historically, bioindicators have been developed and applied primarily since the 1960s. As the years have unfolded, there is expanded use of bioindicators to study aquatic and terrestrial environments, as well as using all major taxonomic groups.

Regardless of geographical location, environment, disturbance type, or organism, good bioindicators share several characteristics.

- A measurable response to disturbances and stresses.
- The response reflects the entire population, community, and ecosystem.
- Response proportional to contamination or degradation.
- A sufficient local population density.
- Moderate variation in climate and environment; relatively stable.
- Ecological and life-history well understood.
- Stable and well-documented taxonomy.
- Taking a survey is simple and inexpensive.
- Harvesting species for other purposes already.
- People are aware of or interested in these species.

All biological processes, species, and communities cannot serve as bioindicators, however. Environmental factors such as light, temperature, substrate, competition vary based on the physical, chemical, and biological environment. The adaptation of populations to a particular range of environmental factors (i.e., fitness) occurs over time. An individual's physiological and/or behavioral responses may alter if their environmental targets are not met. This can lead to a decrease in fitness. In turn, a reduction in fitness can alter community dynamics. Species that act as bioindicators can be effective indicators of environmental change due to their tolerance to changing environmental conditions.

The biotic response of a rare species (or assemblage of species) may go unrecognized because it is too impacted by environmental change, or too rarely sighted. A species (or assemblage of species) that is tolerant of environmental changes has a lower chance of being affected by them. But bioindicators are not restricted to one species or a narrow range of environmental tolerance.

A "biotic index" or a "multimetric" approach can be used to assess environmental conditions by evaluating entire communities, considering a wide range of tolerances for environmental conditions. Bioindicators can also be derived from biological processes within individuals. Coldwater streams in the West of the United States are home to cutthroat trout. In general, most individuals have a thermal tolerance limit of 20°–25°C, so they can be used as a biomarker for water temperature. Cutthroat trout can detect human-related disturbances at many different biological scales such as pasture burning, logging, and livestock grazing. Thermal pollution causes an immediate response in cutthroat. As a result of thermal stress, a protective protein called heat shock protein (HSP) is synthesized to protect vital cell functions. Heat stress can be measured in cutthroat trout by quantifying HSP levels and evaluating changes in the environment (Holt & Miller, 2010).

Bioindicators can be used in the following three situations:

1. In cases where the environmental factor cannot be measured, for example, climatic change study in paleo biomonitoring, where environmental factors are reconstructed.
2. Conditions in which the factor is tough to measure, such as pesticides, their residues, or highly contaminated effluents containing multiple chemical interactions.
3. When the environmental factor is easy to measure, but difficult to interpret, for example, whether the observed changes are ecologically significant.

Bioindicators can be classified differently based on their perspectives. Typically, three types of bioindicators (Figure 8.2) are distinguished by their aim (Gerhardt):

1. compliance indicators;
2. diagnostic indicators;
3. an early warning indicator.

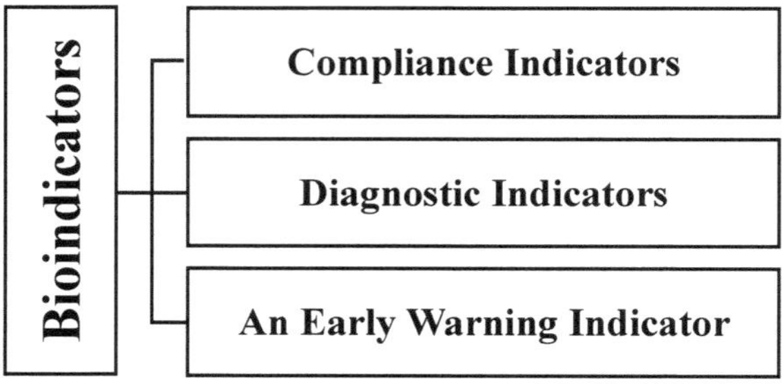

FIGURE 8.2 Types of bioindicators.

A compliance indicator could be an attribute of the fish population based on its size, ecosystem, or community level and focus on issues such as sustainability. Diagnosis and early warnings are based on individual or sub organismal (biomarker) levels. Early warnings are focused on fast and very sensitive responses to changes in the environment. Toxic effect bioindicators, with their effects examined at different levels of bio-organization, are distinguished from accumulation bioindicators (such as mussels, mosses, lichens).

Three categories of bioindicators can be distinguished based on their applications:

1. Environment indicators, such as sentinels, detectors, exploiters, accumulators, and bioassay organisms, are organisms that react predictably to disturbances or changes in the environment. Environmental indicator systems are systems of indicators used to diagnose the state of the environment to make environmental policy decisions.
2. Ecological indicator: The species is well known for its vulnerability to pollution, habitat fragmentation, and other stressors. A representative sample of the community responded to the indicator.
3. Biodiversity indicator: An indicator taxon serves as a proxy for the species richness of local communities. The definition of biodiversity

has been expanded to comprise "measurable parameters of biodiversity," including species richness, genetic parameters, endemism, population-specific parameters, and landscape parameters.

8.9 BENEFITS AND DISADVANTAGES OF BIOINDICATORS

Many countries around the world have passed laws mandating the use of bioindicators, and they are part of several international accords. Nevertheless, bioindicators do not come without their flaws. Some bioindicators serve as early warning signals, much like canaries in the coal mine. The use of bioindicators in heterogeneous environments is limited in some cases because it is difficult to distinguish between natural variability and changes due to anthropogenic activities.

As a result, indicator species may be affected not just by disturbances or stresses but also by factors other than those mentioned (e.g., disease, predation, competition, parasitism), which complicates our understanding of the concept and process that cause change. In addition, bioindicators are scale-dependent in terms of their ability to function as indicators. The biodiversity of large vertebrate indicators (such as fish) may not necessarily reflect that of local insects. As a third point, bioindicator species usually differ significantly from other species in their ecosystem concerning habitat requirements. It may be impossible to protect rare species with different habitat requirements by managing an ecosystem based on the habitat requirements of a bioindicator. Thus, the overall goal of bioindicators is to apply a single species, or a limited number of species, to measure the quality of an environment and how it undergoes changes over time, but this can represent a gross simplification of a complex system.

It is a downside of management tools that their flaws must be acknowledged. Bioindicators, however, have more benefits than limitations. An ecosystem's health can be evaluated using bioindicators at a variety of scales, from cellular to ecosystem. Biological, physical, and chemical components of our world determine individual fitness, population density, community structure, and ecosystem processes. From a management perspective, bioindicators enable us to determine what is and is not biologically sustainable. If we do not have moss on the tundra, cutthroat in the mountain stream, or the canary in the coal mine, we will not be able to detect our disturbances until it is too late to prevent any damages.

8.10 SUMMARY

Environmental health has been monitored using biosensors for decades. Anthropogenic activities continuously release xenobiotics that pollute ecosystems, threatening their integrity. Considering this, xenobiotics, especially chemical agents, should be studied urgently. The chapter has presented a novel approach to environmental monitoring, integrating biomonitor and biosensor systems dynamically. The detection and quantification of various contaminants in the ecosystem have become critically significant in recent times as a result of the extensive use they have had in soil and aquatic ecosystems. There has been considerable attention in the ecosystem due to the persistence, accumulation, and toxicity of both organic and inorganic contaminants.

Recent years have seen a rise in interest in the use of biosensors to study environmental health since organisms that can integrate environmental pollutants and respond in predictable and measurable ways are useful for studying chemical pollutants at various ecological levels.

Biological indicators are useful in monitoring the environment because of their resistance to ecological variability. The term "bioindicator" is defined as any source that responds to ecological changes via biotic and abiotic mechanisms. By shedding light on environmental changes, taxa serve as more than mere markers of natural change.

A change in the natural surroundings, as well as positive or negative impacts, can be detected by using them. Additionally, they can detect any changes in the environment caused by pollutants that can negatively affect the environment's biodiversity.

8.11 EXERCISES

Objective Response Questions

1. Biochemical events are converted into measurable signals by ______.

a) rectifier
b) amplifier
c) transducer
d) opamp

2. A ________ is an organism used to determine how healthy an ecosystem is.

a) Predator
b) Bio-remediator
c) Bioindicator
d) Decomposers

3. Lichens serve as good bioindicators for ________.

a) Environmental radiation
b) None of the above
c) Soil pollution
d) Water and air pollution

4. ________ is an organism used to measure the quality of an ecosystem.

a) Decomposers
b) Predator
c) Bio-remediator
d) Bioindicator

5. Bioindicators were used to measure the quality of the water. What is a bioindicator?

a) any living creature
b) eutrophication
c) An organism is sensitive to its environment
d) a turbidity tubes

6. A biosensor is an analytical tool that uses _____.

a) biological specificity
b) Physical specificity
c) Electrical Specificity
d) None of the above

7. Molecular biosensors are _____.

a) Biomolecules
b) Proteins
c) Recombinant plasmids
d) All the above

8. Luciferase is an enzyme responsible for producing _____.

a) Bioluminescence
b) Chemiluminescence
c) Xenobiotics
d) None of the above

9. A green fluorescent protein was obtained from _____.

a) *Aequorea victoria*
b) *Aequorea japonica*
c) *Aequorea riccia*
d) *Aequorea pinnata*

10. How many classes of bioindicators?

a) 3
b) 4
c) 2
d) 5

Practice Work

Qu 1. What are the advantages of using bioindicators?

Qu 2. Explain different components of biosensors.

Qu 3. Describe the applications of biosensors in environmental monitoring.

Qu 4. What are molecular biosensors? Explain various reporter systems used in it.

Qu 5. Explain the principle of the whole-cell bioreactor.

Qu 6. Describe the role of cell microarrays in monitoring.

Qu 7. What is a bioindicator? Explain different species used as bioindicators.

Qu 8. Explain the advantages and Disadvantages of Bioindicators.

Qu 9. Explain different types of bioindicators.

Qu 10. What is immobilization? How it is applied for biosensor technology.

REFERENCES

Ben-Yoav, H., Melamed, S., Freeman, A., Shacham-Diamand, Y., & Belkin, S., (2011). Whole-cell biochips for bio-sensing: integration of live cells and inanimate surfaces. *Crit. Rev. Biotechnol., 31,* 337–353.

Cai, L., Friedman, N., & Xie, X. S., (2006). Stochastic protein expression in individual cells at the single-molecule level. *Nature, 440,* 358–362.

Cass, A. E. G., (1995). Biosensors. In: Meyers, R. A., (ed.), *Molecular Biology and Biotechnology: A Comprehensive Desk Reference* (pp. 110–114). VCH Publishers; Cambridge, UK.

Elad, T., Lee, J. H., Gu, M. B., & Belkin, S., (2010). Microbial cell arrays. In: Belkin, S., & Gu, M. B., (eds.), *Whole Cell Sensing Systems, I: Reporter Cells and Devices* (pp. 85–108).

Garcia-Balboa, C. C. E., & Lopez, R. V., (2012). Microalgae biosensors for the detection of environmental contaminants: a review. *Rev. Comp. Sci. Vet., 6,* 51–67.

Gau, J. Jr., Lan, E. H., Dunn, B. C., Ho, M., & Woo, J. C. S., (2001). *Biosensors & Bioelectronics, 16,* 745.

Gerhardt, A., (2011). *Bioindicator Species and Their Use in Biomonitoring Environmental Monitoring A* (Vol. I). Encyclopedia of Life Support Systems (EOLSS) LimCo International, Ibbenbüren, Germany. http://www.eolss.net/sample-chapters/c09/e6-38a-01-07.pdf (accessed on 01 April 2023).

Gronow, M., Mullen, W. H., Russell, L. J., & Anderton, D. J., (1988). Biosensors. In: Walker, J. M., & Gingold, E. B., (eds.), *Molecular Biology and Biotechnology* (pp. 323–347).

Hastings, J. W., (1996). Chemistries and colors of bioluminescent reactions: A review. *Gene, 173,* 5–11.

Holt, E. A., & Miller, S. W., (2010). Bioindicators: Using organisms to measure environmental impacts. *Nature Education Knowledge, 3*(10), 8.

Lee, J. H., Mitchell, R. J., Kim, B. C., Cullen, D. C., & Gu, M. B., (2005). A cell array biosensor for environmental toxicity analysis. *Biosens. Bioelectron., 21,* 500–507.

Meighen, E. A., (1991). Molecular biology of bacterial bioluminescence. *Microbiol. Rev., 55,* 123–42.

Morise, J. G., Shimomura, O., Johnson, F. H., & Winant, J., (1974). Intermolecular energy transfer in the bioluminescent system of Aquorea. *Biochemistry, 13,* 2656–2662.

Tecon, R., & Van, D. M. J. R., (2006). Information from single-cell bacterial biosensors: What is it good for? *Curr. Opin. Biotech., 17,* 4–10.

Zhang, F., & Keasling, J., (2011). Biosensors and their applications in microbial metabolic engineering. *Trends Microbiol., 19,* 323–329.

PART II

Phytoremediation

CHAPTER 9

Concept of Phytoremediation

9.1 PHYTOREMEDIATION: INTRODUCTION

Plant interactions with polluted sites to remediate the toxic effects of pollutants. Various mechanisms are involved in phytoremediation. Noxious heavy metals and radionuclides are mainly removed by phytoextraction, phytotransformation, or sequestration mechanism. Willow and alfalfa are mostly used for phytoremediation. Phytoremediation depends on the toxicity of pollutants, plant survival ability, and its adaptability in contaminated environmental conditions, and plant growth rate (Lee, 2013).

By using all of the bioremediation technologies, microbial detoxification may provide new alternatives for removing heavy metals or contaminants. If toxic discharges are not treated, wastewaters and soil are increasingly contaminated, these pollutants may accumulate in living organisms, and become toxic or carcinogenic for them. Various strategies have been developed for solving the environmental pollution problems caused due to toxic pollutants. Bioremediation is safe for all living organisms and is environment-friendly.

Phytoremediation comes from the Greek word phytos, meaning plant, and remedium, which means to correct or remove evil. Research in environmental biotechnology has recently focused on the practice of using vegetation for in situ treatment of contaminated soils, sediments, and water using phytoremediation. Heavy metals contamination sites were first treated with this technique in 1983.

It is an economically cheap, eco-friendly, and widely accepted technology for the clean-up of heavy metal contaminated sites (Neetu Malik and Biswas, 2012).

Phytoremediation is a technology having an esthetic, green, and pleasant approach. It is most appropriate for developing countries and cities.

Bioremediation and Phytoremediation: Technologies for Toxic Pollution
Ashwini A. Waoo, PhD (Author)

Phytoremediation is the engineered approach to the remediation of polluted sites using native plants to remove or deliver environmental contaminants like heavy metals in soil or water. Phytoremediation can be applied to remediate soils contaminated with inorganic contaminants. Phytoremediation focused on the natural capacity of some plants called hyperaccumulators to accumulate, degrade contaminants in water, air, and soil. Contaminants like heavy metals, pesticides, organic waste, solvents, crude oil, and its derivatives are also remediated by phytoremediation. So, it can be known as green, clean, low cost, effective, and ecological suitable technology as other clean-up methods.

Technology for phytoremediation is based on the physiological mechanisms found in higher plants and microorganisms, i.e., transpiration, photosynthesis, mineral nutrition, metabolism, etc. Roots can absorb organic and inorganic compounds, stabilize substances on their surface areas, and interact with microbes in the rhizosphere. Plant cells and tissues transport, store, and convert these substances. (Nelson et al., 2006) These chemicals are released from aerial parts of the plant.

In phytoremediation technology, two approaches are used:

1. To use hyperaccumulator plant having a low amount of biomass.
2. To use plants that generate a large amount of biomass but are not as efficient in accumulating target elements (Tlustos et al., 2006).

According to some research, ideal plants for phytoremediation should possess characteristics such as a rapid growth rate, high biomass, the ability to tolerate various heavy metals, and the capacity to accumulate them in their aerial parts. Hyperaccumulator plants can tolerate high concentrations of toxic metals because they produce molecules (phytochelatins) that bind metals into complexes that can be compartmentalized, so as not to invade sensitive plant tissues.

Phytoremediation uses natural bioprocesses of plants. Biological processes such as water uptake, plant metabolism, waste excretion into the soil, and physicochemical action of plant roots. Sunlight converts water and CO2 into carbohydrates and oxygen during photosynthesis. Plant roots are responsible for extracting macro- and micronutrients effectively. Chemically, contaminants and nutrients are similar, so plants accumulate contaminants rather than nutrients (Pierzynski et al., 1994).

A phytoremediation's success depends on the selection of native plants having hyperaccumulator properties and being able to grow on contaminated soil that contains heavy metals in high concentrations.

Phytoremediation plants should have the following characteristics:

- Ability to accumulate and tolerate high concentrations of heavy metals.
- The high growth rate on marginally contaminated soil.
- The capacity of producing high biomass production (David E et al., 1995).

9.2 MECHANISMS OF PHYTOREMEDIATION

Phytoremediation is the eco-friendly practice in which contaminated land is ameliorated by growing plants that can accumulate the contaminating chemicals. The principles in phytoremediation include phytodegradation, phytovolatilization, phytostabilization, and rhizofiltration as shown in Figure 9.1. The plant and microorganism association in the rhizosphere can enhance the removal of the contaminants.

Based on the process used by the plants to remove or reduce the toxicity of contaminants from the soil, the phytoremediation mechanisms can be categorized into five domains.

Mechanism of Phytoremediation

1. Phytoextraction
2. Phytotransformation
3. Phytostimulation
4. Phytostabilization
5. Phytovolatilization

FIGURE 9.1 Mechanism of phytoremediation.

9.2.1 PHYTOEXTRACTION

Metals or contaminants are eliminated from the soil by accumulating them in the aerial parts of accumulator plants. This technology involves using plants to absorb contaminants from soil or water, and then translocate and accumulate these contaminants in their upper biomass. It is an important technology for soil remediation of heavy metals and metalloids. Heavy metals can be permanently removed from polluted soil using phytoextraction. Hence, it is more suitable for commercial use.

The phytoextraction process involves the following steps:

- Mobilization of a contaminant in soil.
- Contaminant uptake by plant roots.
- Plants translocate nutrients from their roots to their aerial parts.• Sequestration and compartmentalization of contaminants in plant tissues (Ali et al., 2013).

The efficiency of phytoextraction relies on a few factors such as plant selection, plant performance, heavy metal bioavailability, soil, and rhizosphere properties. Therefore, the strategies to improve phytoextraction efficiency are developed in light of those aspects and are discussed below.

The plant species must be selected appropriately for phytoextraction.

Mostly, the plant has the following characteristics that are more suitable for phytoextraction.

- High tolerance to the toxicity of contaminant.
- High extraction ability.
- Hyperaccumulation of a contaminant in aerial parts.
- High biomass production, with fast growth.
- Extensive root and shoot system.
- Easy cultivation and harvest.
- Highly resistant to diseases and repulsive to herbivores to avoid entering into the food chain (Ali et al., 2013).

Effective hyperaccumulators are very important for the successful phytoremediation of contaminants like heavy metals. Up till now, more than 450 plant species belonging to 45 angiosperm families have been identified and studied as metal hyperaccumulators (Suman et al., 2018), such as *Brassicaceae, Euphorbiaceae, Fabaceae, Asteraceae, Scrophulariaceae*, and *Lamiaceae*, families.

The phytoextraction process has some advantages. Phytoextraction, where the contaminants are permanently removed from the soil, is very inexpensive for any contaminated site. The contaminant can be recycled from the contaminated plant biomass after incineration. Despite these advantages phytoextraction show some limitations, i.e., hyperaccumulator plants are generally slow-growing and have small biomass. After appropriate plants must be harvested properly to collect the contaminant. Harvesting and removing plant biomass is followed by metal recovery.

9.2.2 PHYTOTRANSFORMATION

Plant tissues degrade complex organic contaminants, either in full or in part. Plants break down organic contaminants sequestered by their metabolisms, or with the help of enzymes, a process called phytodegradation. Plants can easily degrade or transform organic contaminants into simpler compounds, which promote their growth. Plants undergo phytotransformation when they take up contaminants and accumulate them in their parts. Various factors contribute to the direct uptake of contaminants through the roots of plants, as shown here.

- Uptake efficiency.
- Transpiration rate.
- The concentration of the chemical in soil water.
- Chemical speciation.
- Physical/chemical properties.
- Plant characteristics.

Transpiration is a complex process and will depend on plant type, leaf area, environmental conditions like temperature, wind, moisture, humidity, etc. Organic contaminants after uptake get either translocated and stored in aerial parts of the plants or they may be completely converted into CO2 and water. Volatile contaminants are released into the atmosphere via the plant by the same transpiration process. It is called phytovolatilization, which is also a kind of phytotransformation. But in this case, complete remediation does not occur. Phytovolatilization may be desirable in case of prolonged soil exposure and the low risk of groundwater contamination.

Phytotransformation can be applied for the remediation of organic compounds. Plant enzymes can break down and convert a variety of organic compounds such as chlorinated solvents, ammunition wastes, and herbicides. A petrochemical site or storage area, landfill leachate, fuel spill, or agricultural soil can also be treated with this technology. Some phytotransformation treatments may be combined with other remediation techniques as polishing treatments.

9.2.3 PHYTOSTIMULATION

This process is characterized by the stimulation of microbial degradation by releasing the plant exudates and enzymes into the rhizosphere. Phytostimulation is also called enhanced rhizosphere biodegradation or rhizodegradation, or Bioremediation. In the rhizosphere or root zone, it consists of microbial activity that breaks down organic contaminants. Rhizosphere bacteria can be stimulated in a variety of ways, including:

- Compounds secreted by native microorganisms around the roots, like carbohydrates, amino acids, acetates, enzymes, etc. secreted by the roots along with native microbe populations;
- Root systems provide oxygen to the rhizosphere, for aerobic transformations;
- Rhizosphere organic carbon was increased by hairy root biomass;
- Mycorrhizae fungi located in the rhizosphere can degrade organic pollutants in conjunction with bacteria;
- Plants provide an environment for greater microbial abundance and activity.

Several enzyme systems have been discovered in soils and sediments. The dehalogenase enzyme is used for dechlorination reactions of chlorinated hydrocarbons. Nitroreductase – for the degradation of nitroaromatic compounds. Peroxidase – for oxidation; laccase – for breaking aromatic ring structures of organic compounds; nitrilase – oxidation reactions (Miller, 1996).

Phytostimulation is in situ practice with no disturbance. In this process, complete mineralization of the contaminant can occur. It also has low installation and maintenance cost but in phytostimulation development of an extensive root system is required that takes time, thus the duration of remediation is very long. Soil structure determines the maximum root depth, so phytostimulation occurs only up to that point.

9.2.4 PHYTOSTABILIZATION

This is a method that involves plants to prevent the migration of organic or inorganic pollutants to groundwater or in the food chain by reducing their mobility in the rhizosphere. Phytostabilization is a very less invasive and low-cost technology that uses green plants for the stabilization of the toxic metal ions within or near the rhizosphere. This method is capable of decontaminating contaminated soils and restoring their physicochemical and biological characteristics.

In phytostabilization deactivation or immobilization of the potential toxic ions takes place, so that they get stabilized and do not enter the food chain. The term phytostabilization refers to the mechanical stabilization of contaminants through the plants. Phytostimulation prevents various harmful consequences such as airborne transport, bulk erosion, and the leaching of pollutants (Fitz and Wenzel, 2002). Phytostimulation depends on root structure and root's ability to stabilize contaminant and bioavailability of contaminant in the soil. It is very important to select an appropriate plant for a phytostabilization purpose. There are some selection criteria of a plant species as mentioned here:

- Plants should be tolerant.
- It must grow quickly to set up an extensive root system.
- The establishment and maintenance of plants under field conditions should be easy.
- Plants must have a long life or they can be self-propagated (Mendez and Maier, 2008).

There may be some sort of soil amendments that can be done for enhanced phytostabilization. Soil amendments can facilitate metal immobilization and ultimately phytostabilization occurs this is called aided phytostabilization. These amendments in soil include phosphate, liming material, compost, zeolite, biosolids, and aluminosilicates, fly ash, etc. (Mench et al., 2010). The rooting medium will be improved with these amendments in the soil and, facilitate plant establishment in a contaminated site. Organic and inorganic amendments can be done in this process. Organic amendments include compost, biosolids, and fly ash which provides various beneficial effects such as Increased soil pH, Improved physical structure of the soil, complexation of heavy metals, and slow-release of the nutrient source. In the case of inorganic supplements, NPK fertilizers are added to increase the nutrient supply and lime for the increase in soil pH (Figure 9.2).

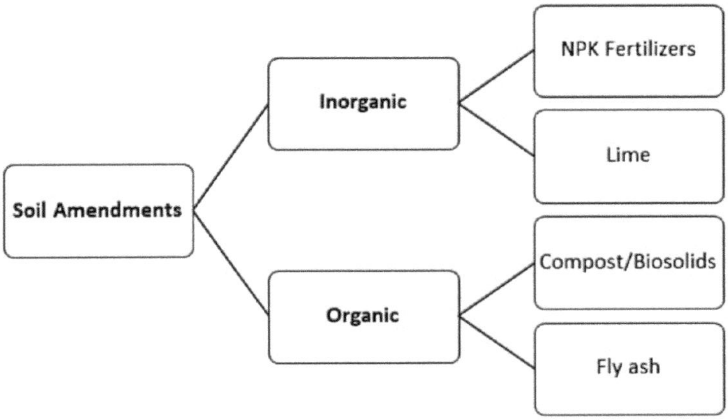

FIGURE 9.2 Soil amendments in aided phytostabilization.

Phytostabilization is applicable to treat contaminated soils with the establishment of a vigorous plant cover that prevents the spread of pollutants by erosion, water flow or percolation, leaching, as well as wind dispersal of toxic pollutants. There are various advantages of phytostabilization such as follows:

- Polluted biomass does not need to be disposed.
- Reduction in leaching and bioavailability of toxic pollutants.
- Soil organic matter, microorganisms, and texture is not affected.
- Low-cost technology.
- Vegetation cover improve the physical stability of the soil.
- Vegetation enhances ecosystem and esthetic sense.

However, phytostabilization does not completely clean up the contaminated soil, but only changes the mobility of toxic elements by reducing their concentrations in the soil. Although phytostabilization, has become more popular and accepted nowadays, further research is required in the field of testing new soil amendments and selection of specific plant species for the process.

9.2.5 PHYTOVOLATILIZATION

This technology is useful for only volatile organic carbons, and a few inorganic compounds (Pilon- Smits, 2005). A plant interacts with its surroundings throughout its entire life cycle, by absorbing chemicals from the soil, water, and air, and releasing them into nature in various ways.

Plant systems manage water and nutrient uptake, translocation, and release, during this process many unintended or the plant can transport undesirable compounds.

The phytovolatilization process involves plants taking up contaminants from the soil, releasing them into the atmosphere as volatiles through transpiration. It occurs as growing plants absorb water and organic contaminants through roots. As water moves upward from the roots to the leaves, some contaminants are transported to the leaves and then evaporated into the atmosphere.

This technology can be used to remove mercury; here mercuric ion is converted into less toxic elemental mercury. Phytovolatilization has been widely employed for a variety of contaminants, including inorganic and organic. Volatile forms of Se, As, and Hg like inorganic compounds can be volatilized by using a plant system. The thrust area of the research domain is the physicochemical domain where phytovolatilization occurs.

Phytovolatilization can work in two different forms: direct and indirect phytovolatilization. Direct phytovolatilization is a more spontaneous and well-studied process when contaminants are taken up by the plant and translocated, eventually resulting in the volatilization of the compound from the leaves and aerial parts of the plant, this process is called "phytovolatilization" and this pathway resembles the transpiration process using the vascular pathway of water.

Indirect phytovolatilization includes the increase in volatile contaminant uptake from the subsurface resulting from plant root activities. As plants move large quantities of water, they explore large volumes or surfaces of soil. This process can cause deep changes in subsurface chemical fate and transport. The plant roots can increase the flux of volatile contaminants through some mechanisms such as; decreasing the water table, gas fluxes caused by water table fluctuations, improved soil permeability, Chemical transport through transpiration force, etc. (Limmer and Burken, 2016).

Plants may be used as an effective pump-and-treat system for mobile contaminants and volatile contaminants like carbon tetrachloride (CCl_4) and ethylene dibromide (EDB). Nowadays considerable interest develops in finding economical remediation processes through plants.

Bioavailability, soil texture, and soil physicochemical properties all affect the mechanisms and efficiency of phytoremediation. In plants, contaminants are absorbed through the root system. Root systems provide a large surface area for absorbing and storing essential components, such as water and nutrients, and non-essential contaminants (Raskin and Ensley, 2000).

9.3 FACTORS AFFECTING PHYTOREMEDIATION

Environmental factors that affect the process and efficiency of phytoremediation include soil texture, organic matter content, pH, oxygen availability, moisture, fertility, temperature, solar radiation, and weathering. These factors affect the properties and bioavailability of hydrocarbons, germination, and productivity of plants, and the colonization and growth of microorganisms in the rhizosphere (Gaskin, 2008). Various factors affect the process of phytoremediation; plant species, medium, rhizosphere, uptake in plant parts, etc.

- ***Characteristics of Soil***

Soil properties, i.e., pH, texture, chemical composition, the addition of fertilizers and chelators affect the process of phytoremediation. Various agronomical practices are applied to enhance these parameters of the soil.

- ***Soil pH***

The pH of the soil is important because most microbial species can only survive within a certain pH range. Also, soil pH affects nutrients. The best pH for biodegradation of petroleum hydrocarbons is 7 (neutral); the acceptable range is 6–8 (US EPA, 2006).

- ***Soil Moisture***

Moisture is necessary for all soil microorganisms to grow and function. Nutrients are diffused into and out of cells by water. Despite this, excess moisture, as in saturated soil, will reduce the amount of oxygen available for aerobic respiration. Biodegradation is slowed by anaerobic respiration (in comparison to aerobic respiration). According to the US Environmental Protection Agency (2006), soil moisture content of 45 to 85% of the soil's water-holding capacity (field capacity) is optimal for petroleum hydrocarbon degradation. Loamy soils with a soil water content of 60% are ideal for the degradation of hydrocarbons.

- ***Soil Composition and Quality***

The quality of the soil affects plant growth and health as well. Poor physical conditioning of heavily contaminated soils is detrimental to the vigorous growth of plants and rhizosphere bacteria. These limitations include poor moisture-holding capacity, low permeability, insufficient aeration, and nutrient deficiencies. The water-holding capacity of contaminated soil can be increased by adding organic amendments such as sewage sludge, aged manure, compost, mulch, or straw. By adding lime and sulfur to the soil, pH can be increased and decreased, respectively (Kamath et al., 2007). Plant germination is improved by adding organic matter high in carbon, such as sawdust, but the yield is decreased due to an increase in the C: N ratio (Amadi et al., 1993). Plants require different soil textures and organic matter contents for optimal germination and growth (Evans et al., 1977).

When screening plants for phytoremediation those species naturally adapted to the soil texture at the contaminated site will likely be more successful than those adapted to different soil textures. Clay and organic matter content also affect microbial populations via their ability to form soil aggregates (Paul and Clark, 1989).

- ***Soil Oxygen***

Soil contaminated with hydrocarbons may have low oxygen availability (Cunningham et al., 1996). Lack of oxygen reinforces seed dormancy of some plants, preventing growth in contaminated soil (Amadi et al., 1993). As the most effective hydrocarbon-degrading microorganisms are aerobic, a lack of oxygen can negatively affect this process (Eweis et al., 1998). Animal manure increases plants to yield more than inorganic fertilizers, likely due to the binding of hydrocarbons to organic matter (Amadi et al., 1993). In freshly contaminated soils, nitrogen and phosphorus are more limited than in older soils, since they tend to be immobilized by microorganisms soon after contamination and mineralized when the carbon to nitrogen ratio declines (Hutchinson et al., 2001a).

- ***Temperature***

Temperature affects the availability and toxicity of soil, and plant and microorganism growth. Indirectly, high temperatures lead to water stress, which decreases plant productivity (Larcher, 1980). Microorganisms benefit from heat; hydrocarbon degradation rates double for every 10 °C increase in temperature (Eweis et al., 1998).

- ***Fertilizer Requirements***

It is usually difficult to establish healthy, vigorously growing plants in contaminated soils because they are deficient in macro- and micronutrients. Plant growth and germination were also affected by the source of nutrients. Organic nitrogen sources were superior to inorganic nitrogen sources.

- ***Plant Species***

Each plant species has its characteristic that influences how a compound is absorbed

Such as a type of metabolism, biomass production, growth rate, its capacity of tolerating a high concentration of the contaminant.

- ***The Rhizosphere***

The rhizosphere microflora mainly affects the phytoremediation mechanism. Increased root diameter and decreased root elongation also affect the process.

- ***Vegetative Uptake Mechanism***

Several environmental factors affect vegetative uptakes like temperature (Burken and Schnoor, 1996), which affect growth substances and thus root length and root structure. Phytoremediation relies on phytoextraction and a contaminant-specific hyperaccumulator (Tu et al., 2004). Understanding the metabolic outcome of contaminants in plants is the key to proving the use of phytoremediation (Bieby Voijant Tangahu et al., 2011; Mwegoha, 2008).

Phytoremediation is moreover adopted as a sustainable solution for soil reclamation. Significant advances in the efficiency of this technology are necessary to obtain higher levels of environmental sustainability. Current phytoremediation techniques do not always work for expected outcomes in field settings because there is the least understanding of the complex biological interactions. In phytoremediation technology, vigorous growth is expected for a study plant since the removal of pollutants from the environment is necessary. Phytoremediation will become a more efficient and reliable technology as scientific thinking evolves. The phytoremediation strategy is expected to be the most promising for contaminant degradation and bioenergy crop production on contaminated poor land.

9.4 PHYTOREMEDIATION OF HEAVY METAL CONTAMINATED SOIL

Clean-up of heavy metal contaminated soil is yet required for an extensive study. Heavy metals are very hazardous contaminants due to their properties such as non-biodegradable, their mobilization is slow and depends on the physical and biochemical properties of soil, and are very toxic at low concentrations. The selection of an appropriate remediation technique depends upon the parameters such as location, size, site, history of the site, treatment, soil and pollutant properties, and extent of the contamination.

In plants, metals are absorbed by the aqueous phase because of their bioavailability, and it depends on the metal retention time and its interaction with other elements. Moreover, upon binding to the soil, its redox potential, pH, and organic matter will affect the potential of the metal to exist in ionic or in available form. Plants affect the soil composition with their ability to reduce the pH and carry out oxygenation of the sediment, which affects the metal availability, i.e., the bioavailability of the heavy metals through the addition of biodegradable physicochemical factors like micronutrients and chelating agents.

9.5 ADVANTAGES OF PHYTOREMEDIATION

One of the most beneficial features of phytoremediation technology was that it allows the study of toxic concentrations of accumulated contaminants in the living tissues of plants, which represent the number of contaminants in that area over some time (Lovett-Doust et al., 1994). Other advantages such as conserving the physical structure of soils and biological activity and also providing the chance of bio-recovery of metals.

The advantages of phytoremediation are as follows:

i) It is very cost-effective.
ii) Does not need expensive equipment or highly skilled personnel.
iii) Maintenance and monitoring of the plants can be easy.
iv) Bio recovery and re-use of valuable products can possible.
v) It utilizes natural plants and is beneficial to the environment. It can be applied to a wide variety of contaminants (Keshav et al., 2010).
vi) This technique applies to a broad range of contaminants including organic and inorganic.

9.6 RECENT ADVANCES IN PHYTOREMEDIATION

Remediation of heavy metal-contaminated sites is essential to prevent their accumulation of food chain and bio-magnification hazards. Metal-contaminated soils are already being remedied with a variety of physical methods such as solidification, soil excavation, soil washing, incineration and pumping, use of permeable barriers. These methods are expensive and less efficient and may cause damage to the environment (Ghosh and Singh, 2005). Hence it is essential to develop greener technological modifications and search for more efficient methods to clean up. Bioremediation can be the best tool emerging as bioagents to remediate the contaminated environment. It is carried by various organisms like bacteria, fungi, and higher plants, either wild-type or genetically modified. This is an eco-friendly and cheap approach with its esthetic sense. Metal phytoextraction is limited by several factors including rate of metal uptake by roots, metal bioavailability, rate of translocation, the proportion of metal accumulated in the roots, and cellular tolerance to the toxic metal (as shown in Figure 9.3).

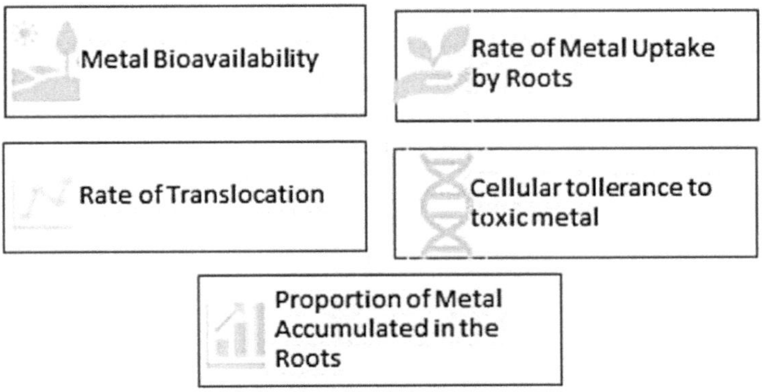

FIGURE 9.3 Factors affecting metal phytoextraction.

Plants have an amazing ability to absorb chemicals from their soil. Extensive root systems and transport tissues of plants are the main functional elements. Phytoremediation is a method of removing pollutants by using green plants. In this technology, biological means are used to remediate organic and inorganic contaminants containing soils, sludges, sediments, and water. Thus, phytoremediation is a natural medicine to clean up our planet. Plants have an extraordinary ability to handle, store and accumulate pollutants from the environment. Accumulation and absorption depend on the type or species of plant Various plant species show a different mechanism for absorption and degradation of the pollutants. Some plants may accumulate heavy metals from the soil and these plants can be used to remediate heavy metal pollution from the industrial dumping site. A similar technique is also used to remediate the river basin by controlling pollutants in the river water.

Biodiversity serves for the maintenance of a balanced ecosystem, also gives economic benefits to the human being. But, the Extinction of many of these species occurred as a result of population explosion and increased human interference in the ecosystem. With the number of contaminated sites on the rise in today's scenario, it has become a major challenge on a global scale. To overcome this, it is necessary to destroy the pollutants or transform them into biodegradable substances. The bioremediation process disposes of various contaminants by using natural biological activity.

Van Dillewijn et al., define "Bioremediation" that it is the process by which various biological agents degrade or render the environmental pollutants into less toxic forms.

A biological remediation agent, a strain of Pseudomonas putida degrading petroleum was first patented in 1974. This technique is low-cost, highly efficient, minimizes chemical as well as biological sludge, is specific to metals, does not require additional nutrients, regenerates biosorbent, and has the option of metal recovery. Based on how waste is removed and transported, bioremediation methods can be classified as in situ or ex-situ.

In situ bioremediation is the remediation of polluted material at the native site. During ex-situ remediation, the polluted material is removed from one site and transferred to another. In 1991, the "phytoremediation" term was introduced. Phytoremediation technologies proved to be interesting as they are cost-effective and environmental-friendly and used for the remediation of organic, inorganic pollutants. These technologies have been widely practiced by academic and industrial researchers all over the world.

Characteristics of suitable remediation method:

1. Make the site biologically productive (Mandelbaum et al., 1995).
2. Should not leave any secondary contaminants (e.g., detergents or solvents) (Ward, 2003).
3. Should be esthetically acceptable to the community (Doick et al., 2009).
4. It should not pollute the surrounding community (Schädler et al., 2011).
5. It should be inexpensive (Grommen and Verstraete, 2002).
6. It would offer some economic incentives to developers or investors (Doick et al., 2009).

Any conventional method cannot clean up a site completely but it requires multiple remediation rounds. Several academic and government institutions should contribute to the creation of sustainable remediation. There are unprecedented pressures on the reservoir of natural resources and the organisms which share them due to exponential population growth. A human being should be aware of these issues and governments, businesses, and communities must work together to remediate the natural resources.

Phytoremediation methods proved to be sustainable but had some practical limitations. Soil type, climate conditions, soil depth, and the severity of contamination can restrict the practical applications of phytotechnology. The time duration required is a significant factor in which phytoremediation projects are questionable. Besides this, conventional remediation methods are more time-efficient than phytotechnology.

Phytoremediation is a thrust area of science and technology which utilizes plants to remediate polluted soil, groundwater, and wastewater. Green plants such as grasses and trees are used to clean up environmental pollutants in soil and water, such as heavy metals, trace elements, organic compounds, and radioactive compounds. Plant-based biological chemical and physical processes are concerned with the uptake, sequestration, and degradation of pollutants, either by plants, microbes, or both.

Plant-based technologies are extensive, self-extending uptake networks enhancing the ecosystem for productive use. Phytoremediation can reduce the threat of dispersal of pollution as no excavation or transportation is required of contaminants. A more polluted site could be treated with it.

Some limitations are dependency on the growth parameters of the plant, e.g., climate, geology, altitude, temperature. To perform remediation, agricultural equipment and knowledge are needed. The tolerance of plant species to the specific pollutant affects the remediation process.

Since 1991, phytoremediation has been used to describe the process of using plants to reduce the mobility, volume, concentration, amount, or toxicity of contaminants in soil and water. It removes pollution from the environment in a very esthetic way.

Phytoremediation research has the potential to find solutions for various types of pollution including metals, explosives, pesticides, cyanides, oil as well as hydrocarbons. The plants having diverse root systems help to prevent carrying pollutants away from the contaminated sites to other areas through running water or wind. Climatic conditions in favor of plant growth and microbial activity enhance the phytoremediation capacity (Zhang et al., 2010).

Researches in the field of phytoremediation are flourished in the last 15–20 years. A very little amount of research funding is available for a research laboratory, greenhouse, and field-level study of phytoremediation plan.

Up till now, more than 500 plant species are identified for having appreciable phytoremediation capacity. Plants also help in preventing soil erosion and maintaining the biodiversity of a particular site, helping in carbon sequestration. Additionally, plants restore natural flora and fauna with esthetic sense and maintain ecological balance.

9.7 GENETIC ENGINEERING IN PHYTOREMEDIATION

The plant species with phytoremediation potential have a few limitations, like slow growth, low capacity of accumulation. These factors inhibit large-scale or field-level applications of such plants for phytoremediation. Sometimes newly introduced plants for phytoremediation purposes do not have adaptation to environmental conditions on contaminated sites. (Gerhardt et al., 2017). Thus, to overcome these limitations, a new strategy is developed with genetic engineering for the improvement of these plant varieties to enhance phytoremediation potential. Transgenic plants created through genetic engineering have improved characteristics of rapid growth rate and more biomass as well as increased hyperaccumulator capacity (DalCorso et al., 2019).

Genetic engineering is a promising technique for enhancing the phytoremediation potential of plants. In this technique a desirable foreign gene from any organism like bacteria, an animal can be introduced into plants. After stable DNA integration, the foreign gene is expressed in plant cells and confers specific character to the plants. This technique requires a very short time.

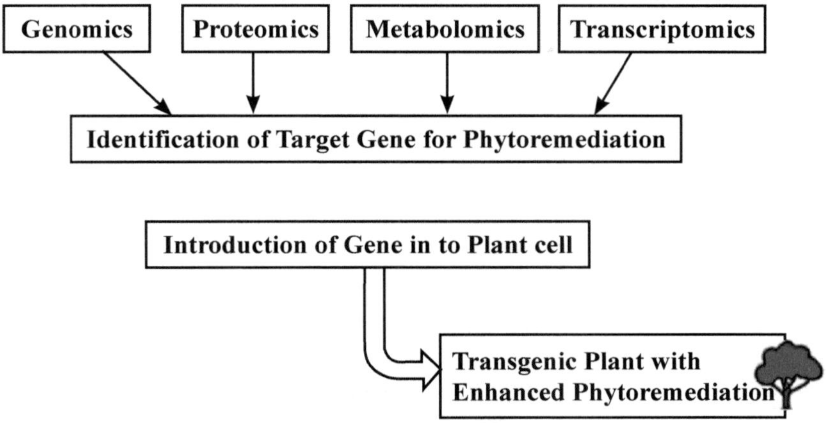

FIGURE 9.4 Omics research for phytoremediation.

Sometimes, gene transfer can also occur within species of sexually incompatible plants. As a result, developing transgenic plants with desired traits using genetic engineering shows promise in phytoremediation. It's more applicable to modify fast-growing, high-biomass species to achieve high metal tolerance and high accumulation capacity than to engineer hyperaccumulators to get high biomass production. Thus, fast-growing, high-biomass plants are often engineered either to enhance heavy metal tolerance or to increase heavy metal accumulation ability, which are the key characteristics of hyperaccumulators. Toxic metal tolerance and accumulation mechanisms in plants should be used to select genes for genetic engineering. The recent research in phytoremediation has proven effective with Clustered Regularly Interspaced Palindromic Repeats (CRISPR)-mediated phytoremediation. An advanced genome-engineering method is an effective genomic tool that enables the augmentation of targeted traits in plants. (Perez Pinera et al., 2013). Studying genomic regions that are associated with contaminant accumulation and plant survival with tolerance is the major focus of research. Researchers can use genomic approaches to genetically manipulate plants to develop their pollutant removal abilities, as illustrated in Figure 9.4. CRISPR systems can help us understand the mechanisms involved in the accumulation, transformation, volatilization, and degradation of contaminants. Thus, phytoremediation can be carried out via targeted engineering. CRISPR allows a specific set of instructions about the genome of the plant. It is an advanced technique of genetic manipulation based on programmable instructions (Basu et al., 2018).

9.8 SUMMARY

Heavy metals and hazardous pollutants are increasingly accumulating in soil due to anthropogenic and natural processes. Toxic wastes such as heavy metals do not biodegrade and remain in the environment for a long time. By biomagnification, they can become part of the food chain by entering agricultural plants. It is therefore imperative to remediate contaminated land. Eco-friendly phytoremediation, a method that can help revegetate polluted soil, can provide a cost-effective and friendly solution. It is critical to better understand the mechanisms underlying heavy metal accumulation and toxicity tolerance in plants to improve the efficiency of phytoremediation. The mechanisms of phytoremediation were discussed in detail in this chapter. In future research, phytoremediation strategies will be updated to improve the efficiency of phytostabilization and phytoextraction processes, through the use of genetic engineering, microbe-assisted, and chelate-assisted approaches, etc.

Compared to conventional soil remediation methods, plants provide several distinct advantages. Weed control and harvesting are marginal costs after planting. Burning food-grown biomass will not release carbon dioxide into the atmosphere because the plants assimilated all of their carbon dioxides during the growing process. The phytoremediation process produces biomass that can be used in specialized facilities.

9.9 EXERCISES

Objective Response Questions

1. **The process of phytoremediation is a way to remove contaminants from soil using plants. Researchers planted Brake Ferns in pots with soil containing lead, iron, arsenic, and copper in equal amounts. The amount of each metal absorbed from the soil after two months of growth was measured. The results indicate that Brake Ferns absorbed more arsenic from the soil than any of the other contaminants. The variable in the above study is?**

   a) the Brake Fern
   b) the container
   c) the soil
   d) the metal

2. **Using plants to degrade pollutants is known as _________.**

   a) Land farming
   b) Composting
   c) Biopile
   d) Phytoremediation

3. **Stabilization refers to changes in the _______ properties of hazardous waste.**

   a) Physical
   b) Chemical
   c) Biological
   d) Physico-chemical

4. **What is the main goal of risk assessment?**

   a) To evaluate hazards and minimize the risks
   b) Remediation of contaminated sites
   c) Hazard management
   d) To find the source of pollutants

5. **Which among the following processes doesn't come under phytoremediation?**

   a) Bio-accumulation
   b) Concentration and excavation
   c) Phytoextraction
   d) Phytostabilization

6. **Which of the following processes uses hydroponics to remove water contaminants?**

   a) Rhizofiltration
   b) Pollution control
   c) Antibiotic development
   d) Manufacturing process

7. **Which of the following is used in the bio degradation of xenobiotics?**

   a) Enzymes
   b) Chemicals
   c) Physical sorting
   d) PAHs

8. This technology uses accumulator plants to eliminate metals or contaminants from the soil by accumulating them in their aerial harvestable parts.

a) Phytotransformation
b) Phytoextraction
c) Rhizofiltration
d) Phytostabilization

9. Identification of specific Gene for Phytoremediation can be braught by _________.

a) Genomics
b) Biofilm technology
c) Quorum sensing
d) All of the above

10. CRISPR, the full form is _________.

a) Clustered regularly interspaced short palindromic repeats
b) Clustered Random interspaced short palindromic repeats
c) Clustered regularly inherited short palindromic repeats
d) None of the above

Practice Work

Qu 1. Explain the mechanism of bioleaching and the microbes involved in that process.

Qu 2. What is phytoremediation? Describe various mechanisms of phytoremediation and their use in the remediation of contaminated soils.

Qu 3. Can phytoremediation be considered part of bioremediation? Justify.

Qu 4. How can one select a suitable plant for the phytoremediation of industrial wastewater?

Qu 5. What are the different mechanisms of phytoremediation?

Qu 6. Explain the concept and history of heavy metals.

Qu 7. Explain the main properties of plants used for phytoremediation.

Qu 8. Describe the steps of the phytoextraction process.

Qu 9. Explain factors that affect the direct uptake of contaminants via the plant root system.

Qu 10. How microbial activity can be stimulated in the rhizosphere.

REFERENCES

Ali, H., Khan, E., & Sajad, M. A., (2013). Phytoremediation of heavy metals-concepts and applications. *Chemosphere, 91*(7), 869–881. doi: 10.1016/j.chemosphere.2013.01.075.

Fitz, W. J., & Wenzel, W. W., (2002). Arsenic transformations in the soil-rhizosphere-plant system: fundamentals and potential application to phytoremediation. *J. Biotech, 99*, 259–278.

Limmer, M., & Burken, J. G., (2016). Phytovolatilization of organic contaminants, environmental science and technology. *American Chemical Society (ACS), 50*(13), 6632–6643. https://doi.org/10.1021/acs.est.5b04113.

Mench, M., Bert, V., Schwitzguébel, J. P., Lepp, N., Schröder, P., Gawronski, S., & Vangronsveld, J., (2010). Successes and limitations of phytotechnologies at field scale: outcomes, assessment and outlook from COST Action 859. *J. Soils Sediments, 10*, 1039–1070.

Mendez, M. O., & Maier, R. M., (2008). Phytostabilization of mine tailings in arid and semiarid environments–and emerging remediation technology. *Environ. Health Perspect., 116*, 278–283.

Miller, R., (1996). *Phytoremediation, Technology Overview Report* (Vol. 3). Ground-Water Remediation Technologies Analysis Center, Series October 1996. https://clu-in.org/download/toolkit/phyto_o.pdf (accessed on 14 May 2023).

Pilon-Smits, E. A. H., (2005). Phytoremediation. *Annual Review of Plant Biology, 56*, 15–39.

Raskin, I., & Ensley, B. D., (2000). *Phytoremediation of Toxic Metals—Using Plants to Clean up the Environment* (p. 304). J. Wiley & Sons, New York.

Raskin, I., & Ensley, B. D., (2000). Recent developments for *in situ* treatment of metal contaminated soils. In: *Phytoremediation of Toxic Metals: Using Plants to Clean Up the Environment.* John Wiley & Sons Inc., New York.

Suman, J., Uhlik, O., Viktorova, J., & Macek, T., (2018). Phytoextraction of heavy metals: a promising tool for clean-up of the polluted environment? *Front Plant Sci., 9*, 1476. doi: 10.3389/fpls.2018.01476.

CHAPTER 10

Rhizosphere

10.1 CONCEPTS OF RHIZOSPHERE

The role of plant-microbe interaction in shaping the phytoremediation mechanism is very promising. The rhizosphere is an interface of soil and plant, here microbial populations are known to affect various factors regarding plant-microbe association and phytoremediation, such as heavy metal mobility and availability of chelating agents, phosphate solubilization, acidification, and redox potential.

The rhizosphere, firstly described in 1904, by Lorenz Hiltner, for many years has focused on agricultural research because of its improved crop yield. The rhizosphere is a region of stimulated or enhanced microbial activity and biomass at the root-soil interface.

Soil contains a diversity of microorganisms and these microbes form a special zone near the plant root system called the rhizosphere, in which "rhizo" means roots and sphere means surrounding zone. This term was first coined by Hiltner (1904), a soil microbiologist from Germany. Perotti and Aureli (1925) suggested that the rhizosphere has two boundaries Edaphosphere and Histosphere.

- *Edaphosphere* means the roots are surrounded by soils on one side, and
- *Histosphere* is surrounded by soil on the other side by the plant tissues.

Further studies revealed the concept of "rhizosphere phenomenon/effect" which focused on the mutual/symbiotic interaction of microorganisms and plant roots occurring in the rhizosphere. Another term "Rhizoplane" denotes the soil particles adhering to the root surface (Clark, 1949).

Bioremediation and Phytoremediation: Technologies for Toxic Pollution
Ashwini A. Waoo, PhD (Author)

10.2 PLANT, SOIL, AND MICROBE INTERACTION

In the rhizosphere, the interaction between plants and microbial communities is complex and provides mutual benefit to both partners. Plants endure large microbial populations in the rhizosphere by secreting various substances such as amino acids and carbohydrates by root cells. Root cap cells, protect the root from abrasion, root cells may be lost to the soil at a rate of 10,000 cells per plant per day. These root cells also secrete a gelatinous substance mucigel which acts as a lubricant for root penetration through the soil during proliferation of extensive root system.

Rhizospheric microorganisms contribute effectively to the successful survival of plants under the toxic stress of contaminated environments as well as nutrient scarcity environments.

Rhizospheric microorganisms contribute effectively to the successful survival of plants under the toxic stress of contaminated environments as well as nutrient scarcity environments. To optimize the interaction between the plants and the associated microorganisms needs to understand the detailed mechanism. In symbiotic interaction, rhizospheres are actively involved in modifying the host and increasing tolerance to a contaminant or any biotic or abiotic stress. Recent research studies have discovered the potential of genetic manipulation through recombinant DNA technology in plants and to improve soil microorganisms for effective plant interaction.

Microbial populations having plant growth-promoting activities inoculated in natural soil for rhizosphere optimization. These microbes proliferate faster with high concentrations of organic nutrients released as root exudates. It has been found that metal-chelating compounds are released by roots of many plant species into the rhizosphere, these root exudates mobilize heavy metals and help in phytoremediation.

The efficiency of phytoremediation is based on sufficient plant biomass growth, dense root proliferation, and root exudates that can help in the formation of microbial consortiums for enhanced phytoremediation in the rhizosphere. A healthy and microbial consortium can benefit the plant. The overall effect of the plant, soil, and microbe interaction, as shown in Figure 10.1, is an increase in microbial biomass efficiency with that of microbial populations in bulk soils. This is called the "rhizosphere effect." It can be expressed quantitatively as, the ratio of the number of microorganisms in rhizosphere soil to the number of microorganisms in non-rhizosphere soil.

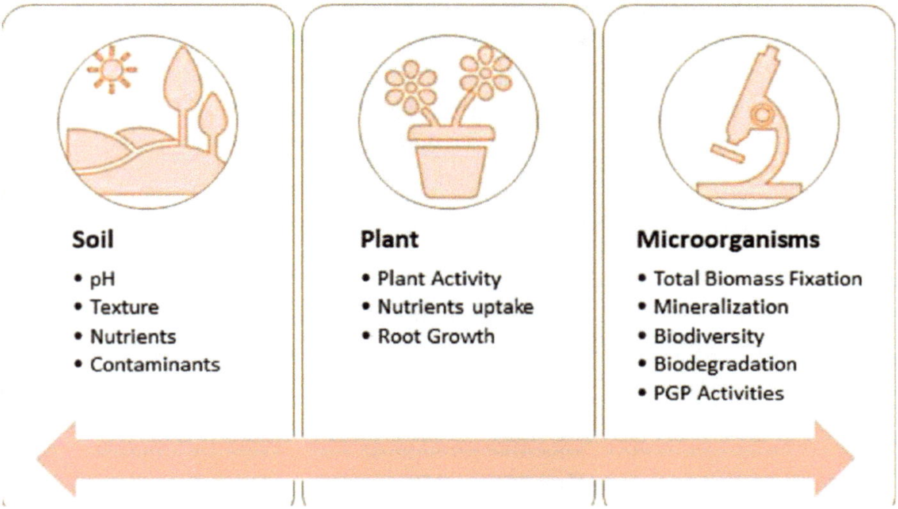

FIGURE 10.1 Interaction of plant, soil, and microbe.

The variety of microbial strains in the rhizosphere play a significant role in trace elements bioaccumulation by the plant. Optimization of rhizosphere microbiome can enhance the potential of phytoextraction and phytoremediation. The efficiency of plant-associated microorganisms is dependent on complex interacting factors: e.g., soil, texture, type and properties, plant-microbe specificity and interaction type, contaminant concentration, type, and chemical nature, microbial diversity, proliferation, and survival rate, etc.

In the rhizosphere, plant roots interact with diverse microbial flora, the potential of phytoremediation is depending on these interactions. This type of interaction affects both the plant-associated microorganisms and the plant. The role of rhizosphere microorganisms is discussed

- Rhizosphere microorganisms carry out the recycling of plant nutrients so that they are readily available to the plant.
- They also help in maintaining soil structure and texture.
- Carry out detoxification of harmful contaminants and promote plant growth (Filip, 2002).
- Rhizosphere microorganisms enhance the remediation capacity of plants and reduce the phytotoxicity of the contaminant.

Plant and microbes have specific interactions which allow the use of unnatural energy sources or carbon sources, thus contaminants in the soil can be degraded biologically. In many nonspecific associations, plants stimulate the microbial population for biodegradation. Plant roots provide root exudates and can increase ion solubility for optimization of the rhizosphere effect. These biochemical processes enhance the remediation activity of microorganisms. The symbiotic relationship between plants and microbes can enhance the adaptation capability and phytoremediation potential of the plant.

10.3 ROLE OF RHIZOSPHERE IN PHYTOREMEDIATION

Successful phytoremediation processes require an understanding of complex interactions in the rhizosphere. Previous literature illustrated that contaminant bioavailability in the rhizosphere plays a crucial role. Limited contaminant bioavailability slower the rate of degradation of the biosphere, this problem can be overcome by the designing of plant–microbial consortia for mobilizing contaminants like metals or metalloids by chemical modification of rhizosphere. Various factors such as pH, ligand exudation, the release of biosurfactants can affect bioremediation or phytoremediation. The selection of plant–microbial associating pairs enhance competitiveness regarding nutrients or habitat sharing, ultimately this can enhance the rate of phytoremediation.

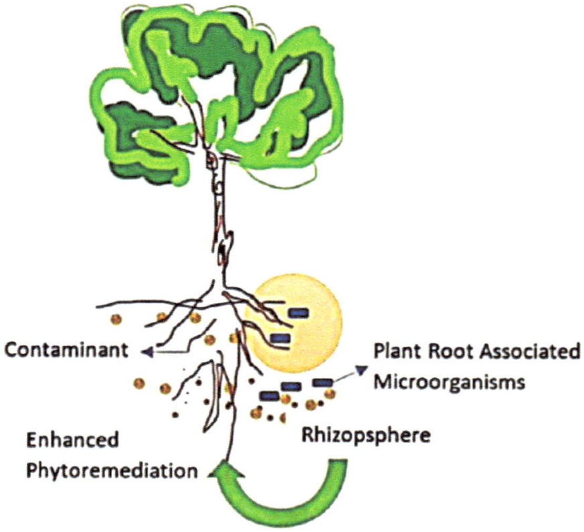

FIGURE 10.2 Role of rhizosphere in phytoremediation.

The rhizosphere is a unique 'remediation unit' to treat contaminated sites, that contains a diversity of microorganisms living with the plant root system.

Physiological and ecological activities of plant roots utilized to remediate contaminated soils, and also accumulate heavy metals, their absorption can take place in association with the microbiome of the rhizosphere. It has been found that the remediation role of the rhizosphere was a significant part of the phytoremediation process, as illustrated in Figure 10.2.

There are various parameters of the rhizosphere that can affect the rate of phytoremediation by promoting plant growth, enhancing the bioavailability of contaminants, stimulating microbial communities. Rhizosphere microorganisms, which are associated with roots, and promote plant growth called Plant growth-promoting rhizobacteria (PGPR) (Glick, 1995). These rhizobacteria can mitigate the toxic effects of heavy metals on plants (Belimov et al., 2004), as shown in Figure 10.3.

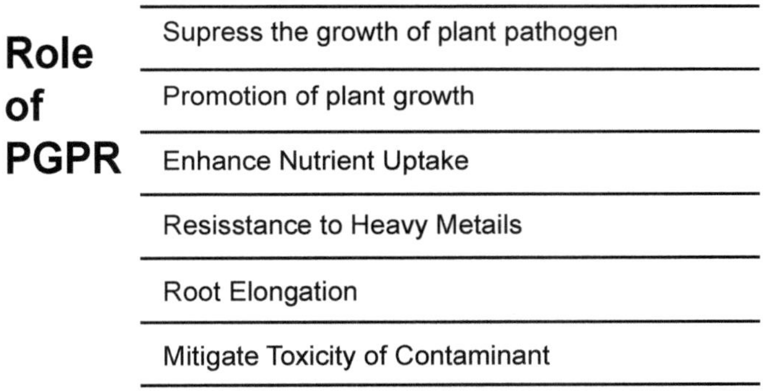

FIGURE 10.3 Role of plant growth-promoting rhizobacteria (PGPR).

High concentrations of toxic pollutants such as heavy metals are also harmful to hyperaccumulator plants, but microorganisms in the rhizosphere optimized the conditions so that plants can sustain there, heavy metals in soil forced some adaptations in the microorganisms. In some cases, high concentrations of heavy metals in soil could decrease the metabolic activity of rhizospheric microorganisms, it can limit their biomass as well as diversity. Some beneficial adaptations can give them the ability to tolerate these adverse conditions.

10.4 PGPR-MEDIATED REMEDIATION OF POLLUTANTS

Rhizobacteria can secrete a variety of compounds like antibiotics, antifungals, some have special mechanisms that are helpful for plants such as phosphate-solubilizing mechanism, secretion of indoleacetic acid (IAA), production of siderophores, absorption of heavy metals All these characteristics ultimately help in promoting vigorous plant growth at contaminated sites (Prakash, 2021).

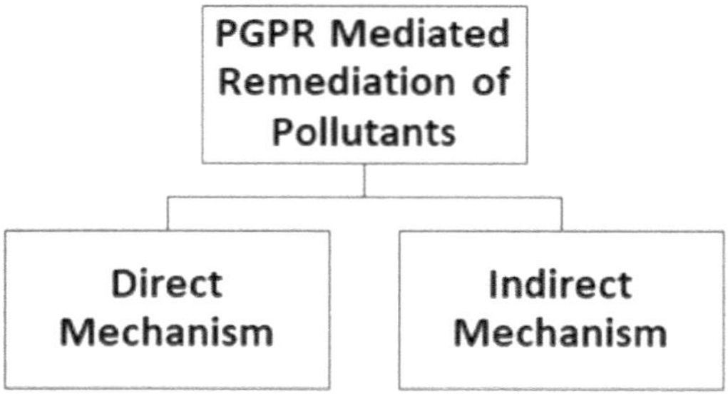

FIGURE 10.4 Types of PGPR-mediated remediation of pollutants.

All the special characteristics of rhizobacteria can be utilized in the remediation of pollutants. These mechanisms can be categorized into direct and indirect mechanisms (Figure 10.4). These mechanisms of PGPR-mediated remediation are illustrated here.

10.4.1 DIRECT MECHANISM

The direct mechanism can be defined as that mechanism by which there is a direct promotion of plant growth this direct mechanism includes production of auxin, nitrogen fixation phosphorus solubilization, etc.

- ***Production of Siderophore***

Siderophores are low molecular weight organic compounds and have a very high and specific affinity to chelate iron (Fe) produced by bacteria and fungi. These compounds are secreted by some bacteria and fungi that are living in low iron conditions (Grobelak and Hiller, 2017). Siderophores are also capable to form complexes with heavy metals such as Cu, Cd, Pb, Zn, etc. This ability can be utilized in phytoremediation. Bacterial siderophores increase the bioavailability of Fe ions from the soil into plants and enhance the phytoremediation process. Some previous researches suggest that plants inoculated with PGPR producing siderophores have no or least symptoms of heavy metal toxicity with an increase in their biomass.

Phytosiderophores act as a good iron source for rhizobacteria. In the siderophore process, most of the heavy metals also accumulated. Heavy metal complexes are comparatively weaker than iron complexes. The bacterial siderophores can mobilize iron to different parts of the plants. Siderophore producing PGPR may decrease the toxicity of some heavy metals by the proper supply of iron. Siderophore producing PGPR help in the overall growth and development of plants in metal contaminated soil (Rajkumar et al., 2010).

- ***ACC Deaminase Activity***

Enzyme 1-aminocyclopropane-1-carboxylate (ACC) deaminase produced by PGPR. It is responsible for the cleavage of the ethylene precursor, ACC, into ammonia and -ketobutyrate in the plant, (Honma and Shimomura, 1978). ACC deaminase-producing organisms can decrease ethylene levels in plants. The hormone ethylene is found in all higher plants, it is a modulator of normal growth and development in plants and act as an indicator of a variety of stresses, illustrated in Figure 10.5.

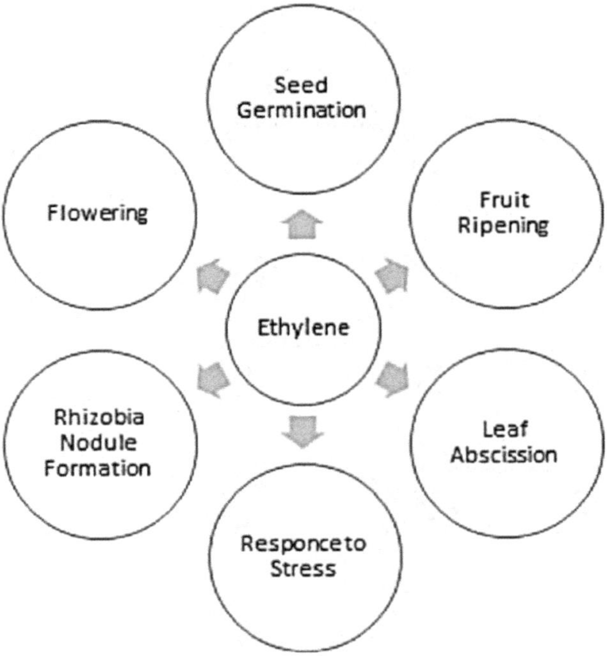

FIGURE 10.5 Role of ethylene in plant growth.

Plant roots have a major role in the phytoremediation process. Vigorous root growth could enhance the rate of hyperaccumulation of inorganic contaminants and also increase the rhizodegradation of organic contaminants, leading to phytoremediation. It has been found that stress induces ethylene production which inhibits root growth and ultimately limits the overall phytoremediation efficiency. Recent researchers found that that ethylene levels in plants regulated by PGPR secreted ACC deaminase enzyme. Plants having ACC deaminase producing PGPR had a low level of Ethelene can form extensive root proliferation (Figure 10.6). This type of extensive root system contributes effectively to phytoaccumulation and phytoremediation. In prospects use of transgenic hyperaccumulator plants that are expressing ACC deaminase will be an excellent tool for phytoremediation of contaminated soil (Arshad et al., 2007).

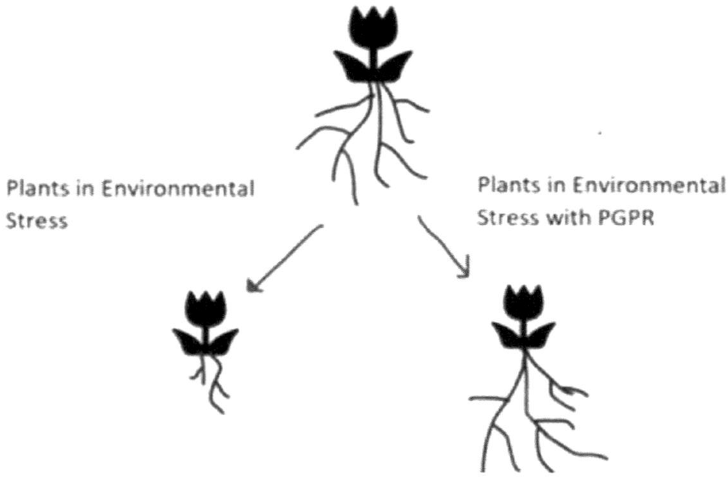

FIGURE 10.6 PGPR-mediated extensive root growth.

- ***Indole-3-acetic Acid Production***

Auxins are significant plant growth regulators mainly used in the agriculture and research field to increase plant growth and biomass. Indole3-acetic acid, i.e., IAA can initiate root elongation and lateral root production. It also activates ATPase in the plasma membrane, affecting the transport of ions across the membrane, those changes can be modulated to accumulation of the heavy metals like contaminants (Tassi et al., 2008).

The phytoremediation potential of any plant is dependent on two major factors:

- Biomass of hyperaccumulator plants and
- The concentration of heavy metals in the harvested plant parts.

IAA helps in increasing biomass as well as the accumulation of heavy metals in plant biomass (Nedjimi,2021). Previous experimental data revealed that the exogenous supply of phytohormones like IAA in the earlier growth stage reduces the toxicity of heavy metals. e.g., the addition of 0.05 M auxin can improve Cadmium tolerance in *Arabidopsis thaliana* with minimum damages (Zhu et al., 2013).

10.4.2 INDIRECT MECHANISM

One interacting mechanism can help the plant indirectly to sustain in adverse conditions like conferring resistance to the plant.

- ***Induced Systematic Resistance***

Rhizobacteria interactions confer disease resistance against some phytopathogens in natural as well as contaminated soils, this mechanism is known as induced systemic resistance, i.e., ISR. When a pathogen attacks the plant, the ISR help in activating the defense system in the plants. Various components are secreted by rhizobacteria such as acetoin, 2,4-diacetylphloroglucinol (DAPG), flagella, 2, 3-butanediol, cyclic lipopeptides, lipopolysaccharides (LPS), siderophores, and other volatile compounds which ultimately activates ISR. There are two types of ISR pathways within the plant: jasmonate and ethylene signaling. These pathways can stimulate plant defense mechanisms against phytopathogens. ISR mechanism also improves resistance against toxic heavy metals and organic pollutants (Gouda et al., 2018).

- ***Secretions from Rhizobacteria***

As mentioned above, rhizobacteria secretion plays a major role in the mechanisms of phytoremediation assisted by rhizobacteria. In the case of indirect mechanisms:

- It prevents phytopathogens from inhibiting plant growth and development, and indirect mechanisms,;
- It promotes nitrogen fixation;

- It promotes the synthesis of siderophores by which plants can solubilize and iron from the soil;
- It induces the production of phytohormones such as auxins and cytokinins, for enhancing plant growth;
- It promotes the solubilization of minerals.

However, there are still many areas where research is needed to manage the rhizosphere for phytoremediation purposes.

- Investigation for the microorganism-induced changes in the rhizosphere;
- Clarifying specific features of microbial-plant and microorganism-soil interactions in the rhizosphere;
- Further research is also required to quantify the effect of rhizospheric processes activated by rhizobacteria;
- Research on heavy metal speciation changes in the rhizosphere;
- Study the mechanisms involved in the mobilization and transfer of heavy metals;
- Optimization of the phytoextraction process.

This kind of research help to understand the role and mechanism of soil rhizobacteria in phytoremediation. To sum up, although the use of rhizobacteria in combination with plants could provide high efficiency for phytoremediation, the microbial ecology in the rhizosphere is not yet fully understood.

10.5 BIOTECHNOLOGICAL TOOLS FOR RHIZOSPHERE

The biotechnological applications of the rhizosphere cover an objective of sustainable agriculture with no or minimal use of chemicals in the form of fertilizers and pesticides; improving crop productivity in adverse conditions like in saline, non-fertile soils, and contaminated soil; and improvement in the active principles in medicinal plants.

In the rhizosphere, microorganisms are the easiest element to manipulate and engineer, since microorganisms native to the rhizosphere play a significant role in plant physiology by directly or indirectly affecting the plant's metabolism. These bacteria perform diverse functions in the rhizosphere such as an increase in nutrient availability in the soil, for better growth of the plant, affecting the plant's hormonal balance and its secondary metabolism which ultimately affect the plant's defense mechanism against a pathogen.

Soils continuously exposed to contaminants like heavy metals such as Cd, Cu, Hg, Ni, Pb, and Zn have high levels of accumulation in plants leading to toxicity. Elevated levels of accumulations of these compounds could pose potential threats to food safety concerns (Ahmadpour et al., 2012; Ali et al., 2013). These metals are potentially toxic to plants and animals causing DNA damage and carcinogenic capacity. Heavy metals persist in soils for a very long time. To minimize these toxic consequences, heavy metals must be prevented from entering the food chain and there must be the employment of cleanup strategies. Metals that are released after the death of bacteria have accumulated metals that are less toxic than their free counterparts. The application of physical or chemical remediation technologies destroys the biotic environment and structure of soils, also the implementation of these technologies is a difficult and expensive task. These toxic elements like heavy metals can be removed from the environment by using this plant-based technology with associated microflora.

In the rhizosphere, the assembly of microbial communities can benefit the plant. But there is also resource competition between plants and microbes. Due to the limiting resources and resource competition, at polluted sites, microbial growth and biodegradation may become limited and for better phytoremediation, it must be enhanced (Unterbrunner et al., 2007). Rhizospheric dynamics also take control over pollutant toxicity. Plants and microorganisms can become adapted to toxic pollutants symbiotically. The rhizosphere is being successfully exploited in alleviating the toxicity by decreasing pollutant concentration. Soil-borne microbial consortia have different genetic and functional activities, that influence soil functions because they are part of fundamental metabolic processes.

Metal-Microbe Interactions are of several types like intracellular accumulation, transformation of metals, and cell wall-metal interaction plant-microbe interactions are controlled by specific molecules responsible for the biogeochemical cycling of nutrients and the maintenance of plant growth and development as well as soil quality.

In phytoremediation, the driving factors such as fast growth rate, and biochemical diversity, metabolize a wide array of natural and xenobiotic chemicals, etc. Ultimately, for successful rhizoremediation, a plant species must have a dense and highly branched root system and a large number of microbial communities. The root is the primary contact point between plants, pollutants, and microbes, it provides a critical factor for the assessment of the phytoremediation efficiency of a particular plant.

Microbe-mediated bioremediation strategies are mostly based on the consortia of a diverse array of organisms indigenous to the contaminated sites having enormous metabolic potential. To isolate and purify such indigenous microorganisms is one of the biotechnological approaches that gives knowledge about microbial metabolites and the pollutant degradation processes (Awasthi et al., 2020).

The recent advances in molecular tools and technologies allow assessing the enormous species of unculturable organisms from natural samples of environments. Molecular approaches such as genomics, transcriptomics, proteomics, metabolomics, etc. help to understand the microbial communities indigenous to a particular contaminated environment.

The beginnings of next-generation sequencing methods and in-silico analyzes enable researchers to solve the problems of unculturable enormous microorganisms simultaneously (Malla et al., 2018). These technologies can produce extensive biological data about microbial communities, their cellular reactions, metabolism, and biomolecules involved in bioremediation.

10.6 METAGENOMICS FOR RHIZOSPHERE

Metagenomics is the new perspective in the field of biotechnology and it is widely used in the rhizosphere approach. Metagenomics involved the molecular level study of the genome of entire communities of microorganisms despite individual species. It can also be known as "communities genomics." This discipline examines the genetic material recovered from environmental samples derived from entire communities of microorganisms. Classical microbiology and genetics, in trying to isolate genes with new and valuable functions, basically are based on the isolation of single microorganisms and the sequencing of their genomes. Metagenomics allows the study of non-culturable, or difficult-to-culture, microorganisms. This approach gives a new estimation of microbial communities independent of their culture in the laboratory (Rolf, 2005).

Rhizosphere metagenomics examines the total DNA extraction from the rhizosphere, the preparation of a clonal library with this DNA, and the screening of the clones to select the genes of interest among the vast genetic reservoir derived from soil microbial communities.

Metagenomics study of the rhizosphere is quite a challenging task. since the rhizosphere is a very complex as well as a dynamic microenvironment Containing a diversity of microorganisms. The rhizosphere is primarily controlled by root exudate root metabolites and other components secreted by different microorganisms. Microbial communities are a very significant part of the rhizosphere. The rhizosphere is said to be a hotspot of plant-microbe interaction. The microbial community is structurally and functionally different and influenced by plant and soil type. Can we say that rhizospheric metabolic output is a function of microbial metabolic reaction? It is essential to study the microbial communities native to any rhizosphere. but due to enormous numbers and diversity, it is a very tedious job Bye every single species of microorganism through individual colony isolation techniques. Soil is one of the major reservoirs of microbial diversity, microorganisms play an important role in maintaining the health of the soil by recycling nutrients in it. The overall rhizosphere is a complex, dynamic, and multi-interactive microenvironment for the soil microbe association.

The culture of microbial species by the traditional method, this approach fails to provide any clear estimate about the total diversity of that soil microenvironment of the rhizosphere but in remediation studies, it becomes necessary to study every species according to its particular function.

10.6.1 ADVANTAGES OF METAGENOMICS

Metagenomics can study the vast majority of genomic data and abundant microflora of the complex and changing environment of the rhizosphere and also have some advantages.

- The metagenome helps in identifying the genomes of diverse groups of soil microorganisms collectively in one step without any culture procedure.
- Directly soil samples can be handled for DNA sequencing studies in metagenomics.
- In a short time, one can assess novel genetic resources from the natural microenvironment.

The use of metagenomics to study soil genomic analysis is a very challenging task and has several difficulties, for example, extraction of humic acid with nucleic acid extraction which interferes with downstream processes. Application of DNA Sequencing methods to soil diverse microbial communities will uncover the hidden resources for example novel genes unknown biomolecules and other significant products which are still and discovered or unknown but have major potential in soil remediation.

The biologically active region of soil near the plant roots is called the rhizosphere. It harbors numerous microbial communities that are significant for soil remediation strategies. Microbial communities involve bacterial algal and fungal species that live symbiotically with the plant root system. Their biological chemical and physical interactions create a dynamic environment in the rhizosphere. Many groups of microorganisms are practically unculturable by the traditional laboratory techniques or standard laboratory conditions. With the help of sequencing techniques like NGS that is next-generation sequencing, it is possible to explore diverse microbial communities at the genome level and reveal their gene structure and function for the study of soil remediation mechanisms.

Initially, these NGS and 16 s-RNA sequencing techniques are utilized for or animal microflora analysis like microorganisms in gut can be identified with metagenomics approach but recently this metagenomic approach is also utilized for plant Research and in remedy, Asian studies soil samples are analyzed for metagenomics and unknown microbial groups having the potential of plant growth promotion, as well as remediation of toxic pollutants, can be identified using metagenomics techniques. Metagenomic approaches also help in comparing the abundance of microbial communities during plant developmental stages and in toxicological studies. The shotgun metagenomics approach is used for exploring the microbial communities in the rhizosphere soil.

Investigation of microbial communities from the rhizosphere in contaminated soils can give knowledge about the microbial species involved in the bioremediation of a, particularly toxic contaminant. Different plant root systems can harbor different types of bacterial communities in their rhizosphere and it results in different compositions of root exudates.

10.6.2 DNA ISOLATION FOR METAGENOMICS

The major objective of the metagenomics study is to obtain good-quality DNA samples from the rhizosphere. DNA isolation protocol can be broadly classified into two types that are direct and indirect DNA isolation methods. All these methods are very according to the degree of sharing purity and in the case of the amount of DNA isolated.

- ***Direct DNA isolation***

Direct DNA isolation technique consists of cell lysis followed by separation of DNA from the cell debris full stop in case of indirect methods celstar extracted from environmental samples before the DNA is extracted are released. Direct DNA isolation methods involve soft and Harsh cell lysis methods. Soft cell lysis technique based on chemical or enzymatic lysis of microbial cells.

The metagenomic analysis is very important to find inaccessible organisms in environmental samples. Natural samples were taken to characterize genomes of all unidentified organisms in the rhizosphere. Some microbial communities or species are in close association with one or another microbial species thus, they cannot be studied by isolating them individually in culture. The metagenomic study can be applied to identify such unknown microbes or microbial communities of natural rhizospheric soil. In this case, the challenging task is to extract the particular sequence from a mixed-species library.

- ***Indirect DNA isolation***

The second strategy, indirect DNA extraction, involves first removing soil from the ground (e.g., 60 grams) and then lysing the soil. Using a density gradient, bacterial and archaeal cells can be separated from eukaryotic cells to some extent (Lefevre et al., 2008). In general, this approach is not the best if eukaryotic sequences are of interest or if interactions between eukaryotes and bacteria or archaea are studied, but it may be useful if eukaryotes should be excluded. The indirect DNA extraction protocols used various methods such as blending and cation exchange (Desai and Madamwar, 2007) or other methods such as silica magnetite nanoparticles for DNA extraction from soil samples (Sebastianelli and Bruce, 2008). A critical aspect of cloning or sequencing is the yield of DNA. This is especially true when vast quantities of DNA are required.

10.7 16S RIBOSOMAL RNA SEQUENCING

Microbe DNA sequencing can often be used to isolate, classify, and quantify microbes within complex biological samples such as environmental samples. The 16S ribosomal RNA is a novel technique for prokaryotic genomes. Approximately 99% of microbes are not easily isolated in culture. The genomic study has been largely driven by this fact. These microbial species are identified by their 16S/18S small subunit ribosomal RNA genes, which are used as phylogenetic markers. Every cellular organism contains these genes.

This study involves DNA extraction directly from an environmental sample such as soil, water, or biofilm. The 16S genes of these community microbes are then amplified using PCR from the mixed genomic DNA sample. After that the PCR products are cloned into vectors and sequenced, this produces rRNA signatures for the microbes present in the sample. These sequences were compared and phylogenetically classified with databases of 16S ribosomal RNA genes. Steps in metagenomics analysis are described here.

- The first step in the metagenomic analysis is to isolate genomic DNA representing the whole microbial community in the target soil sample.
- Then a DNA library is constructed from that isolated DNA.
- After the screening of the available DNA, a library was done for target change identification.

DNA extraction method should be used so that enough yield of DNA can be acquired. This is one of the most crucial steps in metagenomic analysis. The DNA extraction method is selected based on chemical and physical characteristics of soil, it is difficult to standardize the DNA extraction from a particular soil sample because soil represents a very wide and complex texture. Other than soil many substances are extracted with genomic DNA and have any effect on downstream processing and purification of extracted DNA samples for example humic acid fulvic acid, etc.

A DNA library is constructed by fragmenting the isolated DNA into fragments with appropriate sizes and using them for cloning. Restriction enzyme digestion or mechanical shearing can be used for creating DNA Fragments. DNA fragments obtained are cloned into a suitable cloning vector. Usually, plasmid vectors are used for small DNA fragments, and these libraries are called small-insert genomic libraries. Cosmid vectors are used to generate large inserts up to 40 kb in size and BAC vector is used for the inserts with sizes larger than 40 Kb (Simon and Daniel, 2017). Ultimately, a screening assay is applied to identify a gene and its product for a particular function.

There are two different metagenomic approaches used in research:

- The first one is using marker genes like the ribosomal genes 16S rRNA and 18S rRNA to study the composition of the microbial community in a certain environment or to search for specific protein-coding genes having medical or industrial importance, this strategy is called targeted metagenomics.
- The second strategy is shotgun metagenomics, in which there is the use of high-throughput next-generation sequencing methods carried out to

achieve wide coverage of genomic DNA sequences. By this one can assess the entire taxonomic structure or functional potential of microbial communities present in that environment (Quince et al., 2017).

The most difficult aspect in metagenomics is the analysis of such a huge amount of sequence data generated from the constructed library. Various bioinformatics tools have been employed to analyze the metagenomic data and compare it with available online databases.

10.8 PHYLOGENETIC DIVERSITY

Phylogenetic diversity ("PD") is a measure of biodiversity based on phylogeny (tree of life).

The phylogenetic diversity of species is a measure of biodiversity based on phylogenetic differences between species. It is defined and calculated as "the sum of the lengths of all those branches that are members of the corresponding minimum spanning path," in which the 'branch' is a segment of a cladogram, and the minimum spanning path is the minimum distance between the two nodes. In a cladogram, some lines branch out in different directions, ending at a clade, which is a group of organisms that share a common ancestor.

Metagenomics approaches greatly facilitated the identification of novel species and by using this technology, the phylogenetic diversity will be easily assessed and the population structure of the contaminated soil/ water sample. The desired sample may contain several thousand different microbial species including bacteria viruses, etc. Using new metagenomic approaches whole communities of these microbial species can be analyzed for total gene content which can be used for assessing phylogenetic diversity population structure and their trophic interactions can be determined using this data (DeLong et al., 2006). Metagenomics helps to link functional genes of uncultured organisms for finding the identity of that organism; this can be achieved through the identification of phylogenetic marker gene as well as functional gene on the same DNA fragment.

10.9 QUORUM SENSING

Quorum sensing is a process of communication between cells that allows bacteria to share information about cell density and adjust gene expression accordingly. It enables bacteria to make large numbers of energetically heavy processes when their impact on the host and environment will be maximized.

The first quorum sensing in bacteria was observed in *Streptococcus pneumoniae* on genetic competence and in marine Vibrio sp. on bioluminescence. This communication includes specialized auto-inducer signaling molecules (Albuquerque & Casadevall, 2012).

Quorum sensing (QS) of microbes is a stimulus or response that can be visible as a phenotypic change, which is correlated to the microbial population. It is the cell-cell communication phenomenon of prokaryotes. This cell to cell signaling is mediated by autoinducers (AI) in prokaryotes which are produced and secreted by the microbes (Williams, 2007). These auto inducer genes can produce various responses in quorum sensing as shown in Figure 10.7. It is the mechanism of communication between microorganisms based on their cell density. This can affect various behaviors like biofilm formation, virulence factor secretion, bioluminescence competence, and many others in bacteria.

Quorum sensing for example is a process of releasing toxic materials by the pathogens in the host. These toxic chemicals are not abruptly produced by pathogenic microorganisms. When the cell density is low, their virulent genes do not get expressed. But when cell density reached the critical level, these virulent genes get activated and start expression. This helps microbes to ensure their maximum impact. Auto inducer signaling molecules play a key role in the entire process.

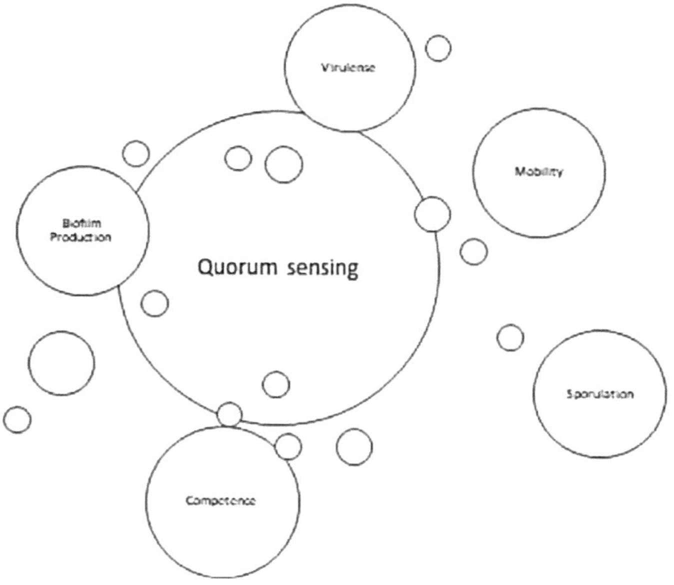

FIGURE 10.7 Responses of auto inducer genes in quorum sensing.

The basic mechanism of QS in bacteria follows three steps;

1. A specialized signaling molecule AI is secreted in the outer space of the cell.
2. Accumulation of molecules takes place with the increase in the cell population density.
3. After reaching the threshold or critical concentration the signaling molecule is sensed by the bacteria and activates the series of regulatory mechanisms (Podbielski & Kreikemeyer, 2004).

10.9.1 QUORUM SENSING IN REMEDIATION TECHNOLOGY

Quorum sensing is mostly used in environmental biotechnology. It may be used for the detection of the pathogenicity of microbes before their use in phytoremediation purposes. Quorum-sensing techniques can easily be exploited for the removal of organic and inorganic pollutants.

10.9.2 QUORUM SENSING FOR HEAVY METAL DETECTION

Quorum sensing is a collaborative approach to communicate between microorganisms for conducting a specific activity in a group. This activity of communicating molecules helps in the formation of various biosensors. This biosensing activity is mostly initiated by a diverse group of microorganisms capable of sensing the bioavailability of various toxic metals.

Biosensors involve the expression of the analyte-induced reporter gene, which may express fluorescence or bioluminescence, or simple colorimetric indications the for detailed information about biosensors refer to Chapter 8. If the concentration of any toxic metal or species increases beyond a critical limit, the AI gene gets expressed forming visible signals (Wang et al., 2013).

Some examples of Biosensors for heavy metal detection

- *Pseudomonas fluorescens* shows the differential gene expression in the presence of the toxic metal cadmium.
- Genetically modified *Vibrio harveyi* is modified to make defective AI1 and AI2, resulting in lead precipitation.
- *Chlorella vulgaris* detects Cadmium, nickel, cobalt, lead, and zinc by conductometric biosensors with gold planar interdigitated electrodes.

Bacterial biosensors have been successfully found to be used for the detection of heavy metals like zinc, cadmium, lead, copper, etc. The whole detection system depends on the activation of gene regulators.

10.9.3 QUORUM SENSING FOR POLLUTANT DEGRADATION PATHWAY

A variety of bacterial species have been identified for their bioremediation abilities for particular toxic pollutants and the majority of them have a quorum sensing approach in their pollutant degradation pathway. *P. aeruginosa* has a wide range of degradation abilities for many contaminants like anthranilate, pharmaceutical, cosmetic, food industry, and oil.

Each degradation pathway is directly related to quorum sensing involving different inducer molecules. Another example is *Burkholderia sp.* DW2–1 can degrade pesticides and wastes present in soil and water. It also has a quorum sensing approach for their degradation metabolism. Supplement of these quorum sensor molecules in the operation site or contamination site, along with the specific degrading bacteria, will result in enhanced, faster, and more effective bioremediation.

10.10 BIOFILM FORMATION

A biofilm is an adherent layer of bacteria that contains many different types of colonies or single cell types. These cells were embedded in extracellular polymeric substances, a matrix made up of DNA, proteins, and polysaccharides. Biofilm formed by the auto-aggregation of the same or multispecies bacteria under specific environmental stimuli such as nutrient availability or stressed conditions, i.e., desiccation, osmotic stress, ultra-violet radiation, or pH change. Exopolysaccharide is the major component of biofilm and other components include protein, lipid, and nucleic acid. Biofilm, generally negatively charged and contains acidic moieties which include uronic acid. The negative charge is due to the abundance of sulfate and phosphate.

A biofilm is an important tool for the remediation of potentially toxic pollutants from water bodies. This innovative technique is helpful for the degradation and removal of heavy metals from polluted aquatic ecosystems. The use of microbial biofilm is a sustainable way of remediating toxic pollutants, pharmaceutical contaminants, synthetic dyes, hydrocarbon pollutants, and plastic waste from polluted water. Biofilm consists of a diversified group of microorganisms mostly surrounded by extracellular polymeric substances. Within the biofilm, all the microorganisms can easily communicate and can do exchange food, materials, and energy. These activities ultimately help the entire community to tolerate various external stresses. Biofilms are continuously exposed to the contaminants in water bodies and then these communities have developed the potential to degrade specific contaminants.

Biofilm forming bacteria are readily adapted for survival in polluted environments and nutrient, as well as oxygen competition, can be seen among themselves. Naturally, biofilms of indigenous microorganisms at polluted sites are found and they are having the efficient potential to survive and manage the stressful and toxic environmental conditions for example pH, temperature, salinity, UV exposure, etc (de Carvalho, 2018).

Biofilm is an organized Cooperative community structure of microorganisms immobilized in the self-secreted extracellular polysaccharides matrix it may contain a mixture of polymers that are excreted by microorganisms play an important role in the flow of nutrients inside the biofilm structure. maybe formed with a single species or multiple species of bacteria, algae, fungi, and archaea. In bioremediation, biofilms can be used as biomarkers for monitoring polluted water.

Biofilm contains cells synthesized matrix; the matrix known as extracellular polymeric substances that is EPS. It protects microbial communities against pollutants, environmental stresses, predatory protozoa. EPS contains various components qualification of proteins, nucleic acids, lipids, humic substances, and surfactants. Water is an important part of biofilm which plays a significant role in nutrients flow inside the biofilm.

EPS has many functions in biofilm maintenance. It affects the density, porosity, water content, charge, mechanical strength of biofilm. It helps in the attachment of biofilm to the substrate it also interacts with the surrounding and via its sorption property. In case of loss of integrity of biofilm then EPS actively supports biofilm failure.

10.10.1 STEPS IN FORMATION OF BIOFILM

Formation of biofilm occurs under the diverse conditions most required moist surfaces. In nature, biofilms are found at all three types of interfaces that is liquid-liquid interface solid-liquid interface, or solid air interface. Biofilm architecture was maintained by EPS components throughout the life of biofilm. Biofilms formed at liquid/liquid interfaces have been applied in hydrocarbon degradation like fuels, oils, and industrial coolants (De Carvalho, 2018).

- Firstly, free-floating, planktonic, bacteria attached to a specific surface.
- Attached bacterial cells produce slimy, extracellular polymeric substances (EPS).
- Then colonization takes place on the surface.
- Cell adhesion occurs with the help of cell adhesion molecules, proteins.
- More EPS production allows the biofilm microbial community to develop a more complex and three-dimensional structure.
- These biofilms can be propagated through the detachment of small or large aggregates of cells that releases individual cells called detachment.

For more detail on biofilm production mechanisms and the role of biofilm in rhizoremediation, please refer to Chapter 11.

10.10.2 BENEFITS OF USING BIOFILMS FOR REMEDIATION OF POLLUTANTS

Biofilm is the collection of the diversity of microorganisms surrounded by EPS. In this biofilm structure, microorganisms can easily communicate and can exchange the food materials and energy inside the biofilm as well as with the environment. These things help the microbial community to tolerate a variety of external stresses and pressures.

These all aspects give a multifaceted and strong mechanism for the self in biofilm to work as an organ system. Thus, microbial communities inside the biofilm develop a potential to tolerate and detoxify toxic contaminants.

10.11 SUMMARY

Rhizosphere soils are "epicenters" of bacterial concentrations because they accumulate a lot of nutrients. Although no attention has been paid to inoculants and soil-type substrates over the years, they are important to bacterial growth. The implications of rhizosphere soil and its associated bacteria for biotechnological applications were assessed in this chapter. Rhizosphere soil is a treasure trove for bacterial community concentration, biodegradation of lignocellulose for biofuel production, the role of rhizosphere soil and its bacteria for bioremediation and biofiltering. Rhizosphere soil, along with its bacteria, can be efficiently utilized in an environmentally friendly way to promote healthy and sustainable environments. There is evidence that plant root exudates and other rhizo-deposits attract beneficial bacteria to the rhizosphere, as well as unwanted ones. Because of variations in the composition of the root exudate, a specific plant microbiota is favored and attracts a specific host-microbiome induced by the host plant.

Globally, biotechnology has a significant impact on food and health, as well as environmental sustainability. Rhizosphere soil bacterial materials were reviewed as a prospective essential tool for utilizing naturally occurring materials in biotechnology. The reason is that they can be used for soil amendment, as well as to produce biofuels, to remediate heavy metals, and to filter gasses. Further critical research is required to maximize the full potential of rhizosphere soil for biotechnological purposes, taking advantage of the metabolic potential of bacteria found in rhizosphere soils. Also, experimental methods that stimulate and improve rhizosphere soil restoration by bacteria should be tested in the future to certify a sustainable application of rhizosphere bacterium materials.

10.12 EXERCISES

Objective Response Questions

1. Which of the following is not a subgroup of Metal-Microbe Interactions?

a) Intracellular accumulation
b) Transformation of metals
c) Cell wall-metal interaction
d) Hydrolysis

2. What of the following does not contribute to the bacterial assimilation of metals?

a) Detoxification
b) Enzyme functions
c) Physical properties of cell
d) Resistance

3. 0.3% of trace metals can be found in *E. coli.*

a) True
b) False

4. Which of these binds to toxic metals?

a) Extracellular ligands
b) Intercellular ligands
c) Intracellular ligands
d) Integral proteins

5. Which metal can be accumulated by a large number of bacteria?

a) Cadmium
b) Water
c) Carbon dioxide
d) Fungus

6. What happens to bacteria that accumulate metals after they die?

a) Metal species released
b) Metal species degraded
c) Free metal species released
d) Free metal species dissolves

7. Biomarkers having high toxicological _______.

a) Relevance
b) Identity
c) Occurrence
d) Quantity

8. Physico-chemical components include which of the following?

a) Cells or tissues
b) Transducer
c) Anti-bodies
d) Enzymes

9. Which of these microbes is not indigenous for bioremediation?

a) *Pseudomonas aeruginosa*
b) *Phanerochaete sordida*

c) *Piscirikettsis salmonis*
d) *Deinococcus radiodurans*

10. Rhizosphere consist of the dynamic interaction of ________.

a) Plant
b) Soil
c) Microorganisms
d) All of the above

Practice Work

Qu 1. Through plant microbiome analysis, what kind of insights do you anticipate.

Qu 2. Explain the term soil microbiome.

Qu 3. What is Quorum sensing? Explain the mechanism in detail.

Qu 4. What is the significance of metagenomics research?

Qu 5. Explain the rhizosphere effect and its implications on bioremediation research.

Qu 6. Discuss the Role of Rhizosphere microorganisms in Nutrition of Plants and their PGPR activities

Qu 7. What is biofilm? Explain in detail the steps involved in biofilm formation.

Qu 8. describe the role of 16S ribosomal RNA Sequencing in rhizosphere research.

Qu 9. Explain PGPR Mediated remediation of pollutants.

Qu 10. What are Siderophores? Explain their significance.

REFERENCES

Albuquerque, P., & Casadevall, A., (2012). Quorum sensing in fungi—A review. *Medical Mycology, 50,* 337–345.

Arshad, M., Saleem, M., & Sarfraz, H., (2007). Perspectives of bacterial ACC deaminase in phytoremediation. *Trends in Biotechnology, 25,* 356–362. 10.1016/j.tibtech.2007.05.005.

Awasthi, M. K., Ravindran, B., Sarsaiya, S., Chen, H., Wainaina, S., Singh, E., Liu, T., et al., (2020). Metagenomics for taxonomy profiling: tools and approaches. *Bioengineered, 11*(1), 356–374. doi: 10.1080/21655979.2020.1736238.

Basu, S., Roel, C. R., Negi, S., & Shukla, P., (2018). Engineering PGPMOs through gene editing and systems biology: A solution for phytoremediation. *Trends in Biotechnology, 36*(5), 499–510. ISSN 0167-7799. https://doi.org/10.1016/j.tibtech.2018.01.011.

Carvalho, D. C. C. J., (2018). Marine biofilms: a successful microbial strategy with economic implications. *Front. Mar. Sci., 5*, 126.

Clark, F. E., (1949). Soil microorganisms and plant roots. In: *Advances in Agronomy* (Vol. 1, pp. 241–288). New York: Academic Press.

DalCorso, G., Fasani, E., Manara, A., Visioli, G., & Furini, A., (2019). Heavy metal pollutions: state of the art and innovation in phytoremediation. *Int. J. Mol. Sci., 20*(14), 3412. doi: 10.3390/ijms20143412.

DeLong, E. F., Preston, C. M., Mincer, T., Rich, V., Hallam, S. J., Frigaard, N. U., Martinez, A., et al., (2006). Community genomics among stratified microbial assemblages in the oceans interior. *Science, 311*, 496–503.

Desai, C., & Madamwar, D., (2007). Extraction of inhibitor-free metagenomic DNA from polluted sediments, compatible with molecular diversity analysis using adsorption and ion-exchange treatments. *Bioresour. Technol., 98*, 761–763.

Filip, Z., (2002). International approach to assessing soil quality by ecologically-related biological parameters. *Agric. Ecosyst. Environ., 88*(2), 689–712.

Gerhardt, K. E., Gerwing, P. D., & Greenberg, B. M., (2017). Opinion: taking phytoremediation from proven technology to accepted practice. *Plant Sci., 256*, 170–185. doi: 10.1016/j.plantsci.2016.11.016.

Glick, B. R., (1995). The enhancement of plant growth by free-living bacteria. *Can. J. Microbiol., 41*, 109–117.

Grobelak, A., & Joanna, H., (2017). Bacterial siderophores promote plant growth: Screening of catechol and hydroxamate siderophores. *International Journal of Phytoremediation, 19*(9), 825–833. doi: 10.1080/15226514.2017.1290581.

Honma, M., & Shimomura, T., (1978). Metabolism of 1-aminocyclopropane-1-carboxylic acid. *Agric. Biol. Chem., 43*, 1825–1831.

Malla, M. A., Dubey, A., Yadav, S., Kumar, A., Hashem, A., & Abd, A. E. F., (2018). Understanding and designing the strategies for the microbe-mediated remediation of environmental contaminants using omics approaches. *Front. Microbiol., 9*, 1132. doi: 10.3389/fmicb.2018.01132.

Nedjimi, B., (2021). Phytoremediation: a sustainable environmental technology for heavy metals decontamination. *SN Appl. Sci., 3*(286). https://doi.org/10.1007/s42452-021-04301-4.

Perez-Pinera, P., Kocak, D., Vockley, C. M., Andrew, F. A., Kabadi, A. M., Polstein, A. M., Thakore, P. I., et al., (2013). RNA-guided gene activation by CRISPR-Cas9–based transcription factors. *Nat Methods., 10*, 973–976. https://doi.org/10.1038/nmeth.2600.

Perotti, R., & Aureli, F., (1925). *Soil Science, 19*(1), 85–86.

Podbielski, A., & Kreikemeyer, B., (2004). Cell density-dependent regulation: Basic principles and effects on the virulence of gram-positive cocci. *International Journal of Infectious Diseases, 8*, 8195.

Prakash, J., (2021). Plant growth promoting rhizobacteria in phytoremediation of environmental contaminants: challenges and future prospects. *Bioremediation for Environmental Sustainability, 2021,* 191–218. http://dx.doi.org/10.1016/B978-0-12-820318-7.00009-5.

Quince, C., Walker, A. W., Simpson, J. T., Loman, N. J., & Segata, N., (2017). Shotgun metagenomics, from sampling to analysis. *Nature Biotechnology, 35*(9), 833.

Rajkumar, M., Ae, N., Majeti, P., & Freitas, H., (2010). Potential of siderophore-producing bacteria for improving heavy metal phytoextraction. *Trends in Biotechnology, 28*(3), 142–149. Epub 2010 Jan 13. PMID: 20044160. 10.1016/j.tibtech.2009.12.002..

Sarkar, D., Poddar, K., Verma, N., Biswas, S., & Sarkar, A., (2020). Bacterial quorum sensing in environmental biotechnology: a new approach for the detection and remediation of emerging pollutants. *Emerging Technologies in Environmental Bioremediation* (pp. 151–164). Chapter 6. Elsevier. ISBN 9780128198605. https://doi.org/10.1016/B978-0-12-819860-5.00006-7.

Sebastianelli, T., & Bruce, S., (2008). Extraction of DNA from soil using nanoparticles by magnetic bioseparation. *Lett. Appl. Microbiol., 46*, 488–491.

Simon, C., & Daniel, R., (2017). Construction of small-insert and large-insert metagenomic libraries. In: *Metagenomics* (pp. 1–12). New York, NY: Humana Press.

Sushanto, G., Rout, G. K., Das, G., Paramithiotis, S., Han-Seung, S., & Patra, J. K., (2018). Revitalization of plant growth promoting rhizobacteria for sustainable development in agriculture. *Microbiological Research., 206*, 131–140, ISSN 0944-5013. https://doi.org/10.1016/j.micres.2017.08.016.

Tassi, E., Pouget, J., Peturzzelli, G., & Barbafieri, M., (2008). The effects of exogenous plant growth regulators in the phytoextraction of heavy metals. *Chemosphere, 71*, 66–73.

Unterbrunner, R., Wieshammer, G., Hollender, U., Felderer, B., Wieshammer-Zivkovic, M., Puschenreiter, M., & Wenzel, W., (2007). Plant and fertilizer effects on rhizodegradation of crude oil in two soils with different nutrient status. *Plant Soil, 300*, 117–126. https://doi.org/10.1007/s11104-007-9394-2.

Van, D. P., Caballero, A., Paz, J. A., Gonzalez-Perez, M. M., Oliva, J. M., & Ramos, J. L., (2007). Bioremediation of 2,4,6 trinitrotoluene under field conditions. *Environ. Sci. Technol., 41*, 1378–1383.

Wang, B., Barahona, M., & Buck, M., (2013). A modular cell-based biosensor using engineered genetic logic circuits to detect and integrate multiple environmental signals. *Biosensors and Bioelectronics, 40*, 368–376.

Williams, P., (2007). Quorum sensing, communication and cross-kingdom signaling in the bacterial world. *Microbiology, 153*, 3923–3938.

Zhang, X., Xia, H., Li, Z., Zhang, P., & Gao, B., (2010). Potential of four forage grasses in remediation of Cd and Zn contaminated soils. *Bioresour. Technol., 101*, 2063–2066.

Zhu, X. F., Wang, Z. W., Dong, F., Lei, G. J., Shi, Y. Z., Li, G. X., & Zheng, S. J., (2013). Exogenous auxin alleviates cadmium toxicity in *Arabidopsis thaliana* by stimulating synthesis of hemicellulose 1 and increasing the cadmium fixation capacity of root cell walls. *J. Hazard Mater, 263*, 398–403.

CHAPTER 11

Rhizoremediation Research

11.1 RHIZOREMEDIATION

Rhizoremediation is the process in which soil contaminants can be degraded by microorganisms in the rhizosphere. Generally, organic pollutants are degraded by this process. Plants are not considered as the direct-acting element for remediation in this process. A suitable niche was created by the plant for the flourishing of rhizosphere microbial communities and these microbial populations can carry out the degradation of pollutants. Plants act as a solar pump for the mobilization of water and pollutants in the soil. Plants also secrete substrates and chemical compounds that are beneficial for microbial growth and survival.

Rhizoremediation systems are based on a harmonious relationship between plants and their root system and associated microbial populations. The roots of plants exude organic compounds, which increase the density and activity of microbes that can degrade hydrocarbons. The bacteria having biodegradation abilities and maintenance of these bacteria in the rhizosphere are important ways for the effective rhizoremediation of contaminants. Thus, rhizoremediation contributes mainly to the removal of these undesired substances from the rhizosphere (Prasad et al., 2011).

Rhizospheric microorganisms can degrade most of the environmental pollutants. These degradation processes can't possible in scarcity of food. The microorganisms get access to the best food material available in the soil, e.g., root exudates. The degradation of a selected pollutant by microbes in rhizosphere and their property of excellent root colonization, this process termed as 'rhizoremediation.' Uptake of high concentration of heavy metals from soils by the plants influences variously in the form of the growth, the yields of crops, biomass, etc. It may be due to the disintegration of cell organelles

Bioremediation and Phytoremediation: Technologies for Toxic Pollution
Ashwini A. Waoo, PhD (Author)

and the membranes, disrupting the physiological process, protein synthesis, respiration, and carbohydrate metabolism by affecting photosynthesis.

Rhizoremediation involves the selection of suitable microorganisms. One can isolate such microorganisms from contaminated soil near the rhizosphere. These bacteria having the ability to survive on the contaminated soil can be further enriched. The enrichment technique can be applied with plants where root colonization can be facilitated. Here a host plant supports a pollutant degrading microbial community in its rhizosphere. Indigenous or wild-type organisms can be selected for this purpose by following strategy (Figure 11.1).

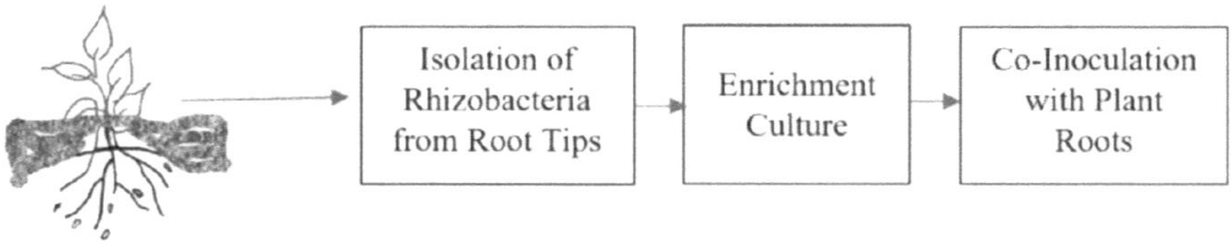

FIGURE 11.1 Strategy of rhizosphere remediation.

Rhizoremediation is an excellent form of bioremediation technique that operates on harnessing light energy through plants and use of that energy to stimulate rhizodegradation of pollutants by the action of native soil microbial diversity. Microorganisms of the rhizosphere can degrade or transform xenobiotics compounds. This rhizoremediation potential gets increased by the increased population of microorganisms and the availability of growth-promoting substances for co-metabolism.

The extensive plant root system provides aeration in the soil layers so that rhizobacteria penetrates through impermeable soil. It also solubilizes the pollutants in soil-water interphase, making them bioavailable for microorganisms.

A variety of microbial communities are found in rhizosphere microhabitats. These microbial communities depend on various factors like the type of plant species, the composition of the root exudates, root type, age of the plant, and soil type. It has been found that gram-negative bacteria, Pseudomonas species predominantly found in the rhizosphere. Rhizosphere bacteria that cause rhizoremediation possess certain special characteristics, such as metal resistance. As well as storing or bioaccumulating metals in their cells, they are also able to transform toxic metals into non-toxic forms using microbial enzymes and exopolysaccharides (Kamaludeen et al., 2008).

Increased availability of nutrients in the rhizosphere offers a diverse population of microorganisms in the rhizosphere which is several times greater than the general soils. However, high concentrations of metals can reduce the microbial population by inhibiting their metabolic activities, biomass. It can cause a permanent change in microbial community structure. The inhibition is mostly seen in newly contaminated soils. After a long time, the diversity and composition of microbial communities contain the tolerant or resistant microbes in rhizosphere soils.

11.2 RHIZOREMEDIATION OF HEAVY METALS

Microbial isolates from the rhizosphere can be effectively utilized for bioremediation of contaminated soils. Rhizospheric PGPR strains (Figure 11.2) are used for bioremediation of metals called rhizoremediation of metals.

PGPR useful metal contaminated soils (Zhuang et.al., 2007)	*Rhizobium leguminosarum*
	Kluyvera ascorbata
	Brevundimonas sp.
	Pseudomonas fluorescens
	Bacillus mucilaginosus
	Azotobacter chroococcum
	Bacillus megaterium
	Bacillus subtilis SJ-101

FIGURE 11.2 List of PGPR associated with plants grown in metal contaminated soils.

The rhizobacteria associated with hyperaccumulator plants such as *Bacillus subtilis*, *Bacillus pumilus Brevibacterium halotolerant*, and *Pseudomonas pseudoalcaligenes* are mostly used in bio- and rhizo-remediation of heavy metal contaminated sites.

Bacteria from the genera *Pseudomonas* and *Bacillus* are mostly nickel tolerant. These isolates can tolerate high concentrations of Ni showing uptake of nickel. Thus, can be used for bioremediation of Ni contaminates sites. Instead of a single strain, a group of bacterial cultures can carry out effective rhizoremediation. Chen and Cutright made use of a microbial consortium to treat heavy metal toxicity in aqueous solutions in 2003. The consortium proved resistant to metal toxicity.

11.3 ENGINEERED RHIZOBACTERIA FOR METAL UPTAKE

A recombinant DNA technology involves modifying the genetic material outside of the organism to enhance and create desirable properties within the organism itself. Various DNA fragments with desirable gene sequences are inserted into an appropriate vector using this method. Microbe plant symbiosis within the rhizosphere is an effective remediation technology for a metal-contaminated site. Recombinant DNA technology can be effectively utilized for engineering rhizobacteria to remediate metal-contaminated sites. There are some examples of previous research, e.g., *Escherichia coli* and *Moraxella* sp. expressing EC20 on the cell surface or intracellularly accumulating up to 25-fold more cadmium or mercury. Recombinant bacteria having the gene phytochelatin synthase are shown to have more Cd accumulation. Inoculation of engineered rhizobacterium with metal-binding peptide Pseudomonas putida has shown a marked decrease in Cadmium phytotoxicity. But the major constraint of this technique is that recombinant bacteria must be capable of facing adverse environmental conditions and competition from indigenous microbial species.

Certain criteria have been suggested for the selection of an appropriate PGPR strain for gene cloning and inoculating into the plant rhizosphere, these are:

- The strain must be stable in cloning.
- The expression of the target gene should be high.
- The PGPR strain must be resistant or tolerant to the heavy metal concentration.
- PGPR strains must associate with a specific plant rhizosphere only (Huang et al., 2004). Although the genetic manipulation for PGPR improvement needs continuous research and development.

Establishing large populations of rhizobacteria is a difficult task for the success of environmental remediation technology. These native microorganisms are successful in the soil due to their high adaptability and their faster growth rate, they have broad biochemical versatility and can metabolize a variety of natural as well as xenobiotic compounds. Generally, the density of rhizobacteria associated with plant roots was found to be up to 10^9 to 10^{11} microbial cells per gram root (Zeng et. al., 2018).

The rhizosphere is the ideal site to modify or engineer microbial populations the process is called bioengineering. Rhizoengineering deals with the strategies favorable for the growth of the desired microbes and enhancing their ability to metabolize toxic or xenobiotic compounds. The general nutrient-limiting conditions in the rhizosphere enabled a nutritional bias that is particularly effective in identifying microbial populations. Rhizoengineering was first successfully applied to opines, an exotic nutrient produced by transgenic plants (Oger, 1997; Savka and Farrand, 1997). As a result, the metabolizing strains grew more rapidly and efficiently than the strains that could not metabolize opines. The success of the above approach is dependent on the general nutrient-limiting conditions which prevail in the rhizosphere. The presence of such conditions in the rhizosphere of many plant species causes root-associated bacterial ("rhizobacterial") populations to starve.

11.4 ADVANTAGES OF RHIZOREMEDIATION

Rhizoremediation results in the decomposition of pollutants through the action of exudates that are derived from the plant. By stimulating the survival and activity of bacteria, these exudates can enhance their survival. Rhizoremediation offers many benefits.

1. It is eco-friendly and cost-effective.
2. No need to add any chemicals for its removal.
3. Little or no side effect.
4. Does not need any investment.

However, chemical degradation processes like chemical oxidation, volatilization, can affect the functioning of the environment. There is no need for anything else without an efficient microbial strain. Researchers must also conduct extensive research on hyperactive accumulators and rhizosphere-microbial variety, as well as rhizoremediation procedures for metal removal from soils using microorganisms. Rhizoremediation has been suggested by many researchers as a means of removing contaminants from the environment.

Rhizoremediation can be accomplished in three ways, i.e., biosurfactant production, organic acid production, and biofilm formation. Further research works oriented towards isolation of different microbial strains from various locations including rhizosphere and any contaminated site, characterize them, and then use them for inoculation at rhizoremediation site (Figure 11.3).

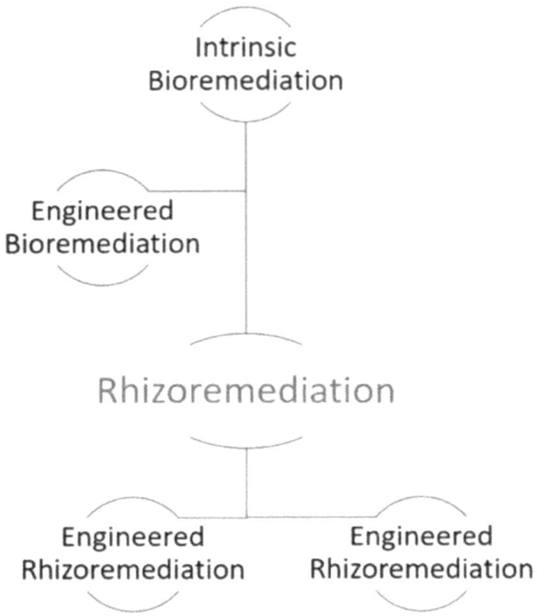

FIGURE 11.3 Strategies of bioremediation.

Rhizoremediation is the process of symbiotic interaction between the rhizosphere of a plant and the soil microbe community. The root exudates provided microorganisms with energy and the plants with protection from contaminant stress along with nitrogen fixation. The root exudate mostly provides energy sources or carbon sources such as glucose, fructose, etc., and can help in the stimulation of metabolic pathways of the microorganisms (Yadav et al., 2018).

The rhizoremediation process offers extra benefits for the entire ecosystem. This process creates the expansion of microbial populations and their diversity. It boosts metabolic activity in the rhizosphere. It also enhances the physicochemical properties of contaminated soil.

11.5 CONTAMINANT REMOVAL MECHANISMS IN RHIZOREMEDIATION

During rhizoremediation, the toxic metals are removed or their toxicity is reduced. Toxic metals are removed from the soil in three significant ways, as shown in Figure 11.4.

1. Biosurfactant production.
2. Biofilm formation.
3. Organic acid production.

11.5.1 BIOSURFACTANT PRODUCTION

Biosurfactants are a special group of surface-active molecules synthesized by microorganisms. They are amphiphilic containing the polar and nonpolar parts in their structural organization. Monosaccharides, polysaccharides, proteins, or peptides group, forms a hydrophobic group and whereas, the hydrophilic part consist of saturated and unsaturated fat, hydroxylated fatty acid, and fatty alcohol (Saharan et al., 2011). These all-specialized surface properties of biosurfactants make them an excellent emulsifying, and foaming agent.

Biosurfactants are secreted in the rhizosphere by the microorganism/ rhizobacteria as extracellular, mixtures of hydrophilic and hydrophobic compounds. As they accumulate between fluid phases, they reduce surface tension and interfacial tension. Micelles are formed between hydrophilic and hydrophobic substances to increase the solubility of hydrophobic contaminants.

This process helps in rhizoremediation.

Currently, biosurfactants gain extra importance in research because of their diversity, eco-friendly nature, suitability for large-scale production, their selectivity and performance under adverse circumstances, and their applications in environmental fortification as well as rhizoremediation.

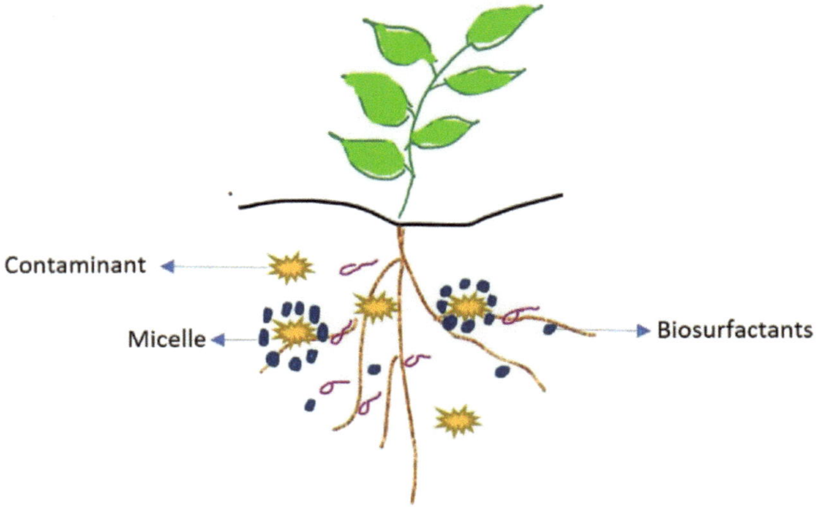

FIGURE 11.4 Microorganisms producing biosurfactants.

Many microorganisms are capable of producing biosurfactants, which can be found in a variety of chemical compositions and differ in their nature and amount of production depending on the type of producing microorganism. Microorganisms such as *Ustilago maydis, Serratia marcescens, Rhodotorula glutinis, R. graminis, Pseudomonas sp., Thiobacillus thiooxidans,* and *Agrobacterium sp.,* can effectively produce different types of biosurfactants.

11.5.2 BIOFILM FORMATION

A biofilm is defined as the collection of microbial species attached to a biological, inert surface and enclosed in a self-synthesized matrix made up of water, carbohydrates, proteins, and extracellular DNA (Figure 11.5). As biofilm contains different microbial species, they are having diverse metabolic pathways, thus consortia of biofilms can capable of degrading different pollutants or a mixture of pollutants, individually or collectively. The application of biofilms is an effective tool for bioremediation because biofilms can easily absorb, degrade or immobilize various toxic pollutants in the environment. Mostly bacterial biofilms exist within native populations at contaminated sites so that all can survive, manage, and tolerate the harsh environmental condition (Sharma et al., 2020)

Biofilm

- DNA
- RNA
- Exopolysachharides
- Protein
- Water
- Ions
- Microbial Cells
- Lipids

FIGURE 11.5 Components of biofilm.

Previous researchers have shown that the biofilm matrix is made up of large amounts of surface proteins that play a significant role in the organization and stability of biofilms. Biofilm proteins are categorized into two subclasses: enzymes and structural proteins.

- ***Mechanism of Biofilm production***

The existence of a surface allows bacterial cells to attach to it, which is an important step in the formation of biofilms. Biofilm formation begins with surface attachment. Most bacteria prefer to colonize on any physical surface in a hydrophobic nature. Some extracellular substances, like exopolysaccharides and lipopolysaccharides, bacterial surface structures, combined with environmental factors and cell signaling, are all involved in the formation of biofilm.

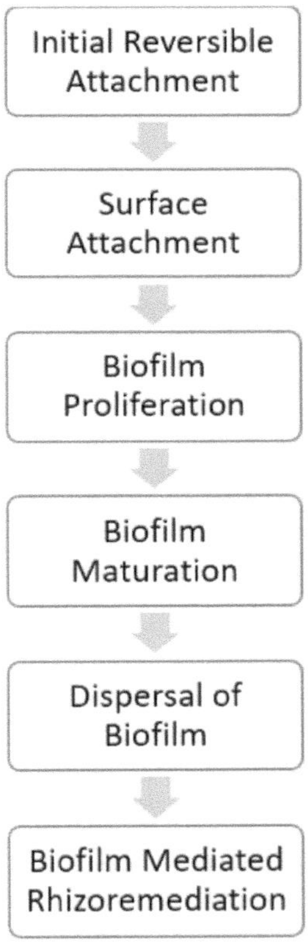

FIGURE 11.6 Stages of microbial biofilm production.

The extracellular polysaccharides act as anion exchangers and prevent the exposure of biofilm to the antimicrobial agents and protect them against different environmental stresses. The initial attachment steps are carried out by external membrane proteins, e.g., calcium-binding proteins, pili, or lipopolysaccharides. Quorum sensing signals are crucial for the development of a mature biofilm. Cell density in biofilm was increased through the cell-cell signaling mechanism.

There is a complex multicellular structure formed by bacteria called a biofilm. There are some general stages involved in biofilm formation (as illustrated in Figure 11.6) such as, attachment of bacteria to a surface, formation of microcolonies, maturation of biofilms, detachment of bacteria, also known as dispersal that can colonize new areas. Bacteria within biofilms, referred to as sessile bacteria, are stationary and show unique phenotypes compared to planktonic bacteria. An exceptional level of resistance to environmental stresses is displayed by bacteria in biofilms. Such biofilms are very significant for pollutant degradation (Figure 11.7).

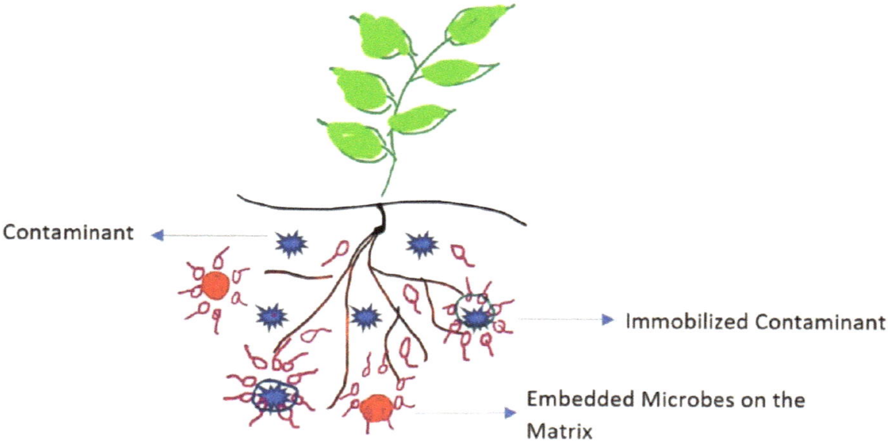

FIGURE 11.7 Biofilm production.

- ***Biofilms for Rhizoremediation of Pollutants***

Biofilm remediation is an environmentally, green, and cost-effective technology to remove pollutants in the environment. Bacterial cells in Biofilm are suitable for survival in harsh conditions by competing for nutrients and oxygen. For the monitoring of environmental pollution and proper planning for remediation, various indicator biofilms can be evaluated based on different parameters such as the composition of species, changes in biomass, photosynthesis, pigment production, and enzymes.

- ***Biofilm Bioreactors for Remediation of Toxic Pollutant***

To increase the rate of degradation of toxic pollutants, biofilm bioreactors can be used as an ex-situ bioremediation strategy, these biofilm reactors can be installed either offsite or near the polluted site. Bioreactors can manage favorable conditions and result in better bioremediation than conventional techniques. Commercial biofilm-based bioreactors are available for cleaning the environment.

Biofilm-bioremediation requires both adequate technical expertise and interdisciplinary approaches from environmental engineering, microbiology, soil science, and civil engineering to be successful. The matrix in biofilm, cell-cell interaction, production of genetically modified organisms to form biofilm as well as efficiently degrade xenobiotics are the future research areas in biofilm Bioremediation.

11.5.3 ORGANIC ACID PRODUCTION

Organic acids for example citric acid, gluconic acid, malonic acid, succinic acid, malic acid, lactic acid, and oxalic acid are produced by the microorganisms present in the rhizosphere. This production of organic acid occurs during their normal growth and development. The organic acids are capable of acting as chelating agents by supplying their protons and organic anions (Figure 11.8). These negatively charged anions can easily bind with positively charged metal ions like Al^{3+}, Ca^{2+}, Fe^{3+} to form complexes and chelate them from the soil (Mardad et al., 2013). These organic acids can also act as a carbon source for the process of Phosphate (P) solubilization and help in increasing the solubility of the contaminants by reducing the pH of the soil of the rhizosphere.

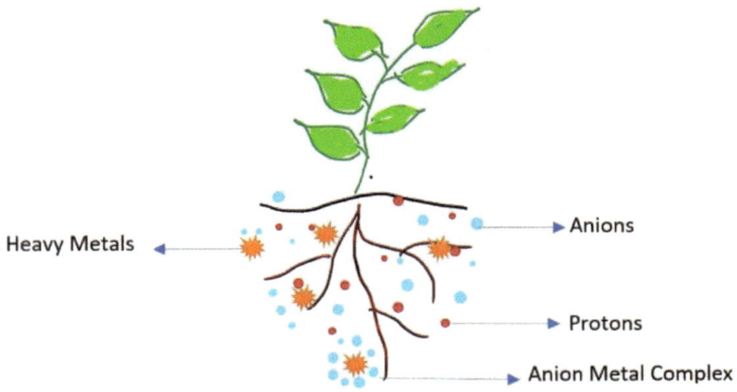

FIGURE 11.8 Mechanism of production of organic acid.

- ***Role of Organic Acids in Rhizosphere***

Rhizosphere organic acids play a very dynamic role (Figure 11.9). They can dissolve insoluble minerals in soil by three major mechanisms, i.e., acidification, chelation, and exchange reactions. In the acidification process, most metals become soluble and released into the soil solution. They help in enhancing mineral acquisition by plants in the case of calcareous soils. The second chelation mechanisms lead to the formation of complexes with metals by the organic acids, that help in facilitating their solubility in the soil. Organic acids have a role in various specific physical and biochemical processes, e.g., nutrient uptake, bacterial chemotaxis, soil structural improvement, anoxia alleviation, metal decontamination, and PAH degradation.

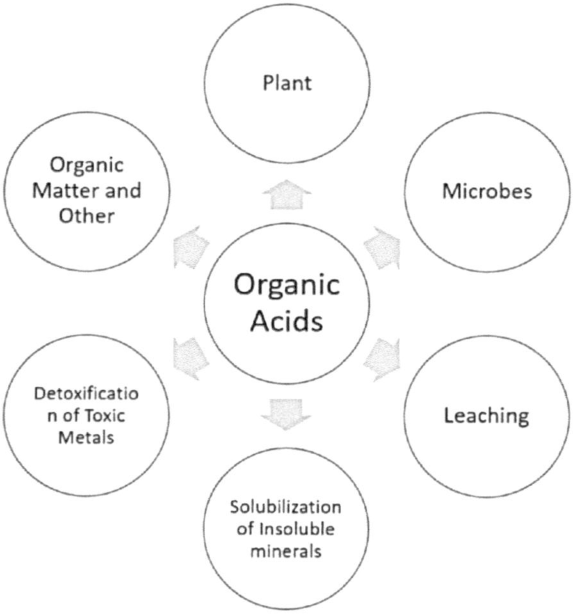

FIGURE 11.9 Role of organic acids in rhizosphere.

The presence of organic acids in the soil can alter or modify the physicochemical properties of the soil in the rhizosphere and help to facilitate the uptake of deficient/ insoluble nutrients.

Various ecological systems have been detected with potentially harmful elements such as Cd, Al, Pb, Zn, Cu, Fe, and Cr (Prasad et al., 2010). The occurrence of these heavy metals in soil is either linked to naturals like Al and Fe or man-made/anthropogenic sources, e.g., Zn Pb, and Cd. Organic acids like citric acid, oxalic acid acts as potential metal chelators in soil. These organic acids detoxify metals by only forming complexes.

There is the involvement of very complex and diverse mechanisms mediated by organic acid for remediation of heavy metals like adsorption, desorption, precipitation, chelation, redox reaction, complexation, respiration, transportation, acidification, and exudation. Several environmental factors are affecting these all processes, such as Soil, pH, Temperature, Microorganisms involved, and pollutant bioavailability as shown in the figure.

- ***Factors Affecting Organic Acid Production***

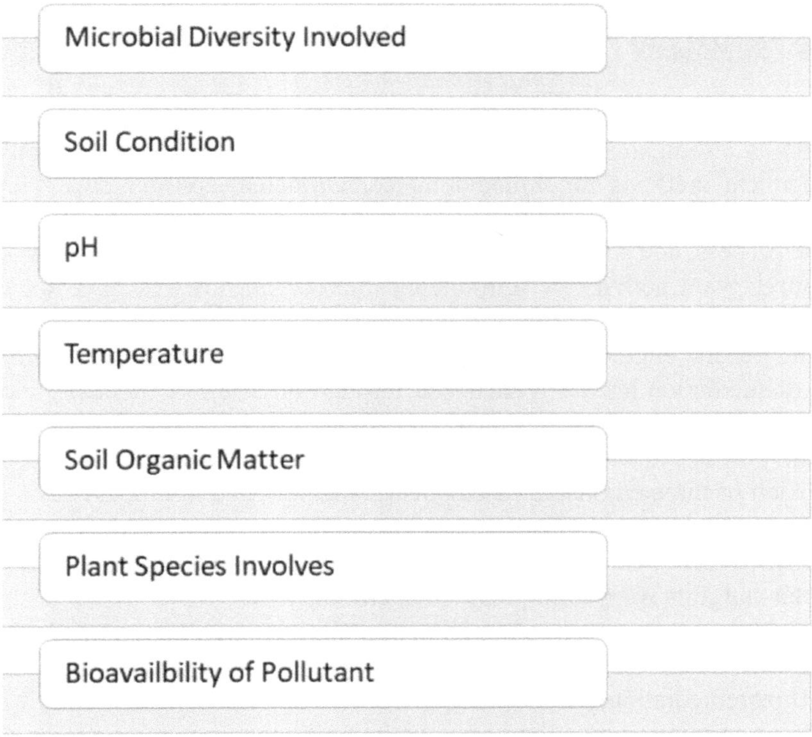

FIGURE 11.10 Factors affecting organic acid production.

The pH, temperature, plant species involved, soil organic matter, and bioavailability of pollutants all affect the production of organic acids in the rhizosphere, as shown in Figure 11.10. In addition to heavy metals, petroleum products, pesticides, herbicides, fly ash, etc. Microorganisms help plants reduce metal toxicity in several ways, including:

- Microbes reduce the toxicity of the metals
- Increases microbial bioavailability and metal uptakes
- Stabilizes the metals in the rhizosphere

Research for the fate and dynamics of organic acids can give insights into the modifications, techniques, and activities involved in rhizoremediation. It also helps to study the properties of the soil ecosystem, Organic acids in the soil may reduce the pH in the rhizosphere leading to the release of unavailable minerals in the soil.

11.6 SUMMARY

Rhizoremediation utilizes symbiotic relationships between root zone plants and soil microorganisms, which is seen as a sustainable alternative to physicochemical methods for remediating contaminated environments. Identifying the right plants and bacteria combinations can enhance the efficiency of the process, and soil amendments can enhance the process. This approach enhances plant activity by introducing microorganisms that have specific catabolic activities in addition to those that already exist in the soil. Introducing genetic engineering to rhizoremediation has made it more effective. Rhizoremediation has been enhanced through the use of genetically modified plants and microbes. A further effect of engineered soil is to enhance the overall process, in addition to genetically modified plants and microbes. The inclusion of three engineered elements (plant, soil, and microbes) has made rhizoremediation more effective. Reviewing the overall work undertaken to improve rhizoremediation for maximum remediation of contaminants, this chapter suggests ways to improve its operation.

Biological remediation is widely recognized as an effective method for restoring contaminated sites.

Rhizoremediation, a technique combining bioaugmentation and phytoremediation, may solve some of the problems encountered with each of the techniques individually. Rhizoremediation has already been shown to contribute to the restoration of polluted sites. Rhizoremediation systems have not been extensively studied, which consist of plants inoculated with bacteria or consortiums with the capability of degrading contaminated soil. Similar microbes and plants can be selected for other pollutants although the studies described above have expressed data for only one pollutant and one plant–bacterium combination. A future study of the selection of suitable rhizosphere bacteria or communities capable of surviving and proliferating on the root system of a plant that is suitable for rhizoremediation or phytoremediation can result in useful novel systems. As a result, bioremediation can be improved and developed further with these systems to become widely accepted.

11.7 EXERCISES

Objective Response Questions

1. **Which method of soil remediation is the most reliable?**
   a) Bio-magnification
   b) Bio-remediation
   c) Lyophilization
   d) Chemical method

2. **Typical contaminants can be degraded using _________ by using its carbon source for growth.**
   a) Microbes
   b) Chemicals
   c) Quorum quenching
   d) Lyophilization

3. **When it comes to soil remediation, which is having the highest potential?**
   a) Bacteria
   b) Plants
   c) Fungi
   d) Pathogens

4. **In which of the following processes are water contaminants removed by hydroponics?**
   a) Rhizofiltration
   b) Pollution control
   c) Antibiotic development
   d) Manufacturing process

5. **The process by which microorganisms degrade contaminants is known as _________.**

   a) Inorganic compounds
   b) Isolated bio-components
   c) Toxic compounds
   d) Mineralization

6. **Decomposition of minerals is caused by __________.**

   a) Other microorganisms
   b) Entering the food chain
   c) Increasing dissolved oxygen
   d) Producing methane gas

7. **What kind of contaminants are removed by immobilization?**

   a) Hydrogen
   b) Nitrogen
   c) Carbon monoxide
   d) Metals

8. **Which of the following is a process of immobilization?**

   a) Bio-accumulation
   b) Bio-remediation
   c) Bio-degradation
   d) Bio-magnification

9. **Bio-remediation is best suited to the degradation of __________.**

   a) Methane
   b) Nitrogen
   c) Organic chemicals
   d) Carbon dioxide

10. **A bio-remediation process can also be applied to destroy ________.**

   a) Methane
   b) Nitrogen
   c) Radionuclides
   d) Carbon dioxide

Practice Work

Qu 1. Briefly describe the mechanism of degradation of various types of poly-halogenated compounds.

Qu 2. What is pathway engineering? How it is useful for Enhanced biodegradation

Qu 3. Write a note on bioconversion of agricultural residue in a suitable product.

Qu 4. Explain bioconversion and utilization of paper wastes.

Qu 5. What is Rhizoremediation? Why it is called a beneficial plant-microbe interaction.

Qu 6. Show how biosensors are used for environmental monitoring and biomonitoring.

Qu 7. Explain the strategy of rhizoremediation research.

Qu 8. Explain the criteria for the selection of an appropriate PGPR strain for gene cloning and inoculation into the rhizosphere.

Qu 9. Describe the advantages of rhizoremediation.

Qu 10. Write a short note on biosurfactants.

REFERENCES

Chen, H., & Cutright, T. J., (2003). Preliminary evaluation of microbially mediated precipitation of cadmium, chromium, and nickel by rhizosphere consortium. *J. Environ. Engrg., 129,* 4–9.

Huang, X. D., El-Alawi, Y., Penrose, D. M., Greenberg, B. M., & Glick, B. R., (2004). Responses of three grass species to creosote during phytoremediation. *Environmental Pollution, 130,* 453–463.

Kamaludeen, S. P. B., & Ramasamy, K., (2008). Rhizoremediation of metals: harnessing microbial communities. *Indian J. Microbiol., 48,* 80–88. https://doi.org/10.1007/s12088-008-0008-3.

Ma, Y., Prasad, M. N. V., Rajkumar, M., & Freitas, H., (2011). Plant growth-promoting rhizobacteria and endophytes accelerate phytoremediation of metalliferous soils. *Biotechnology Advances, 29,* 248–258.

Mardad, I., Serrano, A., & Soukri, A., (2013). Solubilization of inorganic phosphate and production of organic acids by bacteria isolated from a Moroccan mineral phosphate deposit. *Afri. J. Microbiol. Res., 7,* 626–635.

Saharan, B. S., Sahu, R. K., & Sharma, D., (2011). A review on biosurfactants: Fermentation, current developments, and perspectives. *Genetic Engineering and Biotechnology Journal, 2011*, GEBJ-29 2011.

Sharma, A., Jamali, H., Vaishnav, A., Giri, B. S., & Srivastava, A. K., (2020). Microbial biofilm: An advanced eco-friendly approach for bioremediation. *New and Future Developments in Microbial Biotechnology and Bioengineering: Microbial Biofilms* (pp. 205–219). Elsevier, ISBN 9780444642790. https://doi.org/10.1016/B978-0-444-64279-0.00015-3.

Yadav, K. K., Gupta, N., & Kumar, A., (2018). Mechanistic understanding and holistic approach of phytoremediation: A review on application and future prospects. *Ecol. Eng., 120*, 274–298.

Zeng, M., Zhong, Y., Cai, S., & Diao, Y. (2018). Deciphering the bacterial composition in the rhizosphere of *Baphicacanthus cusia* (NeeS) Bremek. *Sci Rep. 8*(1), 15831. doi: 10.1038/s41598-018-34177-1. PMID: 30361644; PMCID: PMC6202335.

Zhuang, X., Chen, J., Shim, H., & Bai, Z., (2007). New advances in plant growth-promoting rhizobacteria for bioremediation. *Environ. Int., 33*, 406–413.

CHAPTER 12

Soil Analysis

12.1 SOIL: A NATURAL RESOURCE

One of the most valuable resources on earth is soil. The soil may be regarded as the natural thin layer on the earth in which life is supported. The complex root systems of plants, as well as other underground components including rhizomes, maize, and bulbs and many soil creatures, all flourish in the soil. Vegetation grows and develops most effectively in soil. It is a natural product of several soil-forming factors (Jenny, 1941). The environment and soil interact to change and breakdown trash into new materials. Soil serves as a filter and traps toxins inside soil particles. Aquifers and rivers receive cleaner water because soil particles trap contaminants. The definition of soil reflects the dynamic and evolutionary nature of soil as a natural body of the earth's surface possessing properties that are dependent on climate, living matter (plants and animals), conditioned by relief (slope) over time (Soil, USDA YearBook, 1957). The soil consists of weathering products of the rocks, mixed with living organisms and the products of their decomposition, moisture and air filling. Different soil types contain different proportions of organic matter, mineral matter, water, and air. Plant growth is dependent on the water and nutrients present in the soil. Beyond this, the soil must provide an environment in which roots can function. By allowing carbon dioxide to leave and new oxygen to reach the root zone, soil also maintains the volume and quality of air. Soil pores have the ability to store water until plant roots require it. The soil frequently acts on a much larger part of the plant's body than does the atmosphere. The physical and biological characteristics of the soil require careful study because the soil is a natural cultivation medium for plants and provides mechanical support. Nazarian et al., (2004) showed that the soil pH, soil moisture, organic carbon, organic matter (edaphic factors), degree, slope, and altitude (topographic factors) from the sea level

Bioremediation and Phytoremediation: Technologies for Toxic Pollution
Ashwini A. Waoo, PhD (Author)

of the region were principal ecological factors. The soil coarse sand, organic matter, water holding capacity, pH, calcium carbonate, and available sulfate and heavy metals (Zn, Cu and Cr) were impacted by industrial emissions and hindered plant growth. The soil of industrial areas has been analyzed by many workers to investigate the effects of soils on plants. Soil texture, maximum water retention capacity, $CaCO_3$, organic material and pH were analyzed by Iqbal. et al., (1983); Shafiq & Iqbal (1987); Mehmood & Iqbal (1995) near industries. The soil was also analyzed for available sulfur (Iqbal & Zaman, 1993); whereas total soluble salts were examined along with moisture content, organic matter, and pH (Wahab & Hashem, 1995) around different factories. The soil around the industrial areas has also been used to detect the accumulation, deposition of metal pollution (Kuo et al., 1983; Hunter et al., 1987; Maxwell, 1991; Wahab & Hashem, 1995). Water and minerals are absorbed by plants from the roots of the soil. during growth. Some workers demonstrated the relation of different soil characters with the plant. Soil pH has an indirect impact on plants as well. (Walter, 1971). Metals effects were recorded on the growth of plants by some workers. Plants directly depend on the soil characteristics and conditions necessary for their successful growth.

Determining the chemical nature of the soil depends in large part on the pH of the soil. Whether the soil is acidic, neutral, or alkaline influences the growth of plants and microorganisms. The application of sewage sludge will also alter soil properties. Figure 12.1 describes the physicochemical parameters of soil which are analyzed by different techniques.

Physiochemical Parameters of Soil

- Soil Color and Texture
- ☐ Soil Temperature
- ☐ Air Temperature
- ☐ Soil Moisture
- ☐ Water Holding Capacity
- ☐ Soil pH
- ☐ Organic Carbon
- ☐ Total Nitrogen, Potassium and Phosphorus

FIGURE 12.1 Physicochemical parameters of soil.

12.2 PHYSIOCHEMICAL ANALYSIS OF SOIL

Composite soil samples were collected from all the three study sites of Raisen District (natural and captive area in Ratapani Wield life Sanctuary and Industrial area). The collection of soil samples was done from each study site for depth (0–15 c.m).

1. ***Soil Color and Texture:*** The soil color was read off directly. The soil texture was determined by the pipette method as described by Piper (1944).
2. ***Soil Temperature:*** Soil temperature was studied with the help of a soil thermometer, which was fixed in the ground at 15 c.m. depth for three locations of the area.
3. ***Air Temperature:*** The air temperature was studied with the help of a thermometer in the three different sites of the study area.
4. ***Soil Moisture:*** Each soil sample was taken at each site within the study area (0–15 cm).Fresh soil was taken immediately packed in polythene bags weight per fresh weight and preserved in an oven. at 105°C till constant weight. The oven-dried soil was weighed again and moisture content the percentage was calculated as follows:

$$Moisture = \frac{\text{Weight of fresh soil} - \text{Weight of oven} - \text{Dried soil}*100}{\text{Dry weight of soil}}$$

5. ***Water Holding Capacity (WHC):*** Water holding capacity or amount of water taken when immersed in water under standardized conditions was most conveniently determined by Keen- Raczkowski box experiment using circular-shaped brass boxes. Whatman number/ filter paper is fixed at the bottom of each brass box. Which are then packed with the air-dried cursed soil. These boxes were in Petri dishes containing water and left aside for 24 hours for the soil to get completely saturated. Then it was taken out wiped dry carefully and weighed. Boxes were then transferred to the oven for 24 hours at 105°C. The oven-dried weight was noted down. Water holding capacity has been calculated as follows.

$$WHC = \frac{b - c - d*100}{c - a}$$

where, a = weight of the unfilled box; b = weight of the unfilled box and saturated soil; c = weight of the can and oven-dried soil; d = average moisture retained by the paper.

Soil pH: The Soil pH was determined by the electrometric method (Mishra, 1968). About 10.2g standard buffer solution of pH (0.05 Potassium biphthlate) was liquified in water and diluted to 1-liter 20 gm soil was transferred to a 125 ml wide mouth bottle of capacity. About 100 ml water was added, the stopper was fixed and shaken mechanically for one hour. The pH meter has been standardized with the buffer solution. Electrodes were removed from the buffer solution and rinsed with distilled water. the bottle was shaken and electrodes were dipped into the soil suspension. The pH was read and recorded.

Organic Carbon: Organic carbon was established by Walkley and Black (19.34) rapid titration method 0.1g of soil was transferred into 500 ml of Erlenmeyer flask. About 10.5 ml potassium dichromate solution added followed by 20 ml of connected sulfuric acid. It was then kept aside for 30 minutes. Then 200 ml of water, phosphoric acid (10 ml) and diphenylamine (1 ml) were added to the mixture. indication were added.

Titration was alone by adding ferrous sulfate until the flashes to green. The 0.5 ml of N potassium dichromate was added and titration was completed by adding ferrous sulfate drop by drop. The carbon percentage of soil was calculated by the expression

$$((V1 - V2)/W) * 0.003 * 100$$

where, V_1 = volume of N potassium dichromate (10.5 ml); V_2 = volume of N ferrous sulfate; W = weight of soil.

The organic matter was obtained by expression organic carbon % * 1.72.

Phosphorous, Nitrogen, and Potassium: For the analysis of exchangeable phosphorous, the solution was prepared by ashing procedure using nitric perchloric and sulfuric acids (Piper, 1944). The known capacity of ammonium molybdate was added to develop blue color and percent transmission of it was measured at 660mm and the amount was calculated from the calibration curve (Jackson, 1962). The exchange of potassium was estimated by Flame Photometer following Jackson (1967).

The total nitrogen of the soil was analyzed by the Kjeldahl method was given by Piper (1994).

12.3 ANALYTICAL TECHNIQUES FOR HEAVY METALS DETECTION

Hazard identification is necessary for human health risk assessment. The process of hazard identification involves determining whether the exposure to two hazards will increase the risk of an adverse health effect, for example, cancer. Hazards may be chemical, physical, or biological and when they are sufficiently exposed to humans can cause illness, disease, or death. Researchers and scientists can assess the effects of hazardous compounds on human and animal health. Characteristics of the hazardous substance should be identified; its potential health problems are characterized.

Humans who are directly exposed to dangerous chemicals can be examined to get the most accurate data on the severity of the reaction. toxicity hazardous substances but there are very few clinical trials on human-animal experiments available for a particular hazardous substance.

A rise in the concentration of heavy metals to levels that are harmful to plants and animals is a concern, even though they are present in the soil, geology, and anthropogenic activities. Metal mining and smelting, burning fossil fuels, using fertilizers and pesticides in agriculture, manufacturing batteries and different metallic products, as well as sewage sludge and municipal waste disposal, are only a few instances of these kind of activities. Plants that flourish on soils that have been contaminated with heavy metals have been recorded to have reduced growth due to changes in physio and biochemical processes. Reduced plant growth reduces yield, which ultimately results in food insecurity. Therefore, heavy metal polluted soils must be remedied as soon as possible. It is possible to remediate metal-polluted soils by physical, chemical, or biological means. Encapsulation, solidification, stabilization, electrokinetic, vitrification, gaseous extraction, and soil washing and flushing are one of the costliest physical and chemical processes. and do not make the soil suitable for planting., on polluted soils, the biological method (bioremediation) supports the establishment/reestablishment of plants. This is an environmentally friendly method because it relies on natural processes. Comparatively to other remediation techniques, bioremediation is also an affordable method. This chapter discusses the properties of heavy metal polluted soils. The growth and performance of plants were examined in these soils. There was an emphasis on biological approaches for remediating heavy metal polluted soils as well.

Heavy metals have been found in soils due to an increase in anthropogenic activities across the globe. These soils reduce the growth, performance, and yield of plants. The remediation of soil polluted with heavy metals is an effective method of treatment. The method can be used for the establishment/reestablishment of crops on treated soils, since it can be used in situ. Microorganisms and plants have distinct mechanisms for the remediation of polluted soils. The use of plants for treating polluted soils is the best way in the process of bioremediation of heavy metal in soils. Combining plants and microorganisms for bioremediation ensures more effective remediation of heavily polluted soils. But, the success of this depends on the species involved.

12.3.1 HEAVY METAL POLLUTED SOILS

Metals have metallic properties such as ductility, cation stability, conductivity, and ligand specificity. Their atomic number is greater than 20 and they have a relatively high density and relative atomic weight. In microgram amounts, organisms need heavy metals like copper, cobalt, manganese, molybdenum, nickel, vanadium, and zinc. Overdoses can however be harmful to organisms. Metalloids (a metalloid but collectively known as heavy metals) including Pb, Cd, Hg, iron, and As do not have any beneficial effects on living organisms, making them the main threat since they are hazardous for living organisms.

Metals can be found separately or in combination. There are many components, including exchangeable ions absorbed on surfaces of inorganic solids, insoluble inorganic metal compounds, metal ions, soluble metal compounds or metal complexes of organic materials, as well as metals attached to silicate minerals are all found in soil solutions. In contrast to metals that are present as separate entities or in very high concentrations in other components, the metals attached to the silicates constitute the background soil metal concentration and don't pose any contamination or pollution problems.

Metal availability varies with soil properties. Generally, soil pH affects metal availability in soil. As soil pH increased, the availability of cadmium and zinc to *Thlaspi caerulescens* roots decreased. By using organic matter and ferric oxide, heavy metals are immobilized. thereby decreasing their availability. Heavy metals have also been found to correlate positively with some soil physical properties, including moisture and water holding capacity. Additional factors that influence soil metal availability include:

- soil pH;
- organic matter;
- moisture content;
- density of soil;
- charge on colloids;
- degree of complexation with ligands;
- the soil's relative surface area.

Therefore, soil colloids play an important role in controlling heavy metal concentrations in natural soils, since they provide large interfaces and specific surface areas. Additionally, High-surface-area soil particles may reduce soluble metal concentrations in polluted soils, though the effect may vary depending on the metal. Hydroxides with a high surface area decrease Cu, Cd, As, Pb and Mo solubilities, while Ni and Zn remain unchanged. Heavy metal availability in soil has also been shown to be influenced by soil aeration, mineral composition and microbial activity.

On the other hand, heavy metals may affect soil properties, especially their biological properties. It is possible to determine the extent of soil pollution by monitoring changes in soil microbiology and biochemistry after contamination since results can be obtained quicker by using these methods rather than monitoring soil physical and chemical properties. The number, diversity, and activity of soil microorganisms are affected by heavy metals. The toxicity of metals to microorganisms depends on several factors, including soil pH, temperature, organic matter, clay minerals, inorganic ions, and metals.

Studies comparing heavy metals' effects on soil's biological properties show discrepancies. Heavy metals can negatively impact soil biological properties, but some soil microbiological properties do not correlate with heavy metal concentrations, according to other researchers. It is possible that there are inconsistencies because some of these studies used soils that were artificially contaminated whereas other studies used soil from polluted areas. Although the origin of the soil used in these experiments is irrelevant, soils need to be studied more deeply to fully understand the effects of heavy metals on their biological properties. It would be better to utilize a wide range of methods when studying the effects of metals on soil properties rather than depend on a single one. Using different methods would provide more comprehensive and conclusive results. In soil contaminated with one heavy metal, plants may be affected. Heavy metals have antagonistic and synergistic behavior. Metal species may also interact with each another.

12.3.2 PROBLEMS OF MAKING MEASUREMENTS IN SOIL

The soil appears to be a homogenous material to a layman. In reality, several factors can affect soil homogeneity. The construction work on the sites can affect the distribution of excavated subsoils on the surface. Hard standing areas can cause soil compaction, which can obstruct water drainage.

Percolating water can lead to finer particles being flushed away from the soil as a result of large flows. This can adversely affect the soil's physical properties. Since the soil is a 3-D matrix, any particular measurement can vary in all three dimensions.

The flow of or surface rainwater can draw pollutants from point sources leading to trails of pollutants across a site where the soil structure acts as a chromatography column packing, and the groundwater is removing the compound through the soil due to their partial solubility. Due to the interval between sampling and results, it is difficult to locate hot spots using off-site laboratory analysis. The most efficient way to utilize on-site screening and off-site lab analysis are by utilizing an on-site screening process that delivers real-time information to the site operator and allows the operator to choose samples for closer examination by an accredited laboratory.

Sampling is crucial to effectively cleaning up a contaminated site, and previous use of the site can provide information about pollution hotspots. For initial sampling, a grid system of sampling areas is being used to determine if a particular region within the grid was previously occupied. To locate the source of pollution, further sampling is performed in areas with high concentrations of pollutants once the initial sampling has been completed and results are obtained. Laboratory analysis differs significantly from on-site analyzes. In the on-site analysis, a few grams of soil are generally used whereas a laboratory sample is taken from a few kilograms of soil which is then homogenized before subsampling and analysis. An evaluation method must be correlated to a laboratory-approved method for confidence in its validity and reliability. Taking subsamples of larger laboratory samples and analyzing them with the onsite equipment is another prudent way to achieve confirmatory results.

12.3.3 ANALYTICAL FIELD TOOLS

The technology offers site investigations the capability of providing rapid results and guiding on-site staff to locate 'Hot Spots.' Due to their cost implications, they enable a much higher number of samples to be analyzed than using only an offsite laboratory. By increasing the density of data, interactive sampling can provide a more accurate characterization of the site and thus is likely to reduce soil removal costs.

Heavy metals bind with organic compounds in the soil, and make them immobile within the soil structure. An extraction process is required to fully release the metals from the soil sample for liquid analysis. For some methods, the sample must be chemically digested to release the metals before analysis can take place, but the XRF method does not require this.

The primary accredited methods for analyzing soil for heavy metals using off-site techniques are XRF, ICP (Inductive Coupled Plasma) and Atomic absorption, However, a nonrepresentative soil sample can negate the accuracy of any of these methods.

12.3.4 X-RAY FLUORESCENCE (XRF)

A portable XRF analyzer provides a very quick and convenient way to quantify significant amounts of elemental concentration in soil quickly. The technique, however, is subject to many sampling and analytical errors and is therefore regarded as only a screening tool. There are two main sources of error in this technique: sampling error due to the heterogeneity of pollutants in the soil, and water content variation in the samples. To ensure data authenticity, samples must be precisely prepared. XRF *in situ* analysis (using an XRF in direct contact with the ground) requires minimal preparation. But it is primarily a screening approach. It can be used for:

- identifying hot spots on a site;
- indicating the extent of contamination;
- identification of potential contaminants;
- contributing to remediation decision-making;
- the screening of hazardous waste.

On-site soil analysis is performed with two versions of the instrument. If the 'gun' is placed close to the soil and a near-instantaneous reading is obtained or if samples are presented to the analyzer in a laboratory. With an appropriate stand, the handheld units can also be used in the laboratory.

Theory of Operation: When a high-energy X-ray from a gun strikes a metal atom, some of the low-energy electrons within the atom are knocked out of their orbits, leaving a "hole" from which more high-energy electrons are absorbed by the distant electron shell , but in doing so, they must lose energy to stay in this orbit. This energy is lost in the form of X-rays. These X-rays are called fluorescent X-rays, and each metal has a series of characteristic emissions that can be used to identify the metal, the magnitude of which provides information about the concentration of the element in the sample of interest.

The advantage of XRF guns is that they are safe to use with minimal training and get results almost immediately. The disadvantage is that X-rays penetrate soil samples to a depth of only a few millimeters, and there are attenuation problems in wet soil or sludge, which can be overcome by drying the samples before analysis.

12.3.5 ANODIC STRIPPING VOLTAMMETRY

The anodic stripping voltammetry (ASV) technique is traditionally used to detect metals at trace levels, while the AdSV technique can detect metals and organic compounds as well. The limits of detection (LOD) are significantly reduced compared to direct voltammetry measurements (sub-nanomolecular versus micromolecular levels) by preconcentrating metal ions onto the electrode surface. Analyzing real (and complex) samples, such as water, biofluids, and tissues, is indeed crucial. Multi-ion analysis can be performed using ASV techniques using a single scan, identifying each (redox-active) ion depends on its redox potential.

It is an electrochemical method that requires the metal ions in solution, so the sample has to undergo wet chemical degradation to release the metal ions and destroy any organics that may be securing them. When the metals have been solubilized, a sample is introduced to an electrolyte, causing a fraction of the ions to be reduced to a metallic state onto the electrode. This process re-oxidizes the metal as well as ions into solution. The oxidation of the metal releases an electron whose flow is measured by another electrode in the solution. Metals are identified by the voltage at which they re-ionize, while the peak height is used to measure their concentration.

This method has the advantage that turbid samples won't interfere with the measurement and analysis can be done quickly once set up. One drawback of this method is, the metal ions are released into the solution only after an extraction process is performed on the sample.

12.3.6 COLORIMETRIC TEST KITS

The analysis of most cations and anions in water is performed using colorimetric systems. Though, the analysis of metals in soils rise few challenges. To measure bound and free metals, Sample digestion is necessary, and since colorimetric measurements often require a very specific pH for chemistry to work properly, this technique is not well suited for analyzing metals in soil. Sample Preparation and Analysis for Heavy Metal Detection

Soil sampling was done with the help of instruments like fawda and khurpi, soil augers, spade, shovel, etc. It depends upon the availability and sometimes specific purpose. Soil augers are generally preferred while collecting depth-wise samples from any area with precision. Generally, for surface samples, khurpi may be used comfortably. The common instruments used for surface soil samplings are spade and khurpi are shown in Figure 12.2.

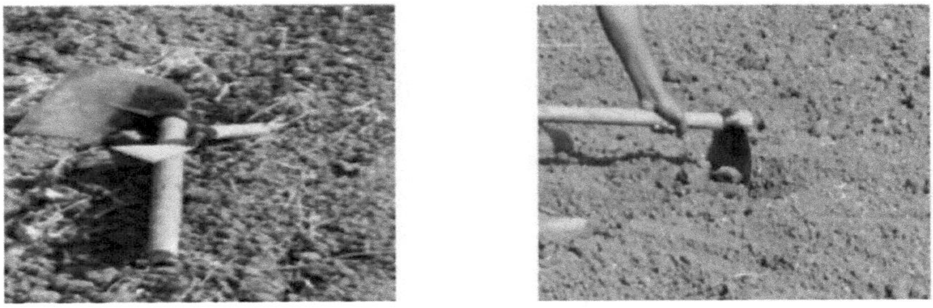

FIGURE 12.2 Surface soil sampling instruments khurpi and spade.

- A soil sample may be taken with the following tools, depending on the purpose and precision needed.
- Soil augers can be tube augers, post hole augers, or screw-type augers, or they may even be spades used to take soil samples.
- A bucket or tray to mix the soil.
- Cloth bags of specific size.
- Pencil for markings and tagging of samples.

An area's soil is like a measuring meter/device that tells us how healthy it is. It is the soil that controls and regulates the interaction between the environment and biomass production. A soil test measures the "plant-available" content of plant-nutrient elements (phosphorus, potassium, calcium, magnesium, sulfur, manganese, copper, zinc, etc.) in the soil.

12.3.7 SOIL SAMPLING METHOD

The soil sample collection procedure varies according to the purpose of the study (Soil testing manual, 2011). This soil sample was carefully selected to make sure it was demonstrative of the whole study area since the purpose was to evaluate the occurrence of metals in the soil. Each representative soil sample was collected from 5–6 points in a zone and mixed thoroughly because an inconsistent or incorrect sampling depth is frequently cited as the major source of soil sampling error. As a result, the quality of soil testing services is dependent on the skill and care with which samples are taken. The single largest source of error in a study is non-representative samples. As part of the soil sample collection, the whole area was divided into four groups according to the zone. An appropriate sampling pattern was chosen. Before obtaining the soil sample, e.g., grass, leaves, etc. were removed from the top surface of the soil. After that using a khurpi the sample was dug in the form of a V-shaped hole in the soil from the desired depth.

The entire soil 'wedge' or slice was collected and similar "wedges" from four spots in the area were taken. The soil from different locations is mixed in a zigzag fashion. All the "wedges" mixed and stones, stalks, leaves were removed. The soil was spread evenly and divided into four quarters. Then two opposite quarters were rejected and mixed the rest of the soil again.

This process was repeated till left with half kg of the soil. The remaining soil was stored in a clean bag and each bag was properly marked for identification of the sample. The clean and contamination-free bag was used for sampling. The soil sample was then condensed to 500–1000 grams and stored in a polythene bag. The bags were clearly labeled as per site so that they can be identified easily at the time of test report is received.

Before sending soil samples to the testing laboratory for heavy metal analysis they were properly identified and marked. All the soil samples were promptly dispatched to the laboratory for further heavy metal analysis only. The general physicochemical characteristics were tested manually.

12.3.8 SOIL SAMPLE PREPARATION FOR ANALYSIS

Drying and post-drying care were used to prepare soil samples for analysis. This section describes in detail how soil samples were prepared. A wooden or enameled tray was used to dry moist samples in the laboratory. It is important to maintain the identity of the sample at all times. Each drying tray was numbered. The soil was allowed to air dry. The trays were also placed in racks within a hot air cabinet whose temperature did not exceed 350°C and relative humidity was between 30 and 60%.

Once the samples were dried, the samples were taken to a preparation room that is separate from the main laboratory. A wooden pestle and mortar were used to crush air-dried samples without destroying the soil particles. Following grinding, the soil was screened by a 2 mm sieve. The remainder of the coarse material on the sieve was returned to the mortar for further grinding. The process was repeated until all aggregate particles could pass through the sieve and only organic residues, concretions, and pebbles remained. During grinding and handling of trace elements, containers made of copper, zinc, and brass were avoided. A cardboard box with wooden drawers was used to store the soil samples after that. The boxes are numbered and arranged in rows in the wooden drawers, which in turn are housed in a cabinet in the soil sample room.

12.3.9 ASH METHOD

Samples are normally collected, washed with tap water, then distilled. This eliminates dust, fertilizers, mud, pesticides, and any pollutants that may be present. Drying the samples at room temperature removes moisture, followed by slicing, weighing, and oven-drying them at 80°C for 72 hours to achieve a constant weight. After being oven-dried, the samples were powdered and sieved through a mesh of 2 mm size. After this, the samples were placed in a desiccator and crucible, then placed in a muffle furnace and ashed at 450°For 12 hours. After this process, the sample is decomposed.

12.3.10 CLOSED FURNACE ANALYSIS OF SOIL SAMPLES AND THEIR DIGESTION PROCESS

A furnace is a closed space where heat is applied to a body to raise its temperature. Fuel or electricity can be used to generate heat. In furnaces, metals, alloys, and sometimes non-metals are heated. Heat is desired for a specific purpose, so temperature and heating rate are determined accordingly. Metals become softer at higher temperature. Consequently, they are susceptible to deformation. The metallic structure can change with or without softening. A lower temperature (below the critical temperature) softens metal by relieving its internal stresses.

The soil samples accumulated from the sphere are very well combined to shape a composite sample; then, the samples had been oven-dried at 105°C for 6 hours till a regular weight is attained. The oven-dried samples are then crushed and sieved through a 2.0 mm wire mesh. The representative samples are then digested thoroughly. In each sample, one gram of ground meat is digested in 10 mL of 1:1 HNO_3 and dried at 95°C, followed by 10 minutes of refluxing. Once the solution has cooled, 5 mL of concentrated HNO3 is added and re-refluxed for 30 minutes until brown fumes are produced. Over and over again, 5 mL of concentrated HNO_3 is added until white fumes appear. On a mantle set to 95°C with a watch glass over it, the solution is vaporized to about 5. To start the oxidation of peroxide, 2 mL of H2O and 3 mL of 30% hydrogen peroxide are added and the solution is placed on the heating mantle until there is no longer any effervescence. Afterward, the acid-peroxide is digested and heated to 5 mL at 95°C. The sample digest is treated with 10 mL of concentrated HCl later, and The solution is heated to 95°C for 15 minutes and then refluxed. The digested material is then filtered and analyzed for heavy metals employing an Atomic Absorption Spectrometer. It is equipped with a specific lamp made of a particular heavy metal.

Analyzing matrices for identification and quantification of elements requires considering the benefits and the drawbacks of certain analytical techniques. Traditionally, metals were determined in liquid and solid samples by titrating and using electrodes. Fortunately, today's technology is more advanced, and new instruments are available in the market. There is sporadic use of electrode techniques such as voltammetry, stripping voltammetry, potentiometry, and polarography. Flow-through coulometry, however, may prove useful in the determination of trace metals, especially when applied to aquatic environmental samples with low concentrations. To determine the number of trace metals in soil extracts, stripping voltammetry is another spectroscopy method.

Anodic, cathodic, and adsorptive stripping voltammetry was used to determine the Pb, Zn, Cu, Co, Ni, Cd, and Cr metals in g.L-1 levels when using glassy carbon as a working electrode. In anodic stripping voltammetry, Pb, Zn, and Cd metals were analyzed using mercury and bismuth modified pencil graphite electrodes. The simplicity, low cost, and commercial availability of pencil graphite make this electrode a popular choice for voltammetric measurements.

The stripping voltammetry technique used for the determination of trace organic species and metal complexes in a variety of media, using either mercury drops or solid electrodes. After specific reagents are added to soil extracts, the spectrophotometric method can be used to determine boron levels. Among the spectrophotometric methods commonly used for boron determination, the Azomethine-H method is probably the most widely used. Despite its reputation as the most sensitive method, the Azomethine-H method has some interference.

The X-ray spectrometer (XRF) is characterized by excellent selectivity and low detectability limits of ppm-ppb. Furthermore, it is a non-destructive technique, which does not damage the sample under study. Neutron activation analysis (NAA) also has this feature. One of the most universal analytical techniques, it allows simultaneous determination of about 40 elements. An additional technique, energy-dispersive XRF, can be used to analyze trace elements in situ at the measurement site, which can shorten the analytical process.

Spectrophotometric techniques were developed later to determine heavy metals and metalloids quantitatively. Electromagnetic waves interact with matter in all its macro- and microscopic forms. Atomic absorption spectrometry (AAS) is the most common technique in this group. Based on the radiation absorbed by free atoms in the thermal plasma generated in an atomizer, it determines the presence of a given line spectrum, such as a hollow cathode lamp (HCD) or electrodeless discharge lamp (EDL). It's not surprising that AAS is becoming more popular in quantitative analysis of elements due to its high sensitivity and selectivity. Furthermore, the technique is characterized by low detectability limits and high precision. The type of atomizer is the basis for a more detailed classification of atomic absorption spectrometry. Typically, flame atomization (F-AAS) was used as a technique, in which air or N_2O was used as an oxidizer, and acetylene was used as a flammable gas. The electrothermal atomizer (ET-AAS) can also be used for metal determinations.

The main type of graphite furnace in which thermal atomization takes place is the GF-AAS. A small hole is placed in the upper part of the furnaces for sample introduction. The furnaces resemble small tubes with a diameter of 4–6 cm. Metalloids (selenium, arsenic, antimonium) are quantitatively determined in surface and ground waters or landfill leachates using hydrogen generation atomic absorption spectrometry. Mercury was determined using the cold vapor AAS technique (CV-AAS). A special type of flow cell (quartz or glass with quartz windows) is used in this technique, and Hg has to be present as an ion or in gas form. The use of atomic fluorescence spectrometry is also beneficial when used in conjunction with the above atomization technique is also beneficial as it enables mercury determination. This chapter addresses the problem of heavy metal determination, spectrophotometric techniques are also mentioned, in which light absorption or emission by the sample is being measured quantitatively.

Spectrophotometric techniques, including ultraviolet and visible light spectrophotometers, are rarely used in the analysis of environmental samples. Often, such trends are caused by the necessity of lowering the sample determination limits regularly, as well as the time required for the

TABLE 12.5 *(Continued)*

analysis. This is why spectrophotometric techniques are becoming less and less competitive. Combining classic atomic emission spectrometry (AES) with a plasma spectrochemical source, ICP-AES, was a breakthrough in the development of this method. Therefore, a sensitive and selective technique was developed and used for the quantitative determination of metals in a variety of environmental samples.

ICP-AES is used in environmental applications for the metal determination of a wide range of materials, including natural waters, seawaters, soils, sediments, biological tissues, and air particulates. U.S EPA method describes the preparation of water, wastewater, and solid samples. Method 200.7 of the US EPA describes ICP-AES measurements of metals and some nonmetals. For trace boron analysis, inductively coupled plasma (ICP) is mostly used.

The ICP technique applies to a wide range of elements with a high degree of reliability. This method applies to the following analytes: Mg, Al, P, K, Ca, Li, Be, B, Na, Ti, V, Cr, Co, Ni, Cu, Zn, Mn, Fe, As, Ag, Cd, Hg, Ce, Sn, Sr, Mo, Se, Sb, Pb, and Ba. Metal and metalloids are analyzed using ICP-AES according to OSHA method ID-125G. In welding fume, it can analyze 13 elements, including Be, Cr, Zn, Co, Ni, Mn, Cu, Cd, Pb, Sb, Fe, and Bi. In the early 1980s, a commercial analytical technique was developed, ICP-MS has been used to determine trace elements, minor elements, and even major elements. Using inductively coupled plasma-mass spectrometry (ICP-MS), environmental samples can be analyzed for traces of metals. ICPMS can detect a variety of elements. To analyze traces of metallic elements, ICP-MS is a perfect instrument. It has a low detectability limit and high selectivity. This makes it a common technique used to determine metal concentrations in water, soil, tissues, and ashes.

12.3.11 REMEDIATION TECHNIQUES OF POLLUTED SOILS

Heavy metal contaminated soil is very difficult to remediate on a large scale. Soil is a mixture of organic and inorganic compounds, different gases present in different proportions, and water. Various minerals are present in soils depending on the parent rock materials. Soil has a wide range of physical, chemical, and biological properties. The movement of soil water in the soil is governed by physical factors such as the soil's structure and texture. Soil moisture influences chemical reactions, solute movement, microbiological activities, salt solubility, and the bioavailability of metal ions. Any phytoremediation program faces the challenge of understanding the variations in soil properties of the specific contaminated site.

Physicochemical treatment of polluted soil involves removing the soil at a toxic waste site, fixing the metals or fixing the chemicals, leaching using acid solutions, and returning the clean soil to the original site (Salt et al., 1995). Further processing may include precipitation microfiltration (Raskin et al., 1996) or flocculation followed by sedimentation, reverse osmosis, and ion exchange. In addition to being very costly, these physicochemical approaches may also have side effects on the ecosystem as well as humans (Raskin et al., 1997).

Two methods of bioremediation exist: (1) microorganisms detoxify metals through valence transformation or volatilization, extracellular chemical precipitation, and (2) using enzymes. Use of specific plants for soil or water decontamination by deactivating heavy metals in the rhizosphere transport them in the upper parts of the plant. Phytoremediation is an auspicious technique for reclaiming contaminated sites that is cheaper than other approaches. The use of green plants for decontaminating soils, water, and the air is botanical bioremediation. It can be used to remove both organic and inorganic pollutants from soils, waters, and the air. Species differ significantly in their ability to accumulate heavy metals, depending on their genetic, physiological, morphological, and anatomical characteristics. In remediation mechanism, phytoremediation can take the forms of phytoextraction, phytofiltration, phytostabilization, phytovolatilization, and phytodegradation. Plants are used to degrade contaminants from the soil through phytoextraction. Metal ions accumulated in the vegetative parts of the plant can be removed or burned for metal recovery. Plant roots remove metals from aqueous wastes through phytofiltration. By preventing leaching, phytostabilization involves the absorption of pollutants from the soil and store them in the rhizosphere. Plants are used to volatilize pollutants from their leaves through phytovolatilization. Plants and associated microorganisms degrade organic pollutants by phytodegradation.

12.5 SUMMARY

Traditionally, soil contamination is measured with reference to threshold values with monitoring results. Agricultural, environmental, and governmental agencies tend to use this approach. It protects the environment from pollution caused by runoff and excess fertilizer leaching, aids in diagnosing plant culture problems, and improves the nutrition of crops. A multivariate statistical approach can help classify, model, and interpret soil monitoring data more easily. By using this method, relationships between specific soil parameters and chemical pollutants can be detected between sampling sites, and pollution can be stratified. Furthermore, pollution sources can be identified.

Several chemical processes are involved in the analysis of soil, which determines the number of nutrients in the soil, as well as its chemical, physical, and biological properties. Chemical analysis of soil determines the levels of basic plant nutrients such as nitrogen, phosphorus, potassium, pH, humus content, total carbon, lime availability, organic matter, sulfur content, trace elements, and other physical parameters such as permeability, density, and pH. The fertility of agricultural soils is of utmost importance for today and in the future as well. To maintain soil fertility, soil analysis and its interpretation are necessary tools to determine whether fertilizer and/or manure applications are necessary, as well as to assess the level of toxic pollution within the soil.

12.6 EXERCISES

Objective Response Questions

1. Which among the following does not lead to process integration?

a) *In situ* or ex-situ
b) Extensive or intensive
c) Biological
d) Physical

2. The pick-and-mix method enables flexibility in integrating the process of land remediation.

a) True
b) False

3. What process does not belong to phytoremediation?

a) Bio-accumulation
b) Concentration and excavation
c) Phytoextraction

d) Phytostabilization

4. Each soil has life in terms of ___________.

a) Geologic time
b) Paleologic time
c) Soil time
d) Geo time

5. The soil texture was determined by the ___________.

a) Pipette method
b) Burette method
c) Flask method
d) Texture analysis

6. Organic carbon was determined by ___________.

a) Rapid titration method
b) Physical method
c) Filtration
d) None of the above

7. The exchange of potassium was estimated by ___________.

a) Flame photometer
b) Spectrophotometer
c) Electrophoresis
d) Galvanometer

8. Which is not the soil remediation method?

a) Encapsulation
b) Solidification
c) Stabilization
d) Dilution

9. XRF stands for ______.

a) X-ray filtration
b) X-ray fluorescence
c) X-ray formation
d) None of the above

10. Cold vapor AAS (CV-AAS) is used in determinations of ___________.

a) Mercury
b) Zink
c) Copper
d) Cadmium

Practice Work

Qu 1. Explain the physicochemical analysis of soil in detail.

Qu 2. Explain Atomic Absorption Spectrophotometer.

Qu 3. How will you detect the heavy metal concentration in a given soil sample?

Qu 4. Explain the preparation of samples for AAS.

Qu 5. Explain various soil remediation technologies.

Qu 6. Explain the process of soil sampling and different instruments used for soil sampling

Qu 7. Describe various analytical techniques used for heavy metal detection.

Qu 8. How will you calculate the water holding capacity of particular soil?

Qu 9. Write the significance of soil as a natural resource.

Qu 10. What are the problems of soil measurements?

REFERENCES

Abedin, M. J., Feldmann, J., & Meharg, A. A., (2002). Uptake kinetics of arsenic species in rice plants. *Plant Physiology, 128*(3), 1120–1128.

Dhir, B., (2017). Bioremediation technologies for the removal of pollutants. In: Kumar, R., Sharma, A., & Ahluwalia, S., (eds.), *Advances in Environmental Biotechnology*. Springer, Singapore. https://doi.org/10.1007/978-981-10-4041-2_5.

Paz-Alberto, A. M., & Gilbert, C. S., (2013). Phytoremediation: A green technology to remove environmental pollutants. *American Journal of Climate Change, 2*(1), 71–86. http://dx.doi.org/10.4236/ajcc.2013.21008.

Salgare, S. A., & Acharekar, C., (1992). Effect of industrial pollution on growth and content of certain weeds. *Journal for Nature Conservation, 4*, 1–6.

CHAPTER 13

Phytoremediation of Heavy Metals Contaminated Soils

13.1 HEAVY METAL TOXICITY

Heavy metals such as iron (Fe), zinc (Zn), lead (Pb), cadmium (Cd), chrome (Cr), and nickel (Ni) are essential nutrients that can become harmful and toxic when their concentration exceeds recommended levels. Heavy metals such as lead and cadmium are not essential. Tiny amounts of them can be toxic.

Contaminants are discharged directly, increasing the concentration of trace elements, which accumulate in sediments (Dunbabin and Bowmer, 1992). Roots and shoots are responsible for absorbing elements (Stepaniuk, 1992; Jackson, 1998) in different species show different behavior about their capacity to accumulate elements in their different parts like root, stem, and leaves.

As a result of direct discharge of contaminants into sediments, such trace elements accumulate (Dunbabin and Bowmer, 1992). A plant's ability to accumulate elements in its different parts like roots, stems, and leaves is different in various species because they absorb elements through roots and/ or shoots.

Therefore, the plant organ absorbing the heavy metals has to be detected (Baldantoni et al., 2004; Cyr and Campbell, 1994). One of the most significant advantages of phytoremediation is its use in determining sub-lethal levels of bioaccumulated pollutants in tissues over a period, which indicates the cumulative net number of pollutants integrated into tissues (Lovett Doust et al., 1994).

Plants can accumulate significant amounts of specific pollutants, even from highly diluted solutions, without obvious toxic effects (Ravera et al., 2003). Depending on the type of organisms sampled, passive bio-monitoring may be performed on native, endemic, or native-like organisms, while active

Bioremediation and Phytoremediation: Technologies for Toxic Pollution
Ashwini A. Waoo, PhD (Author)

bio-monitoring may be performed on introduced organisms (Chaphekar, 1991).

A rigorous study of the metal pollution and its monitoring in plants (Ramdan, 2003) was conducted in the Manzala lake, Egypt. Similarly, Ravera et al., (2003) determined trace metals concentration in macrophytes, and macro and accumulation of microelement in *Phragmites australis* and *Typha angustata* were studied concerning spatial gradients of the pond by Baldantoni et al., (2004; 2005).

Plants are engineered to remove or render harmless the contents of contaminants found in soil and water, such as heavy metals, organic compounds, trace elements, and radioactive substances. During cleaning, phytoremediation is esthetically pleasing to people in the area and is considered a more natural solution than machinery and equipment. Phytoremediation is attractive due to its low cost as it is alternative technologies, in many pollutants (e.g., copper) can be reclaimed and recycled.

It is a global problem to pollute the environment with toxic chemicals, metals, and xenobiotics. Several prevention and control measures are available for it, but most remedial technologies are costly and cause decreases in soil fertility, which negatively impacts the ecosystem.

Plant uptake of heavy metals in soil results in accumulation in plant tissues, phytotoxicity, and changes in plant communities. Electroplating industries create a lot of discharges in the region where the research was conducted. Four to five plant samples and three soil samples were collected for heavy metal analysis. There are homeostatic mechanisms in plants that allow them to hold necessary heavy metal ions in cell and decrease the detrimental effects of excess nonessential metal ions.

13.2 HEAVY METAL DETECTION FROM PLANTS AND SOILS

In the industrial area, soil samples were taken from three different locations to study the heavy metals. In addition, six layers of soil were sampled-0–20 cm, 20–40 cm, 40–60 cm, 60–80 cm, 80–100 cm, and 100–120 cm, and four to five species of plants were selected to represent the dominant species commonly occurring in the area of the study.

Three replicate samples were obtained from every sampling site. At each point, more than 1 kg of soil and plants were collected from each layer. soil samples were air-dried for 7 days, crushed, sieved to obtain 2 mm fractions, and oven-dried stems, leaves, and stems of plants to constant weights in the laboratory.

13.2.1 ANALYSIS OF SOILS FOR HEAVY METALS BY ATOMIC ABSORPTION SPECTROMETRY (AAS)

The soil samples collected from contaminated sites were collected in triplicates and dried to constant weight at 105°C. Microwave-assisted Kjeldahl digesters were used to acid-digest three replicates of 0.25g sediments. A total of 5 ml of concentrated nitric acid, 2 ml of hydrochloric acid, and 1 ml of hydrofluoric acid was added to each microwave extraction vessel. A microwave unit at 800 watts was used to heat the vessels to 210°C for 20 minutes at 40 bar pressure. Atomic Absorption Spectrophotometers using flame atomization were used to analyze the digested samples for heavy metals. The results are expressed as dry weight.

13.2.2 ANALYSIS OF PLANTS FOR HEAVY METALS BY AAS

The plant species frequently grown within the target area have been identified and fresh plant parts of all five varieties selected for experiments were randomly collected from the plants within the target area. After air-drying for seven days, all the samples were examined. To grind the samples to powder, the samples were oven-dried at 60°C to a constant weight. In microwave-assisted Kjeldahl digestion, three dried samples were digested with a mixture (3:1) of concentrated nitric acid and hydrofluoric acid. Six ml of nitric acid and two ml of hydrofluoric acid were added to each microwave extraction vessel along with 0.8 g of plant sample. With a pressure of 25bar and a temperature of 190°C, the vessels were capped and heated in a microwave at 800W for 20 minutes. The digested samples were diluted to 50 ml and analyzed using an atomic absorption spectrophotometer using flame atomization. Results are expressed as dry weights of each component.

13.3 HEAVY METALS ACCUMULATION IN PLANTS

As a cleanup method for hazardous waste sites, phytoremediation is effective and inexpensive. To restore and manage our natural water resources, plants can be used to degrade, absorb, metabolize, and detoxify contaminants in cost-effective and ecologically sound ways.

Indian soils and water bodies need to be decontaminated using phytoremediation technology. This will bring new resources and technology to solve India's environmental problems caused by industries.

Implementing and demonstrating bioremediation's effectiveness requires an interdisciplinary, systematic evaluation strategy at every level. Before any decision to use a particular treatment, the process is made, a detailed characterization of the site needs to be conducted, this would provide balance information on the chemical nature of the contamination.

Plants found in the polluted sites, identified for phytoremediation studies are generally wild annual herbs, with low economic importance and very little biomass. Phytoremediation with crops of economic value or high biomass producers has been also emphasized (Bona et al., 2007).

It is suggested that *in vitro* screening of organs or propagules would be an efficient, simple alternative, and promising method (Mills and Benzioni, 1992) for studies on tolerance and accumulation in plants. *In vitro* culture of the present weeds, plants have been used in investigating maximum responses of plants to toxic metal ions. With the use of *in vitro* techniques, in this research selected plants are tolerant and accumulators to metal ions. Numerous *in vitro* experiments have concentrated on the effects of increasing concentrations of heavy metals on the response of plants tolerant or sensitive to industrial pollution (El-Aref and Hamada, 1998; Ramgareeb, 1999). In vitro or under natural conditions of environmental pollution, plants may be selected.

Additionally, the association of microorganisms and plants present in the rhizosphere seems to enhance the removal of contaminants. Although slow, phytoremediation is green, cheap, requires little equipment or human labor, can be performed quickly, and sites can be cleaned without removing the polluted soil; it is an in-situ method. The hyperaccumulator can also harvest and extract precious metals like gold, and chromium through phytoextraction. In order to be successful, phytoremediation depends on the identification of plants that are tolerant and suitable for each area, ones that can accumulate high concentrations of the required metal. Microbes have evolved a number of mechanisms to tolerate heavy metal ions in metal-stressed environments. Thus, this study demonstrates that plants, bacteria, and fungi can survive in an environment that is heavily polluted by heavy metals. Phytoextraction relies largely on plants themselves, microbial interactions, and heavy metal bioavailability in soils. Chemical chelates can enhance the accumulation of metals by plants (Do Nascimento et al., 2006). Compounds such as these can increase the risk of metal leaching and have negative effects on soil fertility (Blaylock et al., 1997; Kos and Lestan, 2003).

Phytoremediation can remediate the heavy metal polluted soil (Khan et al., 2000; Sekhar et al., 2005; Fischerova et al., 2006). In heavy metal contaminated soil, hyperaccumulator plants may have a slow growth and lower biomass, which may limit their effectiveness (Kumar et al., 1995; Burd et al., 2000). Microbes that are resistant to heavy metals can promote plant growth and improve metal uptake by hyperaccumulator and nonhyperaccumulator plants (Whiting et al., 2001, de Souza et al., 1999, Zaidi et al., 2006, Abou-Shanab et al., 2003).

In contaminated environments, metal-resistant bacteria may survive and accomplish bioremediation by using various mechanisms and can even be selected for their efficacy for bioremediation process (Piotrowska et al., 2005).

13.4 PLANT SPECIES FOR PHYTOREMEDIATION

Metal hyperaccumulation in a variety of plant species has been studied extensively and significant mechanisms were traced out. Different mechanisms of metal accumulation were observed in various plant species, including exclusion and compartmentation. Species classified as hyperaccumulators can accumulate metal concentrations of more than 1000 mg/kg of dry weight (Reeves, 2003). There are several species of *Thlaspi*, Arabidopsis, and *Sedum alfredii* that have been studied extensively. These plants are in the *Brassicaceae* and Alyssum families). *Thlaspi* sp. Can usually hyper accumulate more metals, e.g., *T. ochroleucum* for Ni, Zn (Prasad and Freitas, 2003). *Thlaspi caerulescens* has been extensively studied for remediation of Cd and Zn contaminated soils. Furthermore, it has been shown to absorb up to 2600*10*6 Zn without causing any harm (Brown et al., 1995) and to remove up to 22% of cadmium. Cd is more readily absorbed by its roots due to a specific rooting strategy and high uptake rates in the root membrane caused by Cd-specific transport channels.

Zn is sequestered in soluble form in the vacuoles of epidermal cells of *T. caerulescens* (Frey et al., 2000). Zn was accumulated in the mesophyll cells of *A. halleri* leaves, (Sarret et al., 2002). Cosio et al., (2004) investigated the process of Zinc and Cadmium accumulation in some species and also studied ion compartmentation in mesophyll protoplast. The existence of a regulation mechanism is present on the plasma membrane of mesophyll protoplast.

Various terrestrial plants especially the members of *Brassicaceae* are good heavy metal accumulators (Reeves et al., 2000; Gratao et al., 2005). Some aquatic plants such as *Ceratophyllum demersum, H. verticillata, Pistia stratiotes, Bacopa monnieri,* etc. also have the potential of heavy metal accumulation. Algal species can destroy pollutants. As suggested by Yoon et al., (Yoon et, al., 2006), native plants on a contaminated site have a better potential to tolerate stress conditions than plants from other locations. Thus, native plants must be studied for their use in phytoremediation.

For metal-contaminated environments including soil, water, and sewage sludge Phytoremediation is potentially an efficient detoxification technique.

Plants that accumulate heavy metals can be harvested for phytoremediation. Phytoremediation is a suitable technique when human and environmental health is on priority. Heavy metal-hyperaccumulator plants accumulate more amounts of metals in their aerial parts. There are more than 400 plant species known to be hyperaccumulators, and many can accumulate two or more elements. Out of the known hyperaccumulators, *Noccaea caerulescens* is one of the few model plants used for phytoremediation.

Hyperaccumulators and their threshold of heavy-metal accumulation Plants have developed three strategies to survive on metalliferous soils (Baker et al., 2000; Lasat et al., 2000; Rascio & Navari-Izzo, 2011), and plants can generally be divided into three main groups (van der Ent et al., 2012): metal excluders, metal hyperaccumulators and metal indicators. A metal excluder is a plant that prevents metals from entering its aerial parts over a wide range of metal concentrations.

These plants can accumulate metals in their aerial tissues, which reflect soil metal concentrations. The aboveground tissues of hyperaccumulating plants are enriched with metals in levels far exceeding those of the soil or those of non-accumulating species growing nearby, without suffering phyto-toxic effects (Baker et al., 2000; Lasat et al., 2000; Rascio & Navari-Izzo, 2011). The capacity to accumulate metals in the shoots of these plants can be up to 100 times greater than in non-hyperaccumulators (Lasat, 2002).

Hyperaccumulating plants differ from related non-hyperaccumulating plants in three ways: their enhanced ability to take in heavy metals, their longer root-to-shoot translocation, and their capacity to sequester heavy metals in their leaves (Rascio & Navari-Izzo, 2011). Threshold values are used to determine whether a plant is a hyperaccumulator, and they vary according to the specific heavy metal. The term 'hyperaccumulator' was first used to indicate plants which can accumulate Ni at over 1000 mg kg^{-1} leaf dry weight (DW) (0.1%) (Brooks et al., 1977), whereas Ni content in almost all of the leaves of non-accumulating plants was found in the range 1–10 mg kg^{-1} DW (Marschner, 1988). For Zn, a threshold of 10000 mg kg^{-1} DW was suggested (Reeves & Brooks, 1983), but recently it was proposed to be 30000 mg kg^{-1} (van der Ent et al., 2012).

Most non-accumulating plants have Zn concentrations within the range of 30–100 mg kg-1 (Marschner, 1988). The threshold values for hyperac-cumulators of other metals and metalloids are as follows: 100 mg kg^{-1} for Cd, selenium (Se), and thallium (Tl); 300 mg kg^{-1} for Cu, cobalt (Co), and Cr; 1000 mg kg^{-1} for Ni, arsenic (As) and Pb; and 10000 mg kg^{-1} for Mn (Reeves & Brooks, 1983). The hyperaccumulation of heavy metals has been documented in more than 400 plant species, many of which accumulate two or more metals (Baker et al., 2000). Hyperaccumulators co-sequester metals from soils with low concentrations (Assunção et al., 2003a). A list of hyperaccumulators of Ni, Zn, Cd, Pb, Cu, As, Co, Ni, and Mn are 9 available (Baker et al., 2000; Brooks, 2008; van der Ent et al., 2012), and a large number of these, over 317 species, hyper accumulate Ni. A list of Zn hyperaccumulator plants is also available, 18 species, and they are mainly found growing on soils polluted with Pb, Cd and Zn (Brooks, 2008). Deployment of hyperaccumulating plants for phytoextraction of heavy metals has been assessed and ranked concerning research status, readiness for commercialization, and regulatory acceptance of the technology (Lasat, 2002).

Hyperaccumulation of heavy metals by plants is a complex process that involves several major mechanisms: metal bio-activation in the root zone, metal uptake into root cells and compartments, xylem loading and transport of metals from root to shoot, and distribution and sequestration of heavy metals at the entire plant and at cellular levels (Lombi et al., 2002a; Yang et al., 2005).

The hyperaccumulation capacity of these plants has been explained by a variety of mechanisms, but the mechanisms remain unclear, particularly at the cellular level.

Furthermore, several factors affect the hyperaccumulation mechanisms (Figure 13.1): plant species, properties of the heavy metal, properties of the growth medium, the bioavailability of the metal, environmental condition, and chelating agents (Bieby Voijant, 2011). The plant species is the most influential factor affecting mechanisms of heavy-metal uptake.

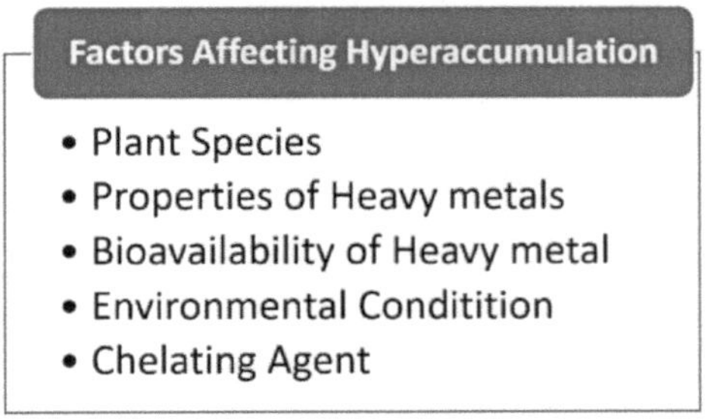

FIGURE 13.1 Factors affecting mechanism of hyperaccumulation.

- ***Plant species***

The production of large amounts of biomass and hyper accumulation of heavy metals is crucial to successful phytoextraction (Baker, 1981; Baker et al., 2000). Plants respond to particular metals in different ways, depending on their genetic background (Baker et al., 2000). Properties of the growth medium are referred to as those agronomical practices that enhance remediation, such as pH adjustment and addition of chelators and fertilizers (Baker et al., 1994a; Monsant et al., 2008).

For instance, nitrogen applied as $Ca(NO_3)_2$ was found to enhance Zn uptake in the hyperaccumulator *Noccaea caerulescens* (Monsant et al., 2010; Monsant, 2011). The root zone is important in phytoremediation because it takes up contaminants and stores them inside the plant (McGrath et al., 1997; Whiting et al., 2001).

- ***Metal bioavailability***

Through collaboration between plant roots and microorganisms in the root zone, proton, organic acids, phytochelatins, amino acids, and enzymes are released in the rhizosphere to enhance metal bioavailability (Yang et al., 2005).

This finding proved that the secretion of protons by roots acidifies the rhizosphere and increases the metal dissolution 10 and enhances root absorption. Root reductases and some plant enzymes radiated from the roots probably contribute to metal transformation in the rhizosphere (Lasat, 2002; Yang et al., 2005).

- ***Environmental conditions***

In other words, environmental factors such as temperature, humidity, and light directly affect plant growth parameters such as shoot development, root proliferation, and root length, thereby altering a plant's uptake mechanism (Li et al., 2009).

- ***Properties of heavy metals***

Understanding the properties of heavy metals is critical for proving the applicability of phytoextraction (Baker et al., 2000). Metals are bound to the soil; therefore, the soil redox potential, pH and organic matter will affect the the metal tendency to exist in an ionic and plant-available form (Bieby Voijant, 2011). Plants can lower the pH of soils and sediments to make metals more bioavailable (Caixan & Zdenko, 2003).

- ***Chelating agents***

A chelating agent is a chemical substance that increases heavy-metal bioavailability, thus enhancing the uptake of metals by plants and stimulating the microbial community to take in metals (Pierzynski, 1993; Bieby Voijant, 2011). The faster the uptake of heavy metals, the cheaper and faster the remediation process will be. The withdrawal of metals from contaminated soil is enhanced by the addition of chelating agents, like EDTA (Ethylene diamine tetra acetic acid) (Gupta and Sinha, 2006).

EDTA-exposed plants showed improved metal translocation into the tissues as well as better phytoextraction (Lombi et al., 2002b). When using synthetic chelating agents, however, you need to consider the possibility of more leaching (Baker et al., 1994a; Alloway, 1995).

13.5 MECHANISMS OF HEAVY-METAL UPTAKE AT THE CELLULAR LEVEL

At the cellular level, plants have a number of mechanisms for detoxification and stress tolerance. According to the study by Hall (2002), the mechanisms include:

- Constraint of metal translocation to roots with mycorrhizae,
- Heavy-metal binding to cell walls and root exudates,
- Reduced influx across plasma membranes,
- Active efflux into the apoplast,
- Chelation in the cytosol by various ligands,
- Repair and protection of plasma membranes under stress conditions,
- Transport of low or high weight proteins into the vacuole by the metal complexes phytochelatin (PC) and metallothionein (MT), and
- Accumulation of metals in the vacuole. Chelation, Transport and sequestration are the key processes of cellular metal homeostasis and tolerance in plants (Clemens, 2001).

These processes determine the uptake, distribution, and detoxification of metal ions at the cellular and whole-plant levels. Hypersensitivity of metals occurs if one of these crucial processes fails. Toxicity is avoided by the synthesis of specific low molecular-weight chelators which prevent excess trace metal(loid)s from binding to physiologically important proteins, and enable them to be transported into the vacuoles (Memon, 2001). According to these studies, PCs and MTs were found to be active and to transport across plant cells some metal ions that support the synthetic enzymes PC synthase and MT synthase. However, PC synthesis was found to be essential for constitutive levels of tolerance to Cd, Hg, and As, but not essential for Copper and Zink tolerance (Assunção et al., 2003b). In the cytoplasm, PCs and MTs are the principal heavy-metal chelators. The binding of cations to cell walls in the root apoplast contributes toward heavy-metal tolerance in plants (Frey et al., 2000; Hall, 2002). Cell walls are composed of a matrix of pectin, lignin, and glycoprotein embedded with cellulose micro-fibrils that include carboxyl groups with a high binding affinity for polyvalent cations, which restrict their movement and absorption into the root.

Metals cross the plasma membrane via transport proteins and intracellular high-affinity binding sites. Following are the several classes of sites of high affinity are involved in the transporttal transport in plants. In plants, metal ion homeostasis and tolerance are regulated by heavy-metal ATPases,

- The natural resistance-associated macrophage family of proteins,
- The cation diffusion facilitator, i.e., CDF family of proteins (Yang et al., 2007), and
- The Zn–Fe-regulated transporter-like (ZIP) family of proteins. However, some studies have indicated that transporters of metal cations are not specific.

There is evidence that non-essential metals enter cells via cation transporters with a range of substrate specificities (Clemens, 2002). Multiple passages exist for metal-ion transport in plants, although which transporters significantly contribute to specific metal-ion uptake in specific plant species is unknown.

Heavy metals are the most hazardous pollutants in ecotoxicology due to their high toxicity and significant concentrations released into the environment. The environment has concerns about heavy metal pollution in terms of water, air and soil pollution. This results in a reduction of crop yield and also causes health hazards. Heavy metal stress induces the antioxidative systems of plants to minimize the damage.

An excessive number of heavy metals blocks many physiological processes like photosynthesis, respiration, transpiration, cell elongation, N-metabolism, and mineral nutrition, decrease in biomass, and consequently plant death. Plant responses to heavy metals include changes in redox potential, the concentration of signaling molecules, membrane permeability, the activity of antioxidant enzymes, cysteine, glutathione (GSH), protein content, and phytochelatin (PCs) contents, etc. This impact of heavy metals depends on their doses (Morkunas et, al., 2018).

Plants have evolved to adapt to a variety of abiotic and biotic stress conditions over time. Constitutive defenses and stress-inducible responses were involved. Important tolerance pathways, such as induction of membrane-localized transporters for compartmentation of ions, osmolyte accumulation, antioxidant enzyme enhancement, and accumulation of elements against pathogens, must all be investigated in this regard. Plants have evolved an ion homeostasis system that allows them to cope with a little amount of heavy metal contamination. Metals that are potentially toxic yet are necessary for regular development must be absorbed and dispersed to various sections of the plant. Chaperons, chelators, and specialized transmembrane transporters have all evolved, according to Hall and Williams. It has been demonstrated that arsenic (As) tolerance may be achieved by phytochelatin-mediated sequestration.

13.5.1 OXIDATIVE STRESS AND REACTIVE OXYGEN SPECIES

"Reactive oxygen species" is derived from molecular oxygen. Various ROS are present transiently in all aerobic organisms, including (a) oxygen-derived non-radicals, e.g., hydrogen peroxide (H_2O_2 organic hydroperoxide (ROOH); and (b) oxygen-derived free radicals, e.g., hydroxyl (HO•), peroxyl (RO·2RO2•), superoxide anion (O·−2O2•-), and alkoxyl (RO•) radicals. ROS are natural byproducts of oxidative metabolism which cause damage to aerobic organisms. Some of them are important signaling molecules that can modulate the activity of some defense proteins. They are unstable and very reactive molecules having a short life and can oxidize lipids, proteins, and nucleic acids resulting in cell structure alteration and mutagenesis.

Chloroplast, mitochondria, and peroxisomes are said to be the main centers of ROS in photosynthetic organism's production. Toxic heavy metals induce more production of ROS in heavy metals interaction with electron transport activities in chloroplast and mitochondrial membrane. According to Anjum et al., (2017) found increased levels of Cr toxicity in maize resulted in high levels of proline, phenolics, and soluble sugars as well as low soluble protein levels. Electrolyte leakage was caused by increases in malondialdehyde and H_2O_2.

13.5.1.1 METAL INDUCED GENOTOXICITY

Genotoxicity associated with heavy metals involves many mechanisms; these are not all well understood or elucidated. Despite this, each metal has unique physicochemical properties and toxicological mechanisms of action.

Genotoxicity mechanisms are complex and yet not well-understood. It has been found that this type of genotoxicity leads to DNA damage which is the result of oxidative stress and the formation of ROS. Heavy metal-induced genotoxicity causes chromosomal aberrations and lowers the cell proliferation rate. Many previous research works showed heavy metal-induced nucleic acid complications in plants like *Allium cepa, Solanum tuberosum, Helianthus annuus, Nicotiana tabacum,* etc. ROS interaction with DNA can lead to base modification, base deletion, strand breaks. There are four sites for metal binding with DNA

(i) the ribose hydroxyls;
(ii) the phosphate oxygen atoms;
(iii) the keto groups;
(iv) the base ring nitrogen.

13.5.2 INTERFERING SIGNALING PATHWAYS

Calcium signaling, hormone signaling, and mitogen-activated protein kinase (MAPK) signaling are the three main signaling components involved in this process. Other regulators, such as microRNAs and transcription factors, also play an crucial role in heavy metal stress regulation. By triggering several signaling pathways, metal stress is known to play a critical role in imparting resistance to environmental stresses. The MAPK pathway contributes to the management of stress response. The activation of signaling pathways influences the expression of genes by activating transcription factors and other cytosolic proteins (Jalmi et al., 2018).

It is not yet clear how this mechanism works. Heavy metals can directly affect cellular signaling or indirectly through the formation of ROS induced by heavy metals. Studies have shown that heavy metal toxicity leads to uncontrolled regulation of the signaling pathway. G-proteins, receptor tyrosine kinases and growth factor receptors, may be the target group. Plants are negatively affected by heavy metal stress due to transcription activation of responsive genes. Heavy metals in plants like Zn, Pb, Cu, and Cd have been shown to affect mitogen kinase signaling pathways.

13.5.3 BIOCHEMICAL AND PHYSIOLOGICAL RESPONSE

Heavy metal stress causes plants to adapt physiologically, biochemically, and molecularly to counteract the toxic effect. It is important to study to understand how plants respond under heavy metal stress. Plants undergo heavy metal stress changes that should be considered for improving their accumulation capacity and usefulness for phytoremediation on heavy metal-contaminated sites. This information is important for engineers designing plants for improvement in terms of accumulation and for the use of plants for phytoremediation.

Heavy metals interact with cellular components of the organisms such as cells, tissue, organs by modulating cell signaling or metabolic pathway which leads to visible injuries. The anti-oxidative defense system of plant activates many enzymes such as catalase (CAT), ascorbate peroxidase (APX), superoxide dismutase (SOD), glutathione reductase, glutathione (GSH), guaiacol peroxidase (GPX), (GR), peroxidase (POX), etc. which play a significant role in scavenging excess ROS.

13.6 HORMESIS

In the hormesis process, a toxic substance or environmental agent is adaptively beneficial at low levels when it is exposed to a low dose whereas it is toxic at higher levels. The evolutionary theory recognizes hormesis as an essential concept. As long as life has existed on earth, its cells have often been exposed to free radicals and harmful substances. Adapting to environmental hazards has enabled organisms to avoid extinction. Hormesis derives from Greek hormesis, literally "rapid motion, eagerness," which is itself derived from hormaein, which means the action of setting into motion, impelling, and urging on. Hormetic is the science of hormesis.

It can also refer to as a biological phenomenon of organisms' defense against heavy metals. Heavy metals cause a reduction of plant growth and also carried out some alteration in structure of root, based on published literature. Thus, heavy metals toxicity leads to a decrease in crop yield. The hormesis phenomenon have occurred in many organisms due to different stress factors, such as chemical, physical and biological. Calabrese and Baldwin reported the concept of chemical hormesis to originate According to the opinion of some authors these effects are observed for growth as well as development, reproduction, the survival of the plant. These hormetic effects may be closely related to various physiological responses such as photosynthesis rate, DNA and RNA synthesis, metallothionein synthesis, oxygen consumption, mitosis, tissue regeneration, stress protein synthesis, seed germination, etc. According to Damelin et al., hormetic activity is a specific stress response to minimize the harmful effects of toxic agents.

13.7 FACTORS AFFECTING THE UPTAKE OF HEAVY METALS

Several factors can affect the uptake of heavy metals, such as plant species, properties of medium, vegetative uptake, and addition of chelating agent.

- ***The Plant Species***

Heavy metals are absorbed differently by different plant species. For phytoremediation to be successful, it must be determined which native plant species can accumulate heavy metals and produce high amounts of biomass.

- ***Properties of Medium***

The use of various agronomic practices such as pH, fertilizers, chelators, etc. increases the rate of remediation. For example, plants' ability to absorb lead was influenced by the organic matter, pH, and phosphorus of the soil.

- ***The Root Zone***

Phytoremediation relies heavily on the Root Zone, which can absorb pollutants and transform them inside the plant. Plant roots secrete enzymes that break down soil contamination.

- ***Vegetative Uptake***

Vegetative uptake of heavy metals is mostly affected by environmental factors. Root morphology and growth factors are affected by the temperature. The availability of contaminant-specific hyper-accumulator plants is crucial to phytoextraction. Water phase bioavailability also affects the uptake of a metal, which is determined by the retention time and the reaction of the metal with other components in the water. Several factors affect the forms of metals available to plants, including metal binding to the soil, redox potential, organic matter and pH.

- ***Chelating Agent***

Heavy metals can be bioavailable through the addition of biodegradable chelating agents and micronutrients can be added to increase the heavy-metal-uptake capacity of the microbiome associated with plants.

Faster uptake of heavy metals can be a cheap remediation method. There is an increased risk of leaching with some synthetic chelating agents. Chelating agents can accelerate the leaching of contaminants into soil. A chelating agent may be necessary for alkaline soils because heavy metals' bioavailability declines above pH 5.5–6. EDTA exposure for a longer period (two weeks) improved metal translocation in plant tissues and phytoextraction performance. The chelating agent EDTA at 5mmol/kg showed positive results. Citrate and oxalate, which are organic acids in root exudates, affect the bioavailability of metals from plant roots.

13.8 SUMMARY

Anthropogenic pressure related to growing populations leads to soil pollution by heavy metals that threaten the environment and food security. There are many anthropogenic activities that contribute to environmental pollution and human exposure to heavy metals, such as industrial production, mining, and the use of metal-containing compounds in domestic and agriculture.

Human health and the environment are seriously threatened by heavy metal contamination. As a result, remediation of land contamination is of the utmost importance. An eco-friendly method of reclaiming heavy metal-polluted soil through phytoremediation could be a cost-effective mitigation measure. Plant tolerance for heavy metals and accumulation of heavy metals is essential for improving phytoremediation efficiency. This chapter illustrated mechanisms that plants use to take up, transport, and detoxify metals.

In agricultural production and food safety, heavy metal pollution is a crucial issue because of its toxic effects and rapid accumulation in the environment. Heavy metal contamination has been prevented or mitigated in a number of ways, including replanting contaminated soil. Phytoremediation can be used to revegetate heavy metal-polluted soils that have been contaminated by heavy metals. Public acceptance is good, and it offers many advantages over other physicochemical methods. To date, hundreds of hyperaccumulators have been recognized and used for phytoremediation of heavy metals.

13.9 EXERCISES

Objective Response Questions

1. **As a result, lead-contaminated soils pose a harm for children living in neighborhoods because ___.**
   a) When children put their dirty hands and objects in their mouths, they ingest dirt and dust
   b) The average child eats more candy than an adult
   c) Children consume more fruit than adults
   d) Children should be encouraged to eat lots of vegetables from their gardens

2. The map displays the pollution levels at a region before and after cleanup. A small area of contamination was left after two months of treatment. Researchers conducted weekly sampling at the site during the two months and recorded their findings. In the contaminated area, scientists were able to determine over time how much pollution there was. Two months after the crisis, locals were allowed to drink the water in most of the affected areas. Accurate measurements are needed by scientists. As a result of these measurements, what did the scientists learn?

a) Who caused the contamination?
b) When the water was clean enough to drink.
c) When more contaminants might be released.
d) How the contaminants were released.

3. Plants absorb heavy metals through

a) Roots
b) Shoots
c) Leaves
d) Flowers

4. This engineered technology uses the green plant to eliminate heavy metals from environment.

a) Mycoremediation
b) Immobilization
c) Fungal remediation
d) Phytoremediation

5. Soil samples taken from contaminated sites for heavy metal analysis were dried at

a) 105°C
b) 95°C
c) 205°C
d) 155°C

6. Which of the following is not involved in the hyperaccumulation mechanism?

a) Metal bio-activation in the root zone

b) Metal uptake into root cells and compartments
c) Xylem loading and transport of metals from root to shoot
d) Gene cloning

7. A key factor for phytoremediation is _________.

a) The identification of a plant suitable for each area
b) Higher concentration of Heavy metals
c) Low Moisture in the soil
d) High soil pH

8. Precious metals, such as gold collected by the hyperaccumulators and can be harvested by _________.

a) Phytoextraction
b) Immobilization
c) Rhizofilatration
d) Phytostbilization

9. Heavy metal contaminated soil may decrease the efficiency of phytoremediation when _________.

a) Fast growth and low biomass
b) Slow growth and low biomass
c) Slow growth and high biomass
d) Slow growth and low water

10. It is better to use the following plants for phytoremediation _________.

a) Non-indigenous
b) Native
c) Foreign
d) Small

Practice Work

Qu 1. Which factors affect the rate of heavy metal uptake?

Qu 2. What are heavy metals? Explain their adverse effect on the environment.

Qu 3. Explain the process of heavy metal analysis from soil and plant.

Qu 4. Describe various plant species used for Phytoremediation purposes.

Qu 5. Describe the factors affecting hyperaccumulation.

Qu 6. Explain the mechanisms of heavy metal uptake at the Cellular Level.

Qu 7. Explain the classes of proteins that help in metal transport in plants?

Qu 8. Explain the phenomenon of metal-induced genotoxicity.

Qu 9. Write an essay on Heavy metal pollution as a vital issue for the environment.

Qu 10. Explain the physiology of plants during heavy metal accumulation.

REFERENCES

Brown, S. L., Chaney, R. L., & Scott, A. J., (1995). Zinc and cadmium uptake by hyperaccumulator *Thlaspi-caerulescens* grown in a nutrient solution. *Soil Sci. Soc. Am. J., 59*(1), 125–133.

Clemens, S., Palmgren, M. G., & Krämer, U., (2002). A long way ahead; understanding and engineering plant metal accumulation. *Trends Plant Sci., 7*, 309–315.

Cosio, C., Martinoia, E., & Keller, C., (2004). Hyperaccumulation of cadmium and zinc in *Thlaspi caerulescens* and *Arabidopsis hallri* at leaf cellular level. *Plant Physiol., 134*(2), 716–725. doi: 10.1104/pp.103.031948.

Dinh, Ngoc., (2016). *Phytoextraction of Zinc by the Hyperaccumulator Noccaea Caerulescens: Accumulation, Distribution, and Multi-Element Interactions.* Thesis at the University of Queensland in School of Chemical Engineering. https://doi.org/10.14264/uql.2016.183.

Frey, B., Keller, C., & Zierold, K., (2000). Distribution of Zn in functionally different leaf epidermal cells of the hyperaccumulator *T. caerulescens. Plant Cell Environ., 23*(7), 675–687. doi: 10.1046/j.1365-3040.2000.00590.x.

Gratao, P. L., Prasad, M. V., Cardoso, P. F., Lea, P. J., & Azevedo, R. A., (2005). Phytoremediation: green technology for the clean-up of toxic metals in the environment. *Braz. J. Plant Physiol., 17*, 53–64.

Jalmi, S. K., Bhagat, P. K., Verma, D., Noryang, S., Tayyeba, S., Singh, K., Sharma, D., & Sinha, A. K., (2018). Traversing the links between heavy metal stress and plant signaling. *Front Plant Sci., 9*(12). doi: 10.3389/fpls.2018.00012.

Kupper, H., Lombi, E., & Zhao, F. J., (2000). Cellular compartmentation of cadmium and zinc in relation to other elements in the hyperaccumulator *Arabidopsis halleri. Planta, 212*(1), 75–84. doi: 10.1007/s004250000366.

Lone, M. I., He, Z. L., Stoffella, P. J., & Yang, X. E., (2008). Phytoremediation of heavy metal polluted soils and water: Progresses and perspectives. *J. Zhejiang Univ. Sci B., 9*(3), 210–220. doi: 10.1631/jzus.B0710633.

Morkunas, I., Woźniak, A., Mai, V. C., Rucińska-Sobkowiak, R., & Jeandet, P., (2018). The role of heavy metals in plant response to biotic stress. *Molecules (Basel, Switzerland), 23*(9), 2320. doi: 10.3390/molecules23092320.

Prasad, M. N. V., & Freitas, H. M. D., (2003). Metal hyperaccumulation in plants—Biodiversity prospecting for phytoremediation technology. *Electron. J. Biotechnol., 93*(1), 285–321.

Reeves, R. D., & Baker, A. J. M., (2000). Metal accumulating plants. In: Raskin, I., & Ensley, E. D., (eds.) *Phytoremediation of Toxic Metals Using Plants Clean Up the Environment* (pp. 193–229). Wiley, New York.

Sarret, G., Saumitou-Laprade, P., & Bert, V., (2002). Forms of zinc accumulated in the hyperaccumulator *Arabidopsis halleri. Plant Physiol., 130*(4), 1815–1826. doi: 10.1104/pp.007799.

Tangahu, B. V., Abdullah, S. R. S., Hassan, B., Mushrifah, I., Anuar, N., & Mukhlisin, M., (2011). A review on heavy metals uptake by plants through phytoremediation. *International Journal of Chemical Engineering* (Vol. 2011, p. 31). Article ID 939161. https://doi.org/10.1155/2011/939161.

Tripathi, R. D., Vajpayee, P., Singh, N., Rai, U. N., Kumar, A., Ali, M. B., Kumar, B., & Yunus, M., (2004). Efficacy of various amendments for the amelioration of fly ash toxicity: Growth performance and metal composition of *Cassia siamea* Lamk. *Chemosphere, 54*, 1581–1588.

Yoon, J., Cao, X., Zhou, Q., & Ma, L. Q., (2006). Accumulation of Pb, Cu, Zn in native plants growing on a contaminated Florida site. *Sci. Total Environ., 368*, 456–464.

CHAPTER 14

Phytoremediation of Wastewater

14.1 INTRODUCTION

The high demand for clean drinking water and the high consumption of water are pressing issues in modern industrial society. Several approaches have been developed to reduce water consumption, but recycling wastewater provides high-quality water in the long run. It seems appropriate to submit alternative water treatment technologies to ensure high-quality drinking water. On the other hand, landfills are causing serious soil and water pollution.

Environmental pollutants such as heavy metals and organic pollutants affect the quality of soil, water, and air. There have been several strategies developed over the past five decades to address polluted water. The use of conventional methods is preferable to the use of aquatic plants. Phytoremediation involves treating water resources with plant matter to eliminate harmful contaminants that can harm the environment with the following objectives:

1. Examine the harmful effects of toxic chemical pollution on plants, animals, and humans;
2. Review of the physicochemical factors that affect the removal of toxic chemicals, such as heavy metals, from aqueous solutions by aquatic plants;
3. Evaluation and comparison of the removal rates of heavy metals and organic contaminants from aqueous solutions by different aquatic plants; and
4. Analyzing aquaculture plants using chemometric models.

Plants are used to partially or substantially remediate contaminated soil, sludge, sediment, ground water, surface water, and wastewater.

Bioremediation and Phytoremediation: Technologies for Toxic Pollution
Ashwini A. Waoo, PhD (Author)

Biological processes and physical characteristics of plants are used in the remediation of sites. Phytoremediation can be described as agro-remediation, phytoremediation, botanic remediation, or vegetative remediation. Phytoremediation involves multiple processes, each of which occurs to varying degrees depending on the conditions, the media, the contaminants, and the plants involved. These processes are referred to in the literature under many different terms.

Using phytoremediation, contaminants can be degraded, removed (through accumulation or dissipation), or immobilized (Pivetz, 2001):

1. **Degradation**
   - that occurs beneath the ground surface by the action of microorganisms is called Rhizodegradation.
   - Phytodegradation term describes contaminant uptake by plants that occur above ground or below ground, within their roots, stems, or leaves.
2. **Accumulation**
   - Phytoextraction: the process of absorbing and accumulating contaminants for removal.
   - Rhizofiltration: trapping and/or removal of contaminants via root adsorption.
3. **Dissipation**
   - Phytovolatilization: uptake and volatilization of contaminants.
4. **Immobilization**
   - Hydraulic Control: regulating ground-water flow by absorbing water through a plant.
   - Phytostabilization: occurs when contaminants are immobilized in soil.

Phytoremediation requires more than simply planting vegetation and assuming that the contamination will disappear. Phytoremediation requires a detailed understanding of the processes involved, the plants chosen, and the steps required to ensure plant growth. There are lots of candidates, but only a few have been examined. To select the most useful plants, screening studies will be necessary. Studies in hydroponics or greenhouses cannot be extrapolated to field conditions without care. There is a need for further field research. Verification of phytoremediation's efficacy and applicability will likely need to take place on an individual site basis.

14.2 FACTORS AFFECTING PHYTOREMEDIATION

Bioremediation is the use of nature's cleansing powers. Several phyto processes are occurring during these growth stages, including the uptake of chemicals and water, metabolism within plants, and releasing exudates into the soil, which results in contamination loss. For plants to grow, light is converted into carbohydrates and oxygen through photosynthesis. Even in soils with high matric potentials and osmotic potentials, roots can extract water from the soil; the water is transported upward by the xylem. Primary transpiration occurs at stomata (potential gas exchange openings in leaves and stems), with secondary transpiration occurring at lenticels (gas exchange points on stems and roots).

For a chemical to be absorbed by plants, it must be in solution, either in water or in soil. The roots absorb water from the soil solution. Contaminants can pass through the epidermis, pass through the Casparian strip, and then pass through the endodermis, where they can be absorbed, bound, or metabolized. In the transpiration stream, chemicals are transported through the endodermis and into the xylem. Furthermore, these compounds can also react with plant tissue and be partitioned into it, in addition to being metabolized by plants or released into the atmosphere through stomata.

The effectiveness of detoxification or remediation by aquatic plants is dependent on the chemical composition of the water, the contaminant, the plant's physiology, and its genotype (Adriano, 2004; Prasad, 2004). Below is a summary of the physicochemistry in this study (Rai, 2008). By manipulating or modifying these parameters, phytoremediation can be enhanced.

- Light — Uptake is light-dependent in some cases
- pH — Lower pH generally increases the uptake/capacity
- Salinity — Salt concentration/toxicity increases with lower salinity
- Divalent Cations — Ca, Mg, Mn, Fe Monovalent Cations lower uptake
- Anions — Increasing divalent cations reduce the uptake
- Organic Acids — Lower uptake and toxicity
- Sediment Fraction — Binds metals and reduces uptake/toxicity
- Heavy Metals — Inhibit uptake/toxicity by binding to metals complex metals

• Sulfate	Insignificant but reduces uptake slightly
• Nitrate(N)	Significantly reduces toxicity
• Polypeptides	Reduces uptake/toxicity by complexation
• Polysaccharides	Chelate metals, Reduces uptake/toxicity
• Sulfur	Reduces uptake and toxicity indirectly
• Extracellular Products	Reduces toxicity

14.3 CONCEPT OF VEGETATED CAPS

Covering contaminated materials like landfill waste or soil with a cover is known as capping. These covers are referred to as "caps." Caps don't destroy or remove contaminants. Instead, they isolate them and keep them in place to prevent contamination. Caps prevent wildlife and people from coming into contact with contaminants.

To minimize exposure pathways and risks, plants were grown in or over-contaminated materials. Vegetated caps (or covers) are self-sustaining, long-term vegetation covers. The primary function of vegetation is to prevent or minimize infiltration of precipitation and snowmelt into contaminated sub-surfaces. A soil with maximum storage capacity is achieved by optimizing evapotranspiration. The evapotranspiration cap or water balance cover is designed to function as a moisture control device. By increasing soil stability, vegetation can also prevent erosion and remove contaminants through rhizodegradation or phytodegradation. The phytoremediation cap is designed to destroy or remove contaminants as well as prevent infiltration.

Vegetated caps can be constructed over landfills, soils, or groundwater that is contaminated. It might be necessary to maintain the cap for a long time, or the vegetation on the cap might be designed to allow a natural succession of plants to maintain its integrity. Utilizing vegetative caps on landfills for evapotranspiration control or contaminant destruction poses significant problems. The issues include the equivalent performance of standard, regulated landfill covers, the possibility of contamination uptake, the possibility of plant roots damaging the cap, and landfill gas production.

14.4 BUFFER STRIPS

Plants placed in buffer strips along waterways or pollution sources (i.e., riparian corridors) protect from contaminants. These strips may only be utilized for compounds that can be easily assimilated and metabolized. With buffer strips and riparian corridors, relatively soluble contaminants such as nutrients and some organics (especially pesticides) have been managed. Riparian corridors and buffer strips have been used to control agricultural runoff. Remediation requires ground water to be below the level of influence of roots. Vegetation can only be established on land that is sufficient for its establishment. The removal of contaminants will need to be monitored.

14.5 ADVANTAGES

According to early estimates of phytoremediation costs, the process is significantly less expensive than traditional technologies. Pilot-scale studies are showing phytoremediation to be a less expensive technology, although its true costs of the routine application are still unclear. Phytoremediation is considered a "green" and low-tech approach to remediation which is perceived to be more environmentally friendly than more active and intrusive methods. This may lead to greater acceptance by the public. Phytoremediation can be applied *in situ* for shallow soils and groundwaters, as well as for surface waters. Phytoremediation is not as destructive as some conventional technologies, which may remove soil nutrients and structure (Greger and Landberg, 1999). It is more likely that, regardless of how much contamination is reduced, plants will improve the soil's overall health. The benefits of vegetation include reducing or preventing erosion and fugitive dust emissions, as well.

14.6 DISADVANTAGES

Phytoremediation has the disadvantage of shallow distribution of roots. In general, phytoremediation should remove contaminants from the root zone of plants. Selecting deep-rooted plants and induced deep rooting can alleviate this disadvantage. Phytoremediation may take longer because it depends on plant growth rates to establish an extensive root system or substantial aboveground biomass. Heavy metal contamination may take up to 20 years to be remedied, for example, depending on the growth rate and biomass of hyperaccumulators. Phytoremediation (such as phytoextraction and rhizodegradation) may not be appropriate if remedial goals must be achieved rapidly or the land will be returned to use soon. Rhizofiltration, for instance, may speed up the process of phytoremediation.

There is a need to properly dispose of contaminated plant materials or analyze risk pathways. As part of phytoextraction and rhizofiltration, as well as other forms of phytoremediation, plant biomass that accumulates heavy metals or radionuclides must be harvested and properly disposed of. Some regulations govern the handling and disposal of biomass, and a suitable disposal facility must be identified.

Some remediation technologies may encounter some uncertainty in achieving remedial goals, such as meeting concentration goals in soil and groundwater and achieving hydraulic containment. Some forms of phytoremediation can take a long time to work, so a bench-scale or pilot-scale test to determine attainment might not be feasible if rapid remediation is desired. Phytotoxic concentrations in the initial contamination phase can prevent plants from growing. Preliminary phytotoxicity studies are likely to be required for screening candidate plants. Several adverse effects are possible. These potential problems should be minimized in part by selecting or managing species.

14.7 PHYTOREMEDIATION APPLICATIONS

There are several disposals or reuse options for treated wastewater, including gravity or pressure-injected irrigation, soil infiltration, graywater reuse, or surface water discharge (Reeed et al., 2006). Waste and wastewater can be treated naturally using macrophytes as part of phytoremediation.

The use of phytoremediation techniques is a low-cost alternative to nutrient recovery and biomass production for waste and wastewater treatment (Srivastava et al., 2008). Residential and municipal wastewater is a significant source of liquid wastes, and their safe, economic, and effective treatment is a challenging problem. Even sanitation services are inaccessible to billions of people worldwide (WHO, 2012). Domestic wastewater consists of a complex mixture of solids and solutes, which include wastewater from the kitchen, bathroom, laundry, and floor drains. The wastewaters also contain high levels of TSS, COD, BOD, nitrogen, phosphorus, total organic carbon, human pathogens, metals (Arroyo et al., 2010), and organic contaminants, which include pharmaceutical and personal care products (PPCP).

Organic matter produced by waste and wastewater is relatively easy to treat. Furthermore, the effluents represent a source of emerging contaminants and pathogens that are more difficult to treat.

14.8 SUMMARY

Utilizing plants as a remediation tool for soils and water can be beneficial. The choice of treatment is affected by climate, soils, effluents, geomorphology, inundations, and applied plants. Native trees and plants have several advantages including their resilience to disturbances, more diverse communities, and esthetic appeal. Phytoremediation of substrates with excess organic matter and nutrients leads to plant overgrowth as one of the most common complications. Thus, the management plan should take into account the planned harvest and the fate of harvested material. These are some of the potential uses, including energy generation, composting, and use as animal feed and fertilizer. A toxicology analysis must be performed before using harvested material, however, to ensure that no pathogens are present.

14.9 EXERCISES

Objective Response Questions

1. **This cleanup approach includes the removal of groundwater or soil from its natural setting to permit bioremediation**

   a) Bioaugmentation
   b) *in situ* bioremediation
   c) ex-situ bioremediation
   d) Phytoremediation

2. **Phytoremediation is the process of using _____ to clear contamination from water, soil, and sludge.**

   a) Green plants
   b) Filters
   c) Microorganisms
   d) Algae

3. **Which phytoremediation technique uses plant roots to take up contaminated groundwater or surface water?**

   a) rhizofiltration
   b) phytodegradation
   c) phytostabilization
   d) Phytovolatilization

4. **Rhizosphere biodegradation depends on relationships between green plants and _____.**

   a) microorganisms

b) algae
c) people
d) Animals

5. BOD stands for _____.

a) British Oxygen Depletion
b) British Oxygen Demand
c) Biological Oxygen Depletion
d) Biochemical Oxygen Demand

6. The purest form of naturally occurring water is _____.

a) river water
b) well water
c) pond water
d) rain water

7. The chemical oxygen demand (COD) measures the _____.

a) amount of oxygen required for the growth of microorganisms in water
b) amount of oxygen that would be removed from the water to oxidize pollution
c) amount of oxygen required to oxidize the calcium present in waste water
d) none of the above

8. The concentration of nitrate in domestic water is generally limited to _____.

a) 10 ppm
b) 30 ppm
c) 15 ppm
d) 45 ppm

9. Minamata is a disease caused by water pollution due to the presence of __________.

a) Lead
b) Mercury
c) Tin
d) Methylisocyanate

10. In this stage of wastewater treatment, methanogenic bacteria are the most prevalent _____.

a) Secondary treatment
b) Primary treatment
c) Sludge digestion
d) Biological oxidation

Practice Work

Qu 1. What are the main mechanisms used in the phytoremediation of water?

Qu 2. Can phytoremediation be considered under bioremediation?

Qu 3. Phytoremediation is possible on moderately polluted soil. However, what are you referring to as moderate pollution?

Qu 4. How will you select plants suitable for the phytoremediation of industrial wastewater?

Qu 5. Considering that aquatic plants have some contaminants inside, how can they be used after phytoremediation?

Qu 6. Phytoremediation is the process of using plants to get rid of contaminants. How do we dispose of the plants after treating or reclaiming them?

Qu 7. Describe a biotreatment technology for industrial wastewater and surface water

Qu 8. What is the concept of Vegetative Cap?

Qu 9. Explain the role of Buffer strips near contaminated water in detail.

Qu 10. Describe various physicochemical factors that affect the phytoremediation of water

REFERENCES

Adriano, D. C., Wenzel, W. W., Vangronsveld, J., & Bolan, N. S., (2004). Role of assisted natural remediation in environmental cleanup. *Geoderma, 122*, 121–142.

Arroyo, P., Ansola, G., & De Luis, E., (2010). Effectiveness of a full-scale constructed wetland for the removal of metals from domestic wastewater. *Water Air Soil Pollut., 210*(1–4), 473–481.

Bruce, E. P., (2001). *Phytoremediation of Contaminated Soil and Ground Water at Hazardous Waste Sites.* Ground Water Issue, EPA/540/S-01/500. 2001.

Greger, M., & Landberg, T., (1999). Use of willow in phytoremediation. *Int. J. Phytoremediation, 1*(2), 115–123.

Prasad, M. N. V., (2004). *Heavy Metal Stress in Plants: From Biomolecules to Ecosystems* (2nd edn., Vol. XIV, p. 462). Narosa Publishing House, New Delhi. https://link.springer.com/book/10.1007/978-3-662-07743-6. e-ISBN: 978-3-662-07743-6.

Rai, P. K., (2008). Heavy metal pollution in aquatic ecosystems and its phytoremediation using wetland plants: An ecosustainable approach. *International Journal of Phytoremediation, 10*(2), 133–160. http://dx.doi.org/10.1080/15226510801913918.

Reed, S. C., Crites, R. W., & Middlebrooks, E. J., (2006). *Natural Wastewater Treatment Systems.* CRC Press Taylor & Francis Group, Boca Raton, FL.

Srivastava, J., Gupta, A., & Chandra, H., (2008). Managing water quality with aquatic macrophytes. *Rev. Environ. Sci. Biotechnol., 7*(3), 255–266.

WHO, (2012). *Informe 2012 Acerca de Los Progresos Sobre el Agua Potable y Saneamiento.* World Health Organization. http://www.who.int (accessed on 01 April 2023).

CHAPTER 15

Plant Tissue Culture Studies for Enhancing Phytoremediation

15.1 PHYTOREMEDIATION USING PLANT TISSUE CULTURE

Plant tissue culture techniques such as micropropagation, callus culture, cell suspension culture, and hairy root culture can be used recently in phytoremediation research as study model plant systems. In vitro experiments offer an opportunity to study and examine the integral metabolic pathways of plant cells and their ability for toxicity tolerance and accumulation. Plant tissue culture involves a set of techniques. The growth of plant cells and tissues in vitro in a contamination-free environment was done in plant tissue culture. Here, dedifferentiated plant cells are cultured as callus or cell suspensions; organs such as roots and shoots can also be cultured in vitro.

Plant tissue culture is a suitable laboratory tool for studies on phytoremediation. It is possible to keep in vitro cultures for indefinite periods and make them available as needed. Compared with whole-plant studies in hydroponics or soil, whole-plant studies in the soil last only a few weeks. Thus, the time required for experimental investigations may be considerably reduced using plant tissue culture techniques than whole plants. There are various advantages of plant tissue cultures also as compared with whole plants. Growing and maintaining plant tissue cultures under controlled conditions allows researchers to study how plant cells differ from microorganisms found in the rhizosphere in the way they respond and function (Lebeau et al., 2008; Chaudhry et al., 2005).

Plant tissue culture is a key tool in phytoremediation research. Tissue cultures also propose experimental convenience, high speed as compared with studies on whole plant systems. Here the intrinsic capacity of the plant to detoxify the pollutants can be studied by eliminating the rhizosphere effect. But it cannot be the whole replacement for studies on soil-cultivated

Bioremediation and Phytoremediation: Technologies for Toxic Pollution
Ashwini A. Waoo, PhD (Author)

plants. A plant tissue culture technique can be used to study the metal tolerance of a plant by exposing it to culture media containing known quantities of a specific heavy metal. By using this method, plants can be identified and screened for phytoremediation on polluted land. The technique can also be used to study the effect of metal on whole plants.

Environmental pollution is an extremely important issue today, affecting all of us in one way or the other. Due to the rapid increase in human population and industrialization, the demand for natural raw materials and sources of energy is increasing day by day in developing as well as underdeveloped countries, the industrial effluents are released directly or indirectly into natural water resources, mostly without proper treatment, thus posing a serious threat to the environment. Many hazardous waste sites in the industrial area in Bhopal will require clean-up.

There is a mixture of contaminants in the groundwater, soil, and/or wastewater. Salts, organics, heavy metals, trace elements, and radioactive compounds may be among them. Researchers sought to identify some interesting accumulators that could link a significant biomass production with the effective absorption and translocation of heavy metals. The current study examines native plants' ability to accumulate and tolerate high concentrations of heavy metals, which are often associated with polluted environments.

15.2 PLAN OF RESEARCH

- To identify the native plant species best adapted for the uptake, sequestration, or degradation of specific contaminants from the related industries and evaluate these species in controlled tissue culture experiments.
- To enhance and optimize their growth and phytoremediation parameters by toxicity and tolerance studies.
- To identify the best suitable species for phytoremediation.
- To demonstrate plant-based clean-up for heavy metals are low-cost, environmentally friendly, and will operate economically at actual contaminated sites.

15.2.1 SELECTION AND IDENTIFICATION OF PLANT FOR PHYTOREMEDIATION

An ideal plant species used in phytoremediation should have the capacity to produce low biomass with high metal accumulation, or a high biomass plant with increased metal uptake capacity. They also have other characteristics such as flexibility of the plant species to tolerate and accumulate multiple heavy metal contaminants and mixtures.

This study aimed to identify some interesting accumulators that can be associated with important biomass production and an effective heavy metal accumulation, absorption, and translocation. Researchers are examining how some native plants can accumulate and tolerate heavy metal concentrations that are often found in polluted areas.

15.2.2 PLANT TISSUE CULTURE PROTOCOL

Plant tissue culture is an emerging tool for plant biotechnology and a potential tool for the propagation of some economically important crops of agriculture, horticulture, forestry rare, and threatened plants. Plant tissue culture is the artificial growth of plant tissues outside an intact plant. It is a technique essential in many areas of the plant sciences. *In vitro* cultures of individuals or groups of plant cells, tissue, or whole organs, help in understanding both fundamental and applied science.

Plant tissue culture is the technique of growing plant cells, tissues, and organs in an artificially, prepared optimum nutrient medium, solid or liquid form, under aseptic conditions. These explants divide and gradually develop into an unorganized mass of cells called "callus" and subsequently differentiate to form plants or directly give rise to shoots or embryos. This technique involves sterilization, inoculation, and regeneration of plant cells, tissues of the organ under aseptic conditions in culture vials containing artificial nutrient medium, the chemical composition of the medium, and the controlled culture conditions like; light, humidity, temperature, aeration, etc. effectively control the expression of any genotypic and phenotypic potential in explants.

Plant tissue culture technique involves steps such as -sterilization of room, washing & sterilization of glassware, preparation & sterilization of medium, selection and isolation of explants, sterilization & inoculation of explants, incubation, and observation. These steps help in the contamination-free culture of plant tissue organs or cells.

Nutrient media for plant tissue culture are designed to allow plant tissues to be maintained in an artificial environment with optimum nutrient concentration for their growth. Various types of plant tissue culture media have been developed, among them very few have found widespread use, e.g., M. S. (Murashige and Skoog, 1962), Gamborg's B5, and SH (Shenck and Hildebrandt). The most commonly used media, devised by Murashige and Skoog (Murashige and Skoog, 1962) by analyzing the inorganic components in tobacco plants and then adding them to medium in amounts similar to those found in the plants.

In-plant tissue culture, for maintenance of the sterilized environment and standard medium conditions various equipment was used such as pH meter, electronic balance, autoclave, and laminar airflow chamber, which are discussed below:

- ***pH Meter***

pH is the negative logarithm of hydrogen ion concentration. The measurement of pH in the pH meter is based on ion exchange in between hydrated layers formed on the glass surface. Change in ion concentration results in voltage difference causing current flow and the current intensity gives the value of pH.

- ***Electronic Balance***

An electronic balance was used for quick weighing and analytical purposes. This is a single pan balance of capacity 100–200 gm; sensitivity 0.1 mg. Digital readout was used for making stock solutions of growth regulators and for another fine weighing.

- ***Autoclave***

The autoclave was mainly used for sterilization of media, glassware, etc., and also for decontamination of contaminated cultures. It is based on the principle of steam under pressure. High pressure provides a high boiling point that contains more heat content and causes coagulation of proteins and cell death. All the culture media are autoclaved at 121°C and 15 lbs. /inch^2 pressure for 20 minutes. Other materials are autoclaved for 1h.

- ***Laminar Airflow Chamber***

It provides an aseptic chamber for culture manipulation. In laminar constant velocity air is passed through the HEPA filters. The pore size of the filter is 0.22 microns. The blower blows the air through the filter at high pressure from one side creating positive pressure inside the chamber. So the entry of any contaminant is prohibited from the open side of the chamber. UV light is also present in the chamber. Before using the instrument, UV is put on for 15 to 20 minutes and the bench is to be cleaned with a 96% alcohol swab to eliminate microorganisms.

- ***Sterilization of Room***

The maintenance of highly aseptic conditions is the most important factor for tissue cultures laboratory. Thus, the room of such laboratory should be first washed with disinfectant followed by wiping with 95% ethyl alcohol. Commercially available disinfectants like dextran, Lysol, zephiran, and roccal are effective disinfectants.

The final sterilization of the room should be done either by an organs mercury lamp of the low pressure or by UV radiations or more recently by ozone generating system, which gives 90–100% disinfection in the transfer area between 8 m3 to 280 m^3. The disinfected area can be used 8 minutes after sterilization is over. Before starting the experiment, it is necessary to ensure all the materials and tools used are sterile. The three most important aseptic techniques that need to be done were autoclaving filter sterilization and surface sterilization.

The importance of maintaining a sterile environment during the culture of plant tissues was necessary. A few simple precautions to avoid contamination will save valuable time in not repeating experiments. The maintenance of highly aseptic conditions is a vital factor for successful tissue culture. The laboratory should be first washed with disinfectant followed by wiping with 2% sodium hypo chloride solution or 95% ethyl alcohol. Commercially available disinfectants like Extrain, Lysol Zephiran, and Roccal are effective disinfectants. The final sterilization of the room should be done either by the Organo mercury lamp of the low pressure or by U.V. radiation or more recently by ozone generating system which gives 90–100% disinfectant in the transfer area between $8m^3$- $280m^3$. The disinfectant area can be used eight minutes after sterilization is over.

Most tissue culture procedures were conducted in sterile operations, such as a laminar flow cabinet. Besides the special design of the gentle flow of sterile air in the cabinet, the aseptic cabinet is also equipped with a germicidal lamp emitting ultraviolet (UV) light. This type of radiation is useful in eliminating airborne contaminants and for surface disinfection. The tissue culture storage room requires some equipment like a temperature controller, an artificial light source with a timer for day-length control, and shelving for culture rocks.

15.2.3 GLASSWARES REQUIRED FOR MEDIA PREPARATION

Various ranges of glassware are used for plant tissue culture according to the requirement. The following were the main glasswares used for the establishment of culture.

- Flasks (5 ml, 100 ml, 250 ml, 500 ml, 1000 ml).
- Volumetric Flasks (500 ml, 1 lit, 2 lit, 3 lit).
- Measuring Cylinders (25 ml, 50 ml, 100 ml, 1 lit).
- Graduated Pipettes.

- Cultures Bottles with Suitable Closure.

Glassware and all the tools used for the tissue culture processes can also cause contamination. It is extremely necessary to autoclave all the material before using it so that all the microbial contaminants are destroyed. Glassware used in our studies was cleaned by boiling in a saturated solution of sodium bicarbonate for 1 h followed by washing in tap water. These were then immersed in a 30% Nitric acid solution for 30 min and were washed thoroughly with tap water. Then also rinsed with double distilled water these were allowed to dry on a draining rack. Tubes and flasks were plugged with absorbent cotton.

All dissecting instruments were either wrapped singly or were put in closed aluminum cans for sterilization by autoclaving. Filter paper pieces of approximately 10x20 cm were kept in a stack alternatively with brown paper pieces of similar size and packed in autoclavable plastic bags with 20–25 pieces in each bag then autoclaved. Dissection and transfer of explants were carried out on these papers under aseptic conditions and disposed of after use. Pipet tips used for aseptic addition by micropipettes were arranged in cases meant for their size, wrapped with brown paper, and autoclaved. Glassware and instruments were sterilized by autoclaving at 121°C for 1 h in 15 lbs./(inch^2). All the glassware used in the tissue culture technique is first washed with hot or warm water containing soap detergent. The soap detergent is removed under running tap water followed by a final washing with double distilled water. After washing, the glassware should be sufficient to over-drying at 70°C and then autoclave.

- ***Selection of Media***

The selection of a medium for achieving a successful result is an essential criterion in the tissue culture technique. When starting with a new system, it is essential to work out a medium that would fulfill the specific requirements of that tissue. The composition of the tissue culture medium may range from the simple salt mixture to a highly complex mixture of organic & inorganic salt. For each plant, there is a need to formulate and standardize the medium and growth hormone concentration according to the requirement of the plant.

- ***Media Ingredients***

A plant tissue culture medium typically contains macronutrients, micronutrients, an iron supplement, vitamins, sugars, and plant growth regulators such as auxins, cytokinins, gibberellins, Abscisic acid, and ethylene.

The preferable energy source, sugar is essential for in vitro growth and development as most plant cultures are unable to photosynthesize effectively for a variety of reasons. The nutrient medium consists of 95% water. Plant cells use the nutrients in the media as building blocks in the synthesis of organic compounds, or as catalysts in enzymatic reactions. Explants grown on semisolid or liquid media contain plant growth regulators (PGRs) at very low concentrations (0.1 to 100 mM) that regulate the initiation and development of shoots and roots. Tissue culture uses auxins and cytokinins primarily as PGRs. The relative ratio among them determines the morphogenesis of cultured tissues. Only trace amounts of the vitamins were required for their catalytic actions in enzyme systems. A vitamin is an organic substance that performs a function as an enzyme or cofactor. Vitamin mixtures may contain amino acids and nitrogenous compounds. Micronutrients are required in tiny amounts (micromolar, M) while macronutrients are needed in millimolar amounts (mM).

Macronutrients were nitrogen, phosphorus, potassium, calcium, magnesium, and sulfur. These macronutrients are also called inorganic chemicals and were essential elements required in relatively large amounts. Micronutrients were traces of certain elements required by all plant cells. Micronutrient elements include iron, manganese, zinc, boron, copper, molybdenum, iodine, cobalt, nickel, aluminum, and chlorine. Meanwhile, vitamins have catalytic functions in enzyme systems and were required only in trace amounts. Water and solidifying material agar were also part of the nutrient medium. Plant tissues have a diverse range of needs, which has led to the creation of very complex culture media.

Plants require a variety of compounds to grow and develop, and also certain compounds can only be made by the entire plant, not by isolated plant tissues. The basal medium is formulated to provide all of these compounds. The most suitable and commonly used basic tissue culture medium for plant regeneration is Murashige and Skoog's (1962) medium (M. S.). Murashige and Skoog (1962) have developed an M. S. medium that consists of ammonium, nutrient, and other minerals include inorganic nutrients that are needed by the plant to establish growth as shown in Table 15.1.

The appropriate composition of the medium is necessary for the success of a tissue culture protocol. Stock solutions of the media ingredients were prepared by dissolving weighed amounts of these salts as given below in distilled water. These solutions were mixed to prepare the media after weighing accurately.

- ***Preparation of Stock Solution***

Stock solutions are the concentrated solution of media ingredients individually or may be prepared in groups well in advance and used subsequently

TABLE 15.1 Murashige and Skoog (M. S.) Media (1962) Used for *in vitro* Culture

Constituents	Weight(mg/L)
A. Inorganic	
Macronutrients	
$MgSO_4 \cdot 7H_2O$	370
KH_2PO_4	170
KNO_3	1900
NH_4NO_3	1650
$CaCl_2 \cdot 2H_2O$	440
Micronutrients	
H_3BO_3	6.2
$MnSO_4 \cdot 4H_2O$	22.3
$ZnSO_4 \cdot 7H_2O$	8.6
$Na_2MoO_4 \cdot 2H_2O$	0.25
$CuSO_4 \cdot 5H_2O$	0.025
$CoCl_2 \cdot 6H_2O$	0.025
KI	0.83
$FeSO_4 \cdot 7H_2O$	27.8
Na_2EDTA	37.3
B. Organic	
Thiamine HCI	0.1
Pyridoxine HCl	0.5
Nicotinic Acid	0.5
Myo-inositol	100
Glycine	2
Sucrose	30
An optimum pH of 5.8 should be maintained; (Agar 10 g/l)	

to prepare several batches of media. The stock solution should be prepared before experimentation because the dissolution of different components during media preparation is time-consuming. It is necessary to dissolve each constituent in the double-distilled water completely before adding another, otherwise, precipitations may occur. The stock solution should be prepared before experimentation because this solution of different components during media preparation is time-consuming. The life of different stock solutions varies. The procedure for the preparation of the stock solution of the most used medium, i.e., Murashige and Skoog's (M. S.) medium (1962) is given in Tables 15.2–15.7. Stock solution after preparation should be placed autoclave chamber colored bottle and store in the refrigerator. They should be examined for any Contamination Precipitation. Stock solutions of Macro-salts, Microsalts, Iron sources, Phytohormones such as Auxins, Cytokinins, Mesoinositol, and vitamins are prepared separately.

- ***Preparation of M. S. Media from Stock Solution***

The preparation of M. S. media from basic chemicals is time-consuming and convenient but it does have the advantage of providing the flexibility to vary individual components of the medium.

For the preparation of a ready use 1 l agar media, 125 ml of stock MA and 20 ml of stock MB were poured into a 1 L beaker, add some ddH2O, and mixed the solution well. Next, 30 g of sucrose (carbon source) was added to the solution. Then add 0.440 gm/l $CaCl_2$ (it added in the last because it produces precipitation in-stock solution). the pH of the solution was adjusted to pH 5.5. Then 4.5 g of agar powder was added. The solution was transferred to a 1 L volumetric flask, and double-distilled water was added to the final volume (1 L). The solution was autoclaved at 121°C for 20 minutes. There was a caution regarding agar media preparation that they might fluctuate in response to changes in autoclaving and pH. Finally, the hormones BAP, NAA, and kinetin were added according to concentrations.

Note that hormones were added after medium sterilization, as the high temperature will denature the hormones and makes them lose their function, as the hormone is not resistant to heat.

Then, the liquid solution was poured into a suitable container, such as Petri dishes, culture tube, culture bottle, or wide mouth conical flask. These tools were being sterilized before use. After pouring, the containers were left in a sterile laminar flow cabinet, to let the agar media solidify. Next, the container closed tightly and wrapped using Parafilm. The solidified agar was then placed in a suitable place for storage.

TABLE 15.2 MA Stock Solution of M. S. (1962) – Macro Salt (x10)

Constituents	Amount (mg/l) present in the original medium	Amount (g/l) to be taken for a stock solution (*100)	Final Vol. of stock solution (ml)	Amount to be used/ liter (ml)
NH_4NO_3	1650	16.5	1000	100
KNO_3	1900	19	1000	100
$CaCl_2.H_2O$	440	4.4	1000	100
$MgSO_4.7H_2O$	370	3.7	1000	100
KH_2PO_4	170	1.7	1000	100

Most tissue culture manipulations were conducted in sterile operations, such as a laminar flow cabinet. They have the facility of the constant flow of sterile air in the cabinet, the aseptic cabinet is also equipped with a germicidal lamp emitting ultraviolet (UV) light.

TABLE 15.3 MB Stock Solution of M. S. (1962) – Inositol (x50)

Constituents	Amount (mg/L) present in the original medium	Amount (g/L) to be taken for stock solution	Final vol. of stock solution (ml)	Amount to be used/liter (ml)
Myo-Inositol	100	5	250	5

TABLE 15.4 MA Stock Solution of M. S. (1962) – Iron Source (x 100)

Constituents	Amount (mg/L) present in the original medium	Amount (g/L) to be taken for a stock solution (*100)	Final vol. of stock solution (ml)	Amount to be used/ liter (ml)
$FeSo_4.7H_2O$	27.8	2.78		
$Na_2EDTA.2H_2O$	37.3	3.73	500	5

TABLE 15.5 MA Stock Solution of M. S. (1962) – Micro Salt

Constituents	Amount (mg/l) present in the original medium	Amount (g/l) to be taken for a stock solution (*100)	Final vol. of stock solution (ml)	Amount to be used liter (ml)
$MnSO_4.\ 4H_2O$	22.3	2230		
$ZnSO_4.4H_2O$	8.6	860		
H_3BO_3	6.2	620		
KI	0.83	83	500	5
$Na_2MoO_4.2H_2O$	0.25	25		
$CuSO_4.5H_2O$	0.025	2.5		
$CoCl_2.6H_2O$	0.025	2.5		

TABLE 15.6 MB Stock Solution of M. S. (1962) – Vitamins (x 100)

Constituents	Amount (mg/L) present in the original medium	Amount (g/L) to be taken for stock solution (*100)	Final vol. of stock solution (ml)	Amount to be used/liter (ml)
Glycine	2	200		
Nicotinic acid	0.5	50		
Pyridoxine HCL	0.5	50	500	5
Thiamine HCL	0.1	10		

TABLE 15.7 Stock Solution of Phytohormones

Phytohormones	Molecular weight	The required amount of stock solution	Amount of solvent required to be dissolved	Amount of water to be added (ml)	The final volume of stock Solution	Final concentration (mg/l)
			AUXIN			
2,4D	221.04	10	1 ml (0.1) NaOH	99	100	0.1
IAA	175.18	10	1 ml (0.1) NaOH	99	100	0.1
NAA	186.2	10	1 ml (0.1) NaOH	99	100	0.1
IBA	203.23	10	1 ml (0.1) NaOH	99	100	0.1
NDA	202.3	10	1 ml (0.1) NaOH	99	100	0.1
			CYTOKININ			
BA	225.2	10	HCI (0.1N) 1 ml	99	100	0.1
Kn	215.21	10	HCI (0.1N) 1 ml	99	100	0.1
Z	219.2	10	HCI (0.1N) 1 ml	99	100	0.1
21 IP	203.3	10	HCI (0.1N) 1 ml	99	100	0.1
			GIBERLIN			
GA3	346.36	10	NaOH (0.1 N) 1 ml	99	100	0.1

- ***Sterilization of Media***

The nutrient media are generally sterilized by autoclaving at 121°C for 15 to 25 min. and 1.06kgs/cm^2. The minimum time required for sterilization depends upon the volume of the medium in the vessel. Prolonged autoclaving may result in the breaking & denaturation of various media ingredients.

The outcome of plant tissue culture studies depends on culture conditions. In vitro cultures offer the prospect to study and identify the intrinsic enzymatic capacity of plant species for phytoremediation without the influence of microbial communities and free of any type of compartmentation of metabolites in organized structures. In vitro studies carried out under controlled conditions often display less variability than that observed using whole plants grown in less easily controlled situations. The results from in vitro experiments are strongly dependent on the culture conditions applied. Experimental parameters are more easily controlled in tissue culture however, it may be difficult to predict results of pollutant metabolism obtained under controlled conditions that may be similar or not to that of plants in the field.

Some components of plant tissue culture media affect the uptake of contaminants such as metals. The high phosphate content of media reduces the solubility of metal ions, which results in low metal uptake by cultured plant cells.

The resistance or susceptibility mechanism in plants towards organic and inorganic contaminants from the environment depends on uptake, localization, and metabolism of the contaminant is carried out and controlled.

15.3 DEVELOPMENT OF TRANSGENIC PLANTS AND GENETIC STUDIES

Tools and techniques of genetic engineering can be used to enhance plants' inherent degradative abilities in the face of xenobiotic compounds and heavy metals, but for that to be possible, the molecular and cellular mechanisms must be well understood. To improve the efficacy of plants for environmental remediation, heterologous genes have been used from various microorganisms, plants, and animals. Using stable genetic transformation to produce recombinant plants is preferable to the use of transient expression systems such as plant viral vectors. The presence of foreign genes in plant cells is usually achieved through the process of plant tissue culture. Upon transformation, plant cell cultures also play a key role in confirming better phenotypic traits before plant regeneration commences. They also determine if successful transformation occurred (Doran, 2009). By employing these techniques, information can be obtained about the interaction and regulation of pathways and processes involved in handling and detoxifying pollutants.

15.4 PLANT TISSUE CULTURE

Heavy metal extraction activities in mining areas generate a large amount of environmental damage with contamination of surrounding soil and water resources with heavy metals such as copper iron and Zinc. Closed or disposed mines remain as a contaminated area with huge amounts of heavy metal pollution in soil with harsh conditions like Extremely acidic pH.

Phytoremediation Technology uses plants for the removal and stabilization of pollutants in the polluted environment. There are various approaches to possible phytoremediation depending on the genetic and physiological characteristics of plants used in this technique plants can absorb metals, metalloids, and radionuclides from soil and can transfer them into different aerial parts. metal absorbing plants are called phytoextractors. These plants can be incinerated after phytoextraction and Kanchan metals recovered for use of industrial purposes.

Some plants are photo stabilizers that can immobilize heavy metals within their root zone by chelation, precipitation, or adsorption. All these activities prevent the migration of contaminants by leaching, erosion, and runoff. Phytoremediation can be employed to detoxify such polluted mining areas. *In vitro* plant tissue cultures can be interestingly applied for remediation research.

There are several advantages of plant tissue culture such as plant cells can detoxify the pollutants and in tissue culture, it is possible to avoid interference from the contaminating microorganisms.

Still very little is known about the metabolic pathways involved in biodegradation or transformation of xenobiotic compounds inside the plant cell and how the plant cell tolerates or detoxifies or accumulates a higher concentration of heavy metals is also a thirst area of research. Thus, plant tissue culture is considered as a basic tool to provide information about plant-pollutant relationships that help to predict plant response against the contamination and one can improve or enhance the plant characteristics for phytoremediation. Plants having all these properties along with the highest biomass production can be selected as suitable plants for phytoremediation.

Glutathione plays a central role in plant defense mechanisms against the heavy metals for xenobiotic compounds. Various chelating compounds derived from glutathione such as phytochelatins and metallothioneins can detoxify the toxic pollutants. Synthesis of these compounds can be seen in the plant subjected to pollutant stress (Clemens et al., 2002).

Some enzymes, such as lactases, nitroreductases, dehalogenases, and peroxidases, are generally involved in the oxidation of xenobiotics, their transportation, and localization or compartmentation processes in plant cells (Sing and Jain, 2003). Understanding the mode of action of these enzymes and pathways of stress compounds can help in applications of plants for phytoremediation purposes.

Recently in vitro cultures are used to reveal the plant-pollutant relationship and response mechanisms. Callus cultures, hairy roots, cell suspensions and shoot tip cultures are mostly used as model systems for the study of the uptake, compartmentalization, translocation, their metabolism, toxicity, and tolerance of contaminants under artificial aseptic conditions. Due to micropropagation and higher multiplication rate, In vitro cultures can be maintained indefinitely, without the use of whole plant material for research. Established culture media, standardized protocols, the addition of growth factors, and specific additives, like promoters and/or inhibitors, can generate a rapid response in culture.

Even though there are lots of advantages discussed above, there are still some problems associated with in vitro studies. It has been found that the response of in vitro plantlets varied from those of naturally growing plants. The environment containing soil and microorganisms makes a different level of response than plants in artificial aseptic conditions. Thus in vitro tissue culture can mostly help in the production and selection of whole plants with enhanced phytoremediation potential.

Plant tissue culture employs the opportunity to study the intrinsic enzymatic capacity of individual plants without the help of soil microflora. With the help of tissue culture, one can study the individual potential of plants for phytoremediation, independent of microbial influences. Various techniques are applied in artificial tissue culture environments to stimulate the native habitat of the specific plant for the assessment of stress responses. Thus, plant tissue culture results can level the least variability but can introduce new experimental parameters related to toxicity and tolerance mechanisms in plant cells.

In vitro Studies of pollutant Uptake

While studying xenobiotic metabolism, the rate of uptake of contaminants can be measured by using tissue culture studies. This type of uptake assay helps to understand the potential of a particular plant to remediate in a specific period. As heavy metals cannot be degraded, only translocated so it's important to study their rate and extent of uptake during phytoremediation research. Appropriate control of all possible parameters is required during experimentation.

The contaminant uptake mechanism is significantly different in the case of tissue cultured plantlets and the whole plant in nature. Does the result obtain from in vitro culture are not the exact representation of intact plants? There are various mechanisms by which chemicals are transported into the plant parts such as simultaneous uptake of chemicals with water through the transpiration stream, transport facility by microorganisms, or transport by diffusion (Arthur et al., 2005). Oman all these mechanisms only diffusion is available in vitro culture because there is an absence of bulk flow of water through transpiration process in plant tissue culture system.

In the case of cultured cells as well as in whole plants, the transport of substances across the cell membrane of individual cells is controlled by a variety of processes such as diffusion, proton pump, and transport by specific membrane proteins. Uptake of heavy metal ions can be done through the lipid bilayer with the help of Transporter protein. Organic Xenobiotics can be radially uptake by direct diffusion into the cell membrane, this process also depends on the physicochemical properties of the contaminant.

It has been generally expected that uptake of contaminants will be greater in plant tissue culture than the whole plants because contaminants can uniformly expose to the plant cell surface. In tissue culture, translocation is not required but still, considerable evidence was not available for this expectation. There must be proper comparison and analysis of contaminant uptake between naturally growing plants and tissue culture plantlets, along with the study of duration of exposure, the concentration of contaminant, and amount of biomass present. All these parameters should be examined simultaneously in phytoremediation studies.

Some contaminants can affect tissue culture experiments by interacting with the components of plant tissue culture media and indirectly affect the uptake of contaminants in tissue culture. For example, in heavy metals, the high phosphate level of tissue culture media reduces the solubility of heavy metal ions and it results in a low level of metal uptake by cultured cells for this phosphate-free medium is used to study heavy metal uptake in phytoremediation research.

Pollutant uptake is drastically influenced by the bioavailability of Ions and organic molecules. Contaminants in soil components are generally much less available to the whole plant but contaminants in the tissue culture environment can readily be available to the plant cell.

Naturally in soil, the presence of heavy metals may be in the form of there inert ferric or magnesium oxide complexes, sulfides, surface mineral precipitates, mineral formulations, or carbonates These are all formulation that makes the uptake of heavy metals a difficult and slow process. Such conditions can be very well controlled in tissue culture experimentation.

15.5 DISTRIBUTION AND LOCALIZATION STUDIES

Differences in the mode of uptake of heavy metal can also affect the localization of organic as well as inorganic pollutants. These differences are not directly reflected in the whole plant. plants have a greater ability to localize the toxic product of metabolism or localize them by storing them into the internal structures or between the organs. These mechanisms can be possible in vitro as well as in vivo conditions.

In the case of the application of xenobiotic chemicals to the plant cell, the primary product of plant metabolism and product conjugates can distribute themselves as extractable and non-extractable factions of biomass. The non-extractable parts bind with the plant cell wall.

It has been reported that these metabolites are incorporated with proteins polysaccharide cellulose hemicellulose pectin, starch, and cutin components of the cell wall with covalent or non-covalent linkage (Sapp et al., 2003). Plants mostly store the conjugated products of the xenobiotic metabolic pathway rather than eliminating them. Xenobiotic conjugates may be transported to vacuum for storage or plants are considered to store the conjugated products of xenobiotic metabolism rather than eliminate them. Bound residues are one mechanism for this; conjugates may also be transported to the vacuole for storage to reduce any residual toxic effects on the cytoplasm. Nevertheless, plant cell cultures contain substantial proportions of specific metabolites.

15.6 PHYTOREMEDIATION RESEARCH STATUS IN DEVELOPING COUNTRIES

In developing countries like India industrial pollution is a result of urbanization and excessive use of natural resources. There is a big problem of population explosion. Land degradation is the major problem of environmental concern. There are various causes of land degradation such as overgrazing over-cultivation, wrong irrigation, and deforestation. Various phytoremediation strategies can be applied for the restoration of polluted soil in India.

1. Tree crop,; wild hyperaccumulators.
2. Metallocrops.
3. Hairy root cultures and transgenic approach.
4. Constructed wetlands.
5. Biosorption with phytomass.
6. Microbial consortia for industrial pollution.

Phytoremediation is the thrust area of environmental science that has great potential to remediate the contaminated environment. There are various research reviews regarding the phytoremediation research on native flora and microorganisms algae cyanobacteria aquatic macrophytes and higher vascular plants.

Due to low cost and lack of technical equipment, phytoremediation has now become very popular in developing countries like India. In India, lots of research focused on identifying indigenous plant species for remediation purposes for the targeted pollutants such as arsenic, pharmaceutical wastes, fly ash, and heavy metals (Prasad, 2007).

In India there are very few reports available regarding phytoremediation research at the field level, most of the studies are only confined to contaminated water bodies, some phytoremediation work is carried out on mines spoil. There is a scarcity of scientific studies on heavy metal contaminated soils in India. Here we highlight some significant research work carried out in India in the field of phytoremediation.

Increasing urbanization poses a serious threat of water contamination which Industrial effluents and heavy metals. Over growing population creates an intensive practice of agriculture and faulty irrigation which ultimately affects the soil quality. The exploitation of water bodies for food production and urban requirements makes this resource. Thus, most of the research work is directed towards the remediation of heavy metal contaminated water bodies in India.

S. acmophylla has been suggested as a bio-monitoring tool for metal contamination (Cu, Ni, Pb) in soil, water bodies, and for phytoremediation (Ali et al., 1999). Kumar et al., (2008) have studied the accumulation of heavy metals in aquatic macrophytes that can be used as biomonitoring tools and for phytoremediation. Researchers found a greater accumulation of heavy metals in *Nelumbo nucifera* and the poor in the *Echinochloa colonum*.

A study by Ghosh (2020) investigated the hyperaccumulation capacity of *Hydrilla verticillata*, a macrophyte that accumulates As and Cd from contaminated water. The *Ipomoea aquatica* is another species that can accumulate Cd and has less capacity to accumulate Pb.

Rai (2008) investigated the phytoremediation capacity of *Azolla pinnata*, a water fern, in the presence of heavy metals. metal contents. Adhikari et al., (2010) revealed the phytoextraction potential of Pb by two wetland plant species, i.e., *Typha angustifolia* L. of *Typhaceae* and *Behaya* plant (*Ipomoea carnea* L.) of the *Convolvulaceae* family and reported that both the plants show promise for the removal of Pb from contaminated wastewater.

According to Chandra and Yadav (2010), Typha angustifolia could be a potential phytoremediator of various heavy metals (Cu, Pb, Ni, Fe, Mn, and Zn).

According to Neetu Malik and Biswas (2012), who discussed various remediation techniques with a focus on phytoremediation and its benefits, phytoremediation is an aesthetically pleasing method of cleaning up contaminated sites. It is also energy efficient and can be used in conjunction with other more conventional remedial techniques at low-contaminated areas.

Many conventional methods of soil remediation offer limited advantages, such as high costs, intensive labor, and changes to soil properties, as well as disturbance of soil microflora. On the other hand, phytoremediation is a better solution (Ali H, Khan E, Sajad MA, 2013). The use of molecular tools for understanding the mechanisms of translocation, uptake, tolerance, and sequestration of metal in plants was suggested.

15.7 SUMMARY

Phytoremediation uses plants' ability to remove pollutants from the environment as a simple and low-cost cleaning technology. There is a lack of understanding about the metabolic pathways involved in the conversion of xenobiotic compounds and plants' ability to tolerate, detoxify, and store high levels of heavy metals. In fundamental studies, plant cell and tissue culture provide information about the relationship between plants and contaminants, enable the prediction of plant responses to environmental contaminants, and improve the design of plants with enhanced characteristics for phytoremediation. To study the interactions between plants and pollutants in an aseptic environment, callus, hairy roots, cell suspensions, and shoot multiplication cultures are used. Most plant species are capable of accumulating and metabolizing a variety of pollutants, but they normally produce little biomass. The rapid growth, extensive root system, and high-water uptake of fast-growing trees make them excellent candidates for phytoremediation.

The understanding of plant metabolism has been significantly enhanced by plant tissue cultures as a component of phytoremediation research. Compared to whole plants, tissue cultures are easier and faster to use for experiments. To obtain a more accurate estimation of the intrinsic capacity of plants to detoxify pollutants, microflora and translocation barriers are eliminated. Chemicals metabolized by plant cells in vitro indicate that the parent plant has the genetic capability to biotransform those substances. The use of plant tissue cultures has been proven effective in developing a better understanding of pollutant metabolism, toxicity, and tolerance, however, they lack the full set of properties that live plants possess that make them appropriate for phytoremediation. In contrast to soil-cultivated plants, they are powerful auxiliary model systems that can be used for getting additional information that can be used to guide subsequent whole-plant experiments.

15.8 EXERCISES

Objective Response Questions

1. Micropropagation most commonly uses solidifying agent such as _______________.

a) agar
b) dextran
c) Mannan
d) all of these

2. How will you define plant tissue culture?

a) The technique of in vivo growing cells
b) The technique of cutting plants
c) The technique of growing plants in gardens
d) The technique of in vitro maintaining and growing cells

3. The tissue culture technique was developed by ______________.

a) White
b) Haberlandt
c) Halperin
d) Skoog

4. What is an explant in plant tissue culture?

a) A part of plant grown under soil
b) Any part of a plant grown in a test tube
c) A specific part of a plant grown in a test tube
d) Leaves grew under test tube

5. The main effect of cytokines in tissue culture systems is this ______________.

a) adventitious root formation
b) shoot elongation
c) induction of somatic embryos
d) adventitious shoot formation

6. Plant culture media do not include this as a basic component ______________.

a) sucrose/sugar
b) amino acids
c) a complex mixture of salts
d) serum albumin

7. Growth of plant tissues in artificial media is ______________.

a) cell hybridization
b) plant tissue culture
c) transgenesis
d) gene expression

8. The modification process of exogenous compounds by plant cells is called ______________.

a) bioconversion
b) Biotransformation
c) both a and b
d) biophytomodification

9. A plant cell without cell wall is called ______________.

a) tropoplast.
b) protoplast.
c) chloroplast.
d) chromoplast.

10. 2, 4-D alone is sufficient for __________________ culture.

a) callus.
b) organ.
c) anther.
d) pollen.

Practice Work

Qu 1. Explain the applications in plant tissue culture in phytoremediation.

Qu 2. What are the advantages of plant tissue culture?

Qu 3. Explain the selection and identification of plants for phytoremediation and tissue culture studies.

Qu 4. Describe the plant tissue culture protocol in detail.

Qu 5. Explain the preparation of M S Media.

Qu 6. Explain the significance of micro salts and micro salt solutions.

Qu 7. What is callus culture?

Qu 8. Explain the composition of MS Media.

Qu 9. Describe in vitro studies of pollutant uptake.

Qu 10. Explain the plan of research for phytoremediation studies through tissue culture.

REFERENCES

Arthur, E. L., Rice, P. J., Rice, P. J., Anderson, T. A., Baladi, S. M., Henderson, K. L. D., & Coats, J. R., (2005). Phytoremediation—An overview. *Crit. Rev. Plant Sci., 24*,109–122.

Doran, P. M., (2009). Application of plant tissue cultures in phytoremediation research: incentives and limitations. *Biotechnol Bioeng., 103*(1), 60–76. doi: 10.1002/bit.22280. PMID: 19309741.

Murashige, T. and Skoog, F. (1962), A Revised Medium for Rapid Growth and Bio Assays with Tobacco Tissue Cultures. Physiologia Plantarum, 15: 473–497. https://doi.org/10.1111/j.1399-3054.1962.tb08052.x

Sapp, M., Ertunc, T., Bringmann, I., Schaffer, A., & Schmidt, B., (2003). Characterization of the bound residues of the fungicide cyprodinil formed in plant cell suspension cultures of wheat. *Pest Manag. Sci., 60*, 65–74.

Sing, O. V., & Jain, R. K., (2003). Phytoremediation of toxic aromatic pollutants from soil. *Appl. Microbiol. Biotechnol., 63*, 128–135.

CHAPTER 16

Novel Insights of Phytoremediation

16.1 HISTORICAL ORIGIN OF THE PHYTOREMEDIATION CONCEPT

The concept of phytoremediation dates back thousands of years. The first plant species to accumulate high levels of metals in leaves were *Thlaspi caerulescens* and *Viola calaminaria* (Baumann, 1885). In 1935, Byers reported that Astragalus plants accumulated up to 0.6% selenium in their dry shoot biomass. Minguzzi and Vergnano (1948) identified plants capable of accumulating up to 1% Ni in their shoots. Plants were used to treat waste-water about 300 years ago (Hartman, 1975). Rascio (1977) reported high Zn accumulation and tolerance in *Thlaspi caerulescens* shoots.

Plant-based remediation ("phytoremediation") (Cunningham et al., 1997; Flathman and Lanza, 1998) describes the use of plants to ameliorate degraded or polluted substrates. Controlling soil and water degradation with plants has a long history. Industrial activities contaminate soil and degrade vegetation. As a result of its ecological and sanitary significance, phytoremediation plays an important role in such polluted lands.

16.2 DEVELOPMENTS IN THE PHYTOREMEDIATION FIELD

In 1990, Baker and Walker classified plants into three groups based on their strategies for growing on soils contaminated with metals; metal excluders, indicators, and accumulators or hyperaccumulators. The term metal accumulator (hyperaccumulator) refers to plant species that accumulate metals in their above-ground tissues at levels higher than those in the soil or of the species nearby that do not accumulate metals. Phytoremediation involves using plants that can extract heavy metals from soils and concentrate them in their shoots.

Bioremediation and Phytoremediation: Technologies for Toxic Pollution
Ashwini A. Waoo, PhD (Author)

There are thousands of hazardous waste sites in the United States, and the EPA (Environmental Protection Agency) is dedicated to finding the most effective and efficient ways to address them. This guide provides technical information and specialized resources related to phytoremediation.

The roots and stems of plants are where metals accumulate preferentially, according to Abdul R. Memon et al., (2001). Roots and leaves bind potential toxic metals to their cell walls and store them in vacuolar compartments, a mechanism of tolerance or accumulation. Many plants absorb and translocate metals, which accumulate in their parts, such as leaves, without causing any toxicity symptoms.

The success of phytoremediation or phytomining generally depends on the availability of plants native to the region of interest that can tolerate and accumulate high concentrations of heavy metals. Baker and Whiting (2002) have found this to be the case. It has been shown that plants can accumulate up to 1000-fold more heavy metals than normal. Metal hyperaccumulation, or the ability to accumulate at least 0.1% of dry leaf weight in heavy metals, is only observed in less than 0.2% of angiosperms.

From the point of view of McIntire Terry (2003) the ability of a plant to remediate a metal-contaminated site depends upon the factors such as the growth rate of the plant and the planting density number of metals accumulated by the candidate plant. The Ideal plant for phytoremediation is that having sufficient tolerance to the contaminated site conditions to and it should be able to accumulate multiple metal contaminants. It must be fast-growing and easy to harvest.

According to Cobbett (2003), some plants can accumulate metals at levels that are 100 to 1000 times higher than levels normally accumulated by most species. More than 45 plant families and species have been identified that accumulate metals such as Zn, Cd, Cu, Co, Ni, Se, and As or specific combinations of these elements. Research on phytoremediation of contaminated soils has only increased significantly in the last 25 years, according to Robinson et al., (2003), even though plants have been used for hundreds of years to reduce soil erosion, human waste, and protect water quality.

Phytoremediation companies have been established around the world to service a global market that is estimated to be around US$ 100 million/year currently.

Several environmental factors, such as redox, pH, cation exchange capacity, dissolved oxygen, and temperature affect the distribution and accumulation of heavy metals in plants, according to Cheng (2003).

According to Weis & Weis (2004), heavy metals accumulated in plants and can cause hazardous effects on plants when they are translocated to above-ground tissues. Heavy metals could not be transported to aboveground tissues through the epidermis of the roots. The endodermis casparian strip blocks the movement of metals into the vascular bundles.

The highest metal concentrations were found in the xylem, followed by the mesophyll and hypodermis. In contrast to intracellular locations, heavy metal concentrations were found to be higher in cell walls.

According to Suresh and Ravishankar (2004), heavy metal tolerance and accumulation in plants are determined by various physiological factors such as phytochelatin production, uptake, and leaching of metal ions by root systems, antioxidative stress, and root cation exchange capacity (CEC).

According to Kramer (2005), the US Government had launched a massive program to develop environmental clean-up technologies, known as Superfund, which has contributed to the development of a new research area known as phytoremediation worldwide.

In 2005, an overview of phytoremediation of heavy metals and usage of their byproducts is given by Ghosh and Singh. In his opinion, phytoremediation is an esthetic, cost-effective, eco-friendly approach most suitable for developing countries like India, even though it is still not a commercially available technology in India. Their studies focused on the mobility, bioavailability, and plant response to soil heavy metals. The chapter discusses the phytoextraction mechanism in plants and the pathway in which metals are absorbed by plants. Numerous techniques have been devised to increase phytoextraction and utilize byproducts since a lot of biomass is generated during this process.

Joseph Devinny et al., (2005) compared exotic and diverse native plant species to control contaminant degradation. The native species perform well; they were the best choice for rhizosphere degradation mechanisms due to their low cost of maintenance and compatibility with local ecosystems.

Wilfried H.O. Ernst (2006) suggested that the evolution of metal tolerances in plants depends on the exposure of a plant species to the metal concentration of the soils. There was a great similarity in the involvement of one or two genes per metal controlling the specific metal tolerance and in its physiological mechanisms. The tolerance mechanism depends upon modifying performance under the higher level of metals and their compartmentalization of the metals in physiologically fewer active cells and tissues.

According to Erakhrumen et al., (2007), phytoremediation is eco-green, sound technology for pollution prevention, control, manage and remediation in developing countries. Some studies utilize the inherent abilities to live plants to clean the environment, yet many parts of the world have yet to adopt this technology for commercial use. As the costs associated with environmental remediation are staggering, the extensive popularity of phytoremediation can be attributed in part to the relatively low cost of the process.

In 2009, Sandra King et al., uses tissue culture technology to develop plants with acid soil, heavy metal tolerance, potentially for hard-rock mine land reclamation. The local native plants were tissue cultured which grow acid in heavy metal polluted soil.

According to Chhotu D. Jadia and M. H. Fulekar (2009), phytoremediation does not have a destructive impact on soil fertility and structure like some more extensive traditional technologies have such as acid extraction and soil washing. This technique can be applied "*in situ*" to remediate soil, groundwater, and surface water. Also, phytoremediation has been proved to be an eco-friendlier and more low-tech as well as active and disturbing remedial method.

Hussain et al., (2009) had reported that plants can bioaccumulate xenobiotics in their above-ground parts, after that they are harvested for removal. Remediation with the help of plants can be done in several ways, by phytodegradation, phytovolatilization, evapotranspiration, rhizoremediation, reducing the leaching of contaminants, and aerating soil.

According to Sarma (2011), the different strategies of phytoremediation with various mechanisms for remediating metal-contaminated soil or water are Phytoextraction, Phytoaccumulation, Phytovolatilization, Phytostabilization, and Phytofiltration.

It appears phytoremediation can be used on a wide range of contaminants, including various organic compounds such as polycyclic aromatic hydrocarbons, chlorinated solvents, nutrients, BTEX (Benzene, toluene, ethylene, and xylene), and surfactants offering a low-cost method of soil remediation technology (Srujana Kathi and Anisa B. Khan, 2011).

Neetu Malik and A. K. Biswas (2012) discussed different techniques of remediation with a special focus on phytoremediation, its advantages and defined that phytoremediation is an esthetically pleasing method of remediating contaminated sites, it is also energy efficient and it can be used in combination of other more traditional remedial methods at the low contaminated area.

Chemical and physical methods are limited by high costs, intensive labor, alteration of soil properties, and disruption of soil native flora. In contrast, phytoremediation is more effective. (Ali, Khan, Sajad, 2013). The use of molecular tools for understanding the mechanisms of translocation, uptake, tolerance, and sequestration of metal in plants was suggested.

Zhaoqing Chuin (2013) focused on searching for new plant species suitable for phytoremediation and the molecular mechanisms of their responses to pollutant stress.

Kaitar Ben Chekroun and Mourad Baghour (2013) suggested different techniques for bioremediation and the ability of some micro and macroalgal species to absorb organic and inorganic pollutants. They found that some algae are useful for phytoremediation because of their rapid growth rate.

Using rhizobacteria to remediate heavy metal toxicity in contaminated soils was reviewed by Rajendra Prasad Bharti et al., (2014). The study examined the mechanisms involved in the transfer and mobilization of heavy metals by rhizobacteria from the rhizosphere of plants growing on metal-contaminated soil.

Biochar and phytoremediation are two environmental technologies that could be used to remediate heavy-metal contaminated soils, according to Paz-Ferreiro et al., (2014). Biochar and phytoremediation are potentially advantageous when combined, as well as the mechanism by which they interact. Phytoremediation and biochar as tools for remediation need to be subjected to specific research to ensure their sustainability.

16.3 KEY RESEARCH FINDINGS ON PHYTOREMEDIATION

The study by Ebong et al., (2007) examined high metal concentrations in municipal and rural dumpsite soils and the rate of mineral accumulation by *Carica papaya* and *Talinum triangulare* in 2007. According to their findings, the Cd levels in dumpsite soil were higher than the standard, while the levels of Cd and Pb in plants were also higher than the standard.

A field experiment was carried out by Zhuang et al., (2007) to evaluate the phytoremediation potential of six high plants such as *Dianthus chinensis, Vertiveria zizanioides, Rumex crispus, Rumex K-1* (*Rumex upatientia x R. timschmicus*), and two populations of *Rumex acetosa* in comparison to metal hyperaccumulators (*Viola baoshanensis, Sedum alfredii*). The experimental paddy fields were contaminated with Zn, Pb, and Cd. Among the high biomass plants, *R. crispus* extracted Zn and Cd of 26.8 and 0.16 kg ha^{-1}, so it could be a candidate species for phytoremediation of Zn, Cd, and Zn from the soil.

According to Schroder et al., (2007), Phytoremediation helps in controlling soil erosion, runoff, water emissions, lower air, and secondary waste production thus making it an attractive technology. It can thus offer a modest way to decrease the effects of global warming by lowering energy usage, CO_2 emission, and waste emission during wastewater treatment.

U. S. Dissanayake et al., (2002) examined the potential of the alien invasive species *Lantana Camara* L. and *Wedelia trilobata* L. for selective removal of ions of heavy metals, Cd, Cr, Cu, Co, Ni, Pb, Fe, Mn, and Zn. Under static conditions, *L. camara* and *A. odoratissima* removed nearly 50% of heavy metal ions. The high level of ion uptake ability by *L. camara* and *W. trilobata, A. odoratissima* can be attributed to the high amount of parenchyma in their tissues.

Pulford and Watson, (2003); Pulford and Dickinson, (2006), reported that different tree species, such as *Salix*, *Poplus*, *Betula*, *Acer*, *Alnus* have the potential of phytoremediation of various heavy metals.

The research of Erika Nehnevajova (2005) involved improving metal extraction using improved phytoextraction from high-yielding oil crops, namely Brassica juncea L. and Helianthus annuus L. having high biomass. Their oil and biomass can be utilized for technical purposes to enhance in situ decontamination techniques. Phytoextraction is enhanced in this way, creating a lasting value.

At the metal-contaminated site in Rafz (Switzerland), a sunflower mutant has been calculated to produce 26 t dry matter per ha and remove 13.3 kg zinc per ha per year through phytoextraction.

Arun K. Shanker et al., (2005) were studied the potential of chromium (Cr) Phyto-accumulating ability of four agroforestry tree species viz., *Albizia amara, Casuarina equisetifolia, Tectona grandis,* and *Leucaena luecocephala* effect of chemical (citric acid) and vesicular-arbuscular mycorrhizal fungi (VAM) was also tried and suggested a view that *Albizia amara* is a potential Chromium accumulator along with citric acid.

Abdelmalek Hakmaoui et al., (2006) had shown the usefulness of the calculation of TI based on aerial growth for metal tolerance test of woody plants and also proved that Cu-tolerant (*S. pedicelata*) and Cd-tolerant (*S. purpurea*) species may be effective in phytoremediation. They measured the photosynthetic parameters and indicated that metal-induced disturbances on growth could not be partially attributed to changes in Chlorophyll content.

Two plant species, i.e., Trifolium repens and Lolium perenne were studied by grown in metal-polluted (Zn/Cd) soil near a smelter by Bidar et al., (2007) for two years. According to bioaccumulation and transfer factors, metals were primarily accumulated in roots, indicating their availability for metal phytostabilization. According to these results, *T. repens* and *L. perennial* are suitable for the phytoremediation of metal-polluted soil.

Luc Van Ginneken et al., (2007) had given the concept of using oil crops such as *Brassica* species, which were used for phytoremediation of soil contaminated with heavy metals. These plants gave a viable option to generate bioenergy and biodiesel production. By adding biodegradable chemicals to rapeseed, maize, and wheat to enhance their bioavailability, and by stimulating microbial communities in the rhizosphere to take up these heavy metals, the uptake ranges of heavy metals can be extended.

J. I. Nirmal Kumar et al., (2008) were carried out a phytoremediation study at Pariyej reservoir, of Gujarat State, an internationally important wetland in the Asian Directory of Wetlands. A study was conducted to assess the accumulation of heavy metals in aquatic macrophytes.

Water samples, surface sediments, and native aquatic plants were analyzed for Cu, Cd, Co, Pb, Ni, and Zn contamination. Nelumbo nucifera had an accumulation of six heavy metals, in the following order: Zn > Cu > Pb > Ni > Co > Cd.

Gonzalez et al., (2008) carried out a second survey to study the effects of Cu metal on plant diversity in Chile. About 90–900 mg kg^{-1} of total Cu were found in soils and a nearby area close to a slag pile of the Ventanas smelter. At least 22 pseudo metallophyte species and common species were found to be able to tolerate concentrations of Cu in the soil that would otherwise be toxic. As the species with the greatest copper accumulation (614 mg kg^{-1}), *Oenothera affinis* is an attractive candidate for phytoremediation initiatives in Chile, and it is a native herb that is easy to propagate, drought-resistant, and produces large biomass.

In 2008, Ian Leslie examined soil samples and plant tissues from a California vineyard to analyze chromium and nickel levels.

Odoemelam and Ukpe (2008) conducted phytoremediation with *Bryophyllum pinnatum* to remove heavy metals from soils. According to the study, maximum levels of metals were accumulated in the 4th month, with the highest levels of Pb (399.90 mg/kg), Ni (11.91 g/kg), Cr (32.48 mg/kg), and V (3.12 mg/kg). Plants such as *pinnatum* are effective metal hyperaccumulators used for the phytoremediation of soil contaminated by heavy metals.

Phyllis Njeri Kimenyu (2008) examined the uptake of heavy metals by plants grown on contaminated soil. In *Helianthus annuus, Amaranthus hybridus, Commelina bengalensis,* and *Zea mays*, the concentrations of

Cu, Zn, Cd, and Pb were determined. According to the results of the study, *Commelina bengalensis* accumulated more Cd, Pb, and Cu in contaminated soil than *A. hybridus* and *Z. mays* and *A. hybridus* and *Z. mays* showed more accumulation capacity for Zn. *H. annuus* was found to remove lead, zinc, and copper the most efficiently. It was shown that the uptake of heavy metals by *Z. mays* and *A. hybridus* from solutions contaminated with single metals was not different from solutions contaminated with a mixture of the metals.

Oyelola et al., (2009) used *Lycopersicum esculentum* (Tomato) as a phytoremediator for soil contaminated with metals. The total metal content of the soils at dump sites before and after harvesting tomatoes differed significantly ($P>0.05$).

After tomato plants were harvested, Zn, Pb, and Cu levels decreased significantly. These findings indicate that tomato plants are remarkably capable of absorbing and accumulating high levels of metals. It showed remarkable tolerance to heavy metals.

Katharina Keilig and Jutta Ludwig-Müller (2009) used *Arabidopsis thaliana* to investigate the effect of flavonoids on heavy metal tolerance. Their study had proven that flavonoids enhanced the heavy metal tolerance in *Arabidopsis thaliana.*

In a study by Migeon et al., (2009), high accumulation properties were reported for poplar and willow species grown on soils with different metal bioavailabilities and chemical characteristics.

A seasonal variation in metal accumulation was also analyzed. Poplar and willow have high bioconcentration factors suggested that the high potential of these species over other woody plants for heavy metal accumulation.

Garg and Kataria (2009) was studied the effect of different concentrations of copper on two varieties of wheat species such as *T. aestivum* L., var. UP 2338 and var. PBW 373 of North India, and also on Mustard (*Brassica juncea* L.) and Radish (*Raphanus sativus* L.) plants. The study aimed to find suitable plants species for use in the remediation of industrially contaminated soil. Results demonstrated that plant species were having a different potential of Cu uptake and translocation. The maximum Cu^{++} ions were accumulated in the roots of the radish plant.

In 2010, Babatunde Saheed Bada and Kamar Adekunle Raji studied on Phytoremediation potential of kenaf (*Hibiscus cannabinus* L.) for the effect of cadmium concentrations.

Selvarathi, P et al., (2010) carried out studies on phytoremediation plant *Datura metel* L. and evaluated the effect of different concentrations (0.5%, 1%, 1.5%, and 2% on 30%) of paper mill effluent on the growth and biochemical characteristics of *Lycopersicon esculentum Mill.* Their study had revealed that physicochemical parameters especially, biological oxygen demand, chemical oxygen demand, electrical conductivity, total dissolved solids, oil, and grease were reduced after the Phyto-treatment of paper mill effluent with *Datura metel* when compared to the untreated effluent.

Elekes et al., (2010) were assessed the Seven different species of perennial grasses grown in the metallurgic industrial area of Targoviste city to evaluate their usefulness for remediation of high heavy metal-polluted soils. *Lolium perenne* accumulated Cr, Zn, Sn, and Sr up to a level that excels the limit of 100%, and showed that *Lolium perenne* and *Festuca pratensis* are good hyperaccumulators for Sn (410.90% and 704% accumulation, respectively).

By applying earthworms to soils, Thi My Dung Huynh et al., (2011) propose a biologically improved method of phytoextraction. An experiment was conducted in a greenhouse to assess the effects of earthworms, *Pontoscolex corethrurus* on the growth and phytoextraction of Lantana Camara in (500 and 1000 mg kg^{-1}) Pb-contaminated soils. In soils contaminated with lead, earthworm activity increased shoot and root biomass. *L. camara* also absorbed more lead by 2.96 and 2.03 factors at two different levels of soil contamination in the presence of earthworms, and positive correlations were found between lead absorption and biomass of *L. camara* in the presence of earthworms.

Sun et al., (2011) had evaluated the phytoextraction and remediation potential of *Mirabilis jalapa* (L.) in soil contaminated with various metals. This plant was most efficient in Zn accumulation in aboveground tissues.

A study was conducted by Israr et al., (2011) to determine the metal uptake ability of *Sesbania drummondii* and found that well accumulated Pb above the threshold level, and suggested the usefulness of it in phytoextraction of Pb.

Hazrat Ali et al., (2012) were carried out phytoremediation studies on Cu, Cd, Zn, and Pb, by plant *Trifolium alexandrinum*. The study has shown that *T. alexandrinum* is an effective plant for the extraction of selected heavy metals from contaminated soil.

Mojiri et al., (2012) studied the phytoremediation of heavy metals from municipal wastewater using Typha domingensis. The results revealed that *T. dominingensis* was capable of accumulating heavy metal ions preferentially in wastewater rather than in soil. The highest levels of metal accumulation were found in rhizomes, roots, and older leaves.

Phytoremediation technology has been tested by Kokyo Oh et al., (2013) using hydroponic and pot experiments. Biofuel crops were assessed for their potential. In their experiment, the sunflower and maize plant accumulated more zinc, copper, and lead than the other two accumulator plants. It has been shown that inoculating wheat and barley with white-rot fungus significantly enhances soil microbial flora, biomass, and removal efficiency, suggesting that biofuel crops can be used for phytoremediation.

Ekwumemgbo et al., (2013) suggested reducing risk through degradation, removal, or containment of a contaminant. The experiment involved cultivating *Bryophyllum pinnatum* in ten different plastic buckets containing heavy metal polluted soil for 20 months. After the first two weeks and monthly, plant and soil samples were collected and analyzed for the total concentration of heavy metals such as Cu, Cd, Ni, Cr, V, Pb, and Zn. Maximum extracted heavy metals from the soil by the plant were Cu (81.01±2.3 mg/kg), Cd (3.12±1.03 mg/kg), Ni (11.91±2.32 mg/kg) Cr (32.48±3.21 mg/kg), V (5.81±0.08 mg/kg), Pb (3MTM Na, K, Ca, Mg, Fe, Mn, Pb, Zn, Cd, Cu, Cr, Ni, and Co concentrations were analyzed, as well as their ability to absorb and accumulate higher metal concentrations.99.90 ±4.32 mg/kg), and Zn (150.51± 0.33 mg/kg) and this occurred in the 4th month of study. This study confirmed *B. pinnatum* as a potential plant for phytoremediation of soil contaminated by heavy metals.

Said Muhammad et al., (2013) examined the phytoremediation potential of wild plant species in the Jijal, Dubair, and Alpuri areas of Kohistan region, and northern Pakistan. Concentrations of MTM Na, K, Ca, Mg, Fe, Mn, Pb, Zn, Cd, Cu, Cr, Ni, and Co were analyzed and the ability of these wild plant species to uptake and accumulate higher metals concentration was studied.

Das S. et al., (2013) studied Cu hyperaccumulating terrestrial and aquatic plants species. The study revealed that members of *Asteraceae, Brassicaceae, Typhaceae, Euphorbiaceae, Poaceae,* and *Solanaceae* family at contaminated sites. The study revealed that the Indian mustard (*Brassica juncea*) is extensively studied for phytoremediation, due to its fast growth, high biomass, and phytoremediation efficiency.

Muhammad et al., (2013) evaluated the accumulation of toxic heavy metals in six plant species collected from the second industrial zone of Riyadh, Saudi Arabia, including Zn, Cd, Ni, Pb, and Cu. The phytoremediation potential of six plants was evaluated using the translocation factor (TF), biological concentration factor (BCF), and bioaccumulation coefficient (BAC) parameters. Plants named Citrullus colocynthis, *Phragmites australis, Malva parviflora, Datura stramonium, Rhazya stricta, and Lycium shawii.* The pattern of metal accumulation in studied plants was: Zn> Cu> Pb> Ni> Cd. Results obtained showed significant ($P \geq 0.05$) accumulation in the above-ground components of Malva parviflora, Phragmites australis, *Rhazya stricta,* and *Lycium shawii* compared with the root. *Phragmites australis, Lycium shawii, Malva parviflora,* and *Rhazya stricta* combined to have a highly positive phytoextraction potential for Cd and Pb. The plants *Datura stramonium* and *Citrullus colocynthis* were found to be suitable for phytostabilising soils contaminated with Ni and Cu.

Studies of soil polluted by lead from a highway were done using hyper-accumulating plants like Hordeum vulgare was done by Zerrouki Dahbia et al., (2013) in Algeria. Heavy metals incorporated into the environment through traffic have been detected in high concentrations at horizons of the soil surface. In their view, lead uptake by *Hordeum vulgare* is influenced by soil bioavailability.

Dipu Sukumaran (2013) evaluated *Salvinia molesta, Typha latifolia, Pistia stratiotes,* and *Eichhornia crassipes* in the laboratory as aquatic weeds. Eichhornia crassipes and Typha latifolia are the best plants for treating effluent. Eichhornia crassipes removed a greater amount of lead than Typha latifolia. However, *Typhalatifolia* mainly removed copper and cadmium. A study by *Parisa Ziarati* and *Somaye Alaedini* (2014) evaluated whether Amaranthus could phytoextract cadmium, lead, copper, zinc, and nickel from soils in Tehran. The metal transfer factors from soil (TFS) of Amaranthus and the effects of adding black and green tea leaves in soil have been studied. The researchers found that the young plants had a greater capacity for absorbing and concentrating heavy metals. Plant accumulation and the soil pH affected the uptake rate. Analyzing the results, it became clear that the best results for uptake of Lead occurred in soils with pH 6.5 among the various samples. By 45 days of plant growth, the pH translocation factor was the highest, while the best Cadmium phytoextract performance was in pH 6.3 soil and the maximum Copper, Nickel, and Zinc uptake rates were achieved in pH = 6.1 and by 30 days of plant growth.

16.4 HEAVY METAL POLLUTION STUDIES

Environmental pollutants can also include heavy metals because they can remain in the environment for long periods, posing a threat to humans, animals, and plants. There have been numerous studies conducted in India to assess the amount of heavy metal pollution in the soil in densely populated, urban areas that generate enormous amounts of industrial waste.

Heavy metals are a unique class of toxic pollutants since they cannot be degraded into non-toxic forms (Jabeen et al., 2009). The concentration of toxic metals has increased dramatically since the industrial revolution (Ana et al., 2009). As soon as heavy metals enter the ecosystem, they remain a potential threat for a long time. Pollution caused by heavy metals is a global issue. Heavy metals are metals and metalloids with densities greater than 5 g cm-3, and they are usually associated with pollution and toxic effects. However, some of these elements are required by organisms at low concentrations (Andriano, 2001). Cadmium (Cd), chromium (Cr), lead (Pb), and nickel (Ni) are the most common heavy metal contaminants (Lasat, 2002). There must be some innovative methods to prevent pollution of the environment, including soil, given the increased awareness of the hazards of pollution.

Govil et al., (2001) worked on Patancheru industrial development area of Andhra Pradesh, which is situated at the north of Hyderabad city. This research evaluated the type of chemical pollution in the soil. The evaluated data showed high contamination, which is a higher level of toxic elements than normal. The study revealed that higher concentrations of heavy metals such as As, Cr, V, Cd, Se, Fe, Ba, Sr, Zn, Cu, and Mo were found to be present above normal distribution in the soil.

Singh et al., (2002) carried out studies on stream sediments from six urban centers of the Ganga River for heavy metal contamination. The study had shown that stream sediments from Kanpur, Lucknow, Agra, and Delhi and was came in highly polluted and dangerous sediments in the country.

Krishna and Govil in 2004 worked in the Pali Industrial area of Rajasthan in the western zone. Their study revealed that the soil in their study area is highly contaminated with heavy elements like Cu, Pb, Sr Cr, Zn, and V. In 2007, again study on the industrial area of Surat city, near the western state of Gujarat was carried out. The data evaluated have shown that that the study area has contamination of much heavy metal such as Cr, Ba, Co, Cu, Ni, Zn, and Sr.

Lokeshwari and Chandrappa (2006) have studied the city of Bangalore, for assessing the heavy metal contamination in vegetation and soil which is mainly due to the use of lake water contaminated with sewage for agricultural use. A higher amount of heavy metals contamination has been found in soil and vegetation samples.

In Varanasi, Sharma et al., (2007) conducted similar research. Heavy metal concentrations were measured in soil samples taken from major irrigation sites. According to the data, there is a high concentration of cadmium in that area.

In this study, heavy metals (As, Cr, Cu, Ni, Pb, Zn) in soil samples were analyzed by Vandana Partha et al., (2011) near hazardous waste disposal sites in Hyderabad, India. As soon as Cr and Pb concentrations were found to exceed the threshold and natural limits. On the other hand, the highest concentrations of Cu, Ni, and Zn exceeded the prescribed threshold limit.

Shiva Kumar and Srikantaswamy (2012) carried out the investigation of heavy metal pollution in the Mysore city, Karnataka by dissolving the extracted soil sample in the tri acid mixture method. The study revealed that the heavy metals concentration is at the nearby maximum level.

Results have shown that iron levels range from 2.5 gm/kg to 6.7 gm/kg. The copper content of food varies from 6.8 mg/kg to 20.3 mg/kg. There are between 6.6 mg/kg and 22.0 mg/kg of chrome found in this study, whereas 66 mg/kg to 121 mg/kg of zinc and from 10 mg/kg to 18.1 mg/kg of nickel were found. Industrialization is growing much faster than any other activity, and pollution of water, air, and soil is steadily increasing.

16.5 STUDIES ON TISSUE CULTURE FOR PHYTOREMEDIATION

Plant tissue culture, or the growth of plant cells outside of a plant, is essential to many fields of plant science. Individual or groups of plant cells, as well as whole organs, can be used to study fundamental as well as applied science.

Modern biotechnology techniques such as plant tissue culture have proven to be useful in medicinal plant improvement programs. These plants can not only be preserved from being lost but can also be developed into a complete plant from a small plant part developed by scientists.

The cells, tissues, and organs of in-plant tissue culture are grown under aseptic conditions in a static, or fluid, artificial medium containing nutrients. Eventually, these explants divide into an unorganized mass of cells called a "callus" and differentiate to form plants or directly generate shoots or embryos.

The mechanism of tissue culture is the ability to produce embryoids, organs, tissues, etc. Murashige (1974) had described plant regeneration accomplished from explants such as leaves, stems, microsporocytes cotyledons, and shoot tips. According to Gamborg et al., (1974), the most important determinant of plant multiplication and quality of regenerated plants is the initial explant used.

Many studies have shown that plant tissue cultures are a valuable tool in phytoremediation research today. Tissue culture is a powerful tool that allows us to grow millions of cells under in vitro conditions and to understand how plant cells respond to stress. Molecular genetics and plant tissue culture have opened up new avenues in plant sciences. With the use of selection and screening at the cellular level, plants have been developed that are more tolerant of disease, heavy metals, salt, heat, herbicides, cold, drought, and insects.

The tissue and cell culture of plants is an important tool for studying plant-contaminant interactions, as well as helping to predict the effects of environmental pollutants on plants and redesign the plants for phytoremediation. Studies of the interaction between plants and pollutants in vitro are conducted using plant tissue cultures.

Tissue cultures are more than just convenient and fast compared to whole plants. By removing the effects of microflora and translocation barriers, a closer approximation of the intrinsic detoxification capacity of plant cells can be obtained.

Nelson Marmiroli et al., (2006) presented an extensive review of European research on phytoremediation. Phytoremediation is categorized based on research suggesting that it can be divided into two main categories, including a basic understanding of mechanisms and implementation of these mechanisms for applications. The basic research focuses on the molecular, genetic, and cellular mechanisms of heavy metal accumulation and tolerance in plants. Research in Europe focused on the enzymology of phytodegradation of TNT, PAHs, herbicides, and dyes, either in vitro or in whole plants.

Phytoremediation research using plant tissue cultures was examined by Pauline M. Doran (2009). In-vitro cultures can be used to study phytoremediation processes and mechanisms best in areas where in-vitro cultures have the greatest potential. However, tissue cultures have some limitations over whole plants for research purposes. In vitro plant cell and organ, cultures are useful and have limitations in phytoremediation studies, as was revealed in this study. A fundamental understanding of plants requires the ability to identify the ability of plants to absorb pollutants and detoxify them. Plant cells are cultured to study their intrinsic metabolic capabilities and tolerance to toxicity. Phytoremediation technology is ultimately developed through plant tissue culture experiments. Pauline M. Doran points out that plant tissue culture is a convenient laboratory tool for phytoremediation studies. Cell suspensions and hairy roots are the two most commonly used tissue culture methods. The in vitro cultures can be propagated and are available on-demand once established.

Kadhim M. Ibrahim and Shaimaa A. Yousir (2009) conducted several experiments to study heavy metal tolerance in tissue cultures and whole plants of *S. grandiflora.*, the callus was grown and maintained on modified

MS, 1962 medium supplemented with (0.5) mg/l benzyl adenine and (2) mg/l 2,4-phenoxy acetic acid. Various heavy metals were added to the culture medium as contaminants, including Cr, Cd, Co, Zn, and Cu. Heavy metals were also assessed for their effect on seed germination. As heavy metal concentrations in the culture medium increased, callus fresh weight decreased. Zn was absorbed by the whole plant at a maximum of (400) mg/kg soil, reaching (2220) ppm. The percentages of heavy metals absorption increased as heavy metal concentrations increased. The study found that *S. grandiflora* callus cultures and the whole plant level are tolerant to heavy metals compared to other plants.

Abdul R. Memon et al., (2008) investigated 30 plant species from mining areas of the Southeastern region of Turkey. Heavy metals such as Cd, Fe, Mn, Cr, Co, Cu, Zn, Pb, and Ni contents were analyzed and discovered four heavy metal accumulator plants, among them *Brassica nigra* was found to be an accumulator of Cu and Cd. Successful regeneration of *Brassica nigra* plants from tissue culture for mass production was also part of the study. The RNA expression experiments by real-time PCR (RT-PCR) showed that two enzymes of glutathione pathway: (1) γ-glutamyl cysteine synthetase (γ-EC) and (2) phytochelatin synthase (PC) were highly expressed when plants were subjected to 500 μM Cu in the media.

The current study examines native plants' ability to accumulate and tolerate high concentrations of heavy metals, which are often associated with polluted environments. An assessment of heavy metals in soils at the industrial waste site revealed that the amounts of some heavy metals are several times higher than in natural areas.

Research and development in phytoremediation are still in their early stages, with many advances to be made. In contrast to the United States, Europe, and Australia, Indian government agencies do not submit large numbers of applications, although universities and research institutes are very interested.

16.6 SUMMARY

Globally, plants are needed to restore the ecological environment. The mechanism of phytoremediation and the parameters for environmental pollution have not been systematically reviewed. This chapter summarized the detailed review on phytoremediation research from a historical perspective and described how plants purify pollutants based on their purification rate, as well as methods for improving remediation rates. Plants are crucial for restoring the ecosystem and the environment through their use. The interaction between microorganisms and plants is how plants repair their ecological environments. In the remediation process, plants' purification factors are influenced by many factors, including light intensity, stomatal conductance, temperature, and microorganisms. Phytoremediation is also dependent on the species-specific metabolism of the plants, including air absorption and photosynthesis, as well as microorganism diversity in the soil and heavy metal uptake.

This review attempts to summarize the state of the science of phytoremediation and provide a few references to publications that provide more detail. Phytoremediation describes plant processes that uptake, translocate, sequester, and degrade organic and inorganic pollutants once an introduction to various types of phytoremediation has taken place. A discussion is also made of genetic engineering and its potential for phytoremediation.

16.7 EXERCISES

Practice Work

Qu 1. Define term phytoremediation.

Qu 2. Explain the historical developments in the phytoremediation field.

Qu 3. What is EPA? Explain its role in environmental protection.

Qu 4. Explain the types of toxicity symptoms on plants.

Qu 5. What are the key research findings on Phytoremediation?

Qu 6. Explain the historical developments in heavy metal pollution studies.

Qu 7. Write a review on tissue culture studies for phytoremediation.

Qu 8. Explain the need for phytoremediation research.

Qu 9. Explain the historical origin of the phytoremediation concept.

Qu 10. Explain the process of evolution of metal tolerances in plants.

REFERENCES

Ali, H., Khan, E., & Sajad, M. A., (2013). *Phytoremediation of Heavy Metals--Concepts and Applications, 91*(7), 869–881.

Ali, M. B., Tripathi, R. D., Rai, U. N., Pal, A., & Singh, S. P., (1999). Physicochemical characteristics of lake Nainital (UP) India. Role of macrophytes and phytoplankton in biomonitoring and phytoremediation of toxic metal ions. *Chemosphere, 39*, 2172–2182.

Andriano, D. C., (2001). Trace element in the terrestrial environment: Biochemistry. *Bioavailability and Risks of Metals* (2nd edn.). New York Springer.

Baker, A. J. M., & Walker, P. L., (1990). Ecophysiology of metal uptake by tolerant plants: Heavy metal tolerance in plants. In: Shaw, A. J., (ed.) *Evolutionary Aspects*. CRC Press, Boca Raton.

Baker, A. J. M., & Whiting, S. M., (2002). In search for the holy grail—Another step in understanding metal hyperaccumulation. *New Phytologist, 155*, 1–7.

Baumann, A., (1885). *Das Verhalten von Zinksatzen gegen Pflanzen und im Boden* (Vol. 31, pp. 1–53). Landwirtsch, Vers.-Statn. https://chemical.journalspub.info/index.php?journal=IJGC&page=article&op=view&path%5B%5D=156 (accessed on 12 May 2023).

Belsky, A. J., Amundson, R. G., Duxbury, J. M., Riha, S. J., Ali, A. R., & Mwonga, S. M., (1989). The effects of trees on their physical, chemical, and biological environments in a semi-arid savanna in Kenya. *J. Appl. Ecol., 26*, 1005–1024.

Bharti, R. P., Shrivastava, A., Soni, N., Tiwari, A., More, S., & Choudhary, J. R., (2014). Phytoremediation of heavy metal toxicity and role of soil in rhizobacteria. *International Journal of Scientific and Research Publications, 4*(1).

Bidar, G., Garcon, G., Pruvot, C., Dewaele, D., Cazier, F., Douay, F., & Shirali, P., (2007). Behavior of *Trifolium repens* and *Lolium perenne* growing in a heavy metal contaminated field: plant metal concentration and phytotoxicity. *Environ. Pollut., 147*, 546–553.

Chekroun, K. B., & Baghour, M., (2013). The role of algae in phytoremediation of heavy metals: A review. *J. Mater. Environ. Sci., 4*(6), 873–880.

Cobbett, C., (2003). Heavy metals and plants – model systems and hyperaccumulators. *New Phyto., 159*, 289–293.

Cunningham, S. D., Spann, J. R., Crowley, D. E., & Anderson, T. A., (1997). Phytoremediation of contaminated water and soil. In: Kruger, E. L., Anderson, T. A., & Coats, J. R., (eds.), *Phytoremediation of Soil and Water Contaminants* (Vol. 664, p. 219). ACS Symposium Series.

Dahbia, Z., Mhamed, M., Benchaban, H., & Mohamed, A. H., (2013). Phytoremediation of contaminated agricultural soil by lead from traffic pollution using a common barley hordeum vulgare. *International Journal of Environment and Resource (IJER), 2*(3), 49.

Das, S., Goswami, S., & Talukdar, A., (2013). Copper hyperaccumulating plants from Barak valley, South Assam India for phytoremediation. *International Journal of Toxicological and Pharmacological Research, 5*(1), 30–32.

Devinny, J., Longcore, T., & Bina, A., (2005). *Phytoremediation with Native Plants.* pp. 1–65. file:///C:/Users/hp/Downloads/Phytoremediation_with_Native_Plants.pdf.

Dissanayake, U. S., Tennakoon, K. U., & Priyan, N., (2002). Potential of two invasive plant species, *Lantana Camara* L. and *Wedelia Trilobata* L. for selective heavy metal uptake. *Cey. J. Sci. (Bio. Sci.), 29*, 1–11.

Doran, M. P., (2009). Application of plant tissue cultures in phytoremediation research: Incentives and limitations. *Biotechnology and Bioengineering, 103*(1).

Ebong, G. A., Akpan, M. M., & Mkpenie, V. N., (2007). *Heavy Metal Contents of Municipal and Rural Dumpsite Soils and Rate of Accumulation by Carica Papaya and Talinum Triangulare in Uyo.* Nigeria.

Ekwumemgbo, P. A., Eddy, N. O., & Omoniyi, I. K., (2013). Decontamination of heavy metals in polluted soil by phytoremediation using *Bryophyllum Pinnatum*, E3S web of conferences. *EDP Sciences, 4*. https://doi.org/10.1051/e3sconf/20130113004.

Elekes, C. C., Dumitriu, I., Busuioc, G., & Iliescu, N. S., (2010). The appreciation of mineral element accumulation level in some herbaceous plants species by ICP-AES method. *Environ. Sci. Pollut. Res., 17*, 1230–1236.

Erakhrumen, A. A., (2007). Phytoremediation: An environmentally sound technology for pollution prevention, control and remediation in developing countries. *Educational Research and Review, 7*, 151–156.

Flathman, P. E., & Lanza, G. R., (1998). Phytoremediation: Current views on an emerging green technology. *Journal of Soil Contamination, 7*(4), 415–432.

Garg, G., & Kataria, S. K. (2009). Phytoremediation potential of *Raphanus sativus* (L.), *Brassica juncea* (L.) and *Triticum aestivum* (L.) For Copper Contaminated Soil. Gautam Buddha University, Greater Noida, India. *Proceedings of the 53rd Annual Meeting of the ISSS – 2009, Brisbane, Australia, 1*(1). Retrieved from https://journals.isss.org/index.php/proceedings53rd/article/view/1123.

Ghosh, M., & Singh, S. P., (2005). Comparative uptake and phytoextraction study of soil induced chromium by accumulator and high biomass weed species. *Applied Ecology and Environmental Research, 3*(2), 67–79.

Ginneken, L. V., Meers, E., Guisson, R., Ruttens, A., Elst, K., Filip, M., Tack, G., et al., (2007). Phytoremediation for heavy metal-contaminated soils combined with bioenergy production. *Journal of Environmental Engineering and Landscape Management, XV*(4), 227–236.

Gonzalez, I., Muena, V., Cisternas, M., & Neaman, A., (2008). Copper accumulation in a plant community affected by mining contamination in Puchuncavi Valley. *Central Chile. Rev. Chil Hist. Nat., 81*, 279–291.

Govil, P. K., Reddy, G. L. N., & Krishna, A. K., (2001). Contamination of soil due to heavy metals in the patancheru industrial development Area, Andhra Pradesh, India. *Environmental Geology, 41*, 461–469.

Hakmaoui, A., Baron, M., & Ater, M., (2006). Environmental biotechnology screening Cu and Cd tolerance in Salix species from North Morocco. *African Journal of Biotechnology, 5*(13), 1299–1302.

Hartman, W. J. Jr., (1975). *An Evaluation of Land Treatment of Municipal Wastewater and Physical Sitting of Facility Installations*. US Department of Army, Washington, DC.

Hussain, S., Siddique, T., Arshad, M., & Saleem, M., (2009). Bioremediation and phytoremediation of pesticides: Recent advances. *Cri. Rev. Environ. Sci. Tech., 39*, 843–907.

Ibrahim, K. M., & Yousir, S. A., (2009). *In vivo* and *in vitro* studies on heavy metal tolerance in *Sesbania Grandiflora* L. *Biotechnology Research Center (special edition), 3*(2).

Ibrahim, M. M., Alsahli, A. A., & El-Gaaly, G., (2013). Evaluation of phytoremediation potential of six wild plants for metal in a site polluted by industrial wastes: A field study in Riyadh, Saudi Arabia, Pak. *J. Bot., 42*(2), 571–576.

Israr, M., Jewell, A., Kumar, D., & Sahia, S. V., (2011). Interactive effects of lead, copper, nickel and zinc on growth, metal uptake and antioxidative metabolism of *Sesbania drummondii*. *J. Hazardous Mat., 186*, 1520–1526.

Jabeen, R., Ahmad, A., & Iqbal, M., (2009). Phytoremediation of heavy metals: Physiological and molecular mechanisms. *Bot. Rev., 75*, 339–364.

Jadia, C. D., & Fulekar, M. H., (2009). Phytoremediation of heavy metals: Recent techniques. *African Journal of Biotechnology, 8*(6), 921–928.
Kathi, S., & Khan, A. B., (2011). Phytoremediation approaches to PAH contaminated soil. *Indian Journal of Science and Technology, 4*(1).
Keilig, K., & Ludwig-Müller, J., (2009). Effect of flavonoids on heavy metal tolerance in *Arabidopsis thaliana* seedlings. *Botanical Studies, 50*, 311–318.
Kimenyu, P. N., (2008). *Phytoremediation of Heavy Metals in Mimics of Chemically Polluted Soils and Wastewater*. Thesis, Jomo Kenyatta University of Agriculture and Technology.
Kramer, M. N., Coto, D., & Weidner, J. D., (2005). The science of recalls. *Meat Science, 71*, 158–163.
Krishna, A. K., & Govil, P. K., (2004). Heavy metal contamination of soil around Pali industrial area, Rajasthan, India. *Environmental Geology, 47*, 38–44.
Kumar, D. S., & Srikantaswamy, S., (2012). Heavy metal pollution assessment in industrial area soil of Mysore City, Karnataka, India. *Int. Journal of Applied Sciences and Engineering Research, 1*(4).
Kumar, N. J. I., Soni, H., Kumar, R. N., & Bhatt, I., (2008). Macrophytes in phytoremediation of heavy metal contaminated water and sediments in Pariyej community reserve, Gujarat, India. *Turkish Journal of Fisheries and Aquatic Sciences, 8*, 193–200.
Leslie, I. Appel, C., Stubler, C., Davani, K., DeValentine, A., Gomez, J., Kim, D., Konwinski, T., Jones, A., & Patterson, K. (2008). *Analysis of Chromium and Nickel in Soil and Plant Tissue Collected from the Experimental Vineyard at California Polytechnic State University, San Luis Obispo, CA*. DigitalCommons@CalPoly. https://digitalcommons.calpoly.edu/ersc_fac/9/.
Lokeshwari, H., & Chandrappa, G. T., (2006). Impact of heavy metal contamination of Bellandur lake on soil and cultivated vegetation. *Current Science, 91*(5), 622–628.
Malik, N., & Biswas, A. K., (2012). Role of higher plants in remediation of metal contaminated sites. *Sci. Revs. Chem. Commun., 2*(2), 141–146.
Marmiroli, N., Marmiroli, M., & Maestri, E. (2006). Phytoremediation and phytotechnologies: a review for the present and the future. *NATO Science Series*, 403–416. https://doi.org/10.1007/978-1-4020-4728-2_26.
Memon, A. R., Aktoprakligul, D., Zdemür, A., & Vertii, A., (2001). Heavy metal accumulation and detoxification mechanisms in plants. *Turk J. Bot., 25*, 111–121.
Memon, A. R., Yildizhano, Y., & Keskin, B. C., (2008). *Phytoremedıatıon of Heavy Metals from Contamınated Areas of Turkey, 4th European Bioremediation Conference* (pp. 1–4). Chania, Crete, Greece. file:///C:/Users/hp/Downloads/B3.PHYTOREMEDIATIONOFHEAVYMETALSFROMCONTAMINATEDAREASOF.pdf.
Migeon, A., Richaud, P., Guinet, P., Chalot, M., & Blaudez, D., (2009). Metal accumulation by woody species on contaminated sites in North of France. *Water Air Soil Pollut., 204*, 89–101.
Minguzzi, C., & Vergnano, O., (1948). *II Content di Nichel Nelli Ceneri di Alyssum Bertlonii Desv* (Vol. 55, pp. 49–77). Atti della societa Toseana di Science Naturali, Mem Ser A.
Mojiri, A., (2012). Phytoremediation of heavy metals from municipal wastewater by *Typha domingensis*. *African J. Microbiol. Res., 6*, 643–647.
Muhammad, S., Shah, M. T., Khan, S., Saddique, U., Gul, N., Khan, M. U., Riffat, N. M. R. N., et al., (2013). Wild plant assessment for heavy metal phytoremediation potential along the mafic and ultramafic terrain in Northern Pakistan. *BioMed Research International*, 2013, pp. 1–9. https://doi.org/10.1155/2013/194765. Article ID 194765.

Murashige, T. (1974) Plant Propagation through Tissue Cultures. Annual Review of Plant Physiology, 25, 135–166. http://dx.doi.org/10.1146/annurev.pp.25.060174.001031

Nehnevajova, E., (2005). *Non-GMO Approach for the Improvement of Heavy Metal Accumulation and Extraction of High Yielding Crop Species for Efficient Phytoextraction of Contaminated Soil*. Swiss Federal Institute of Technology Lausanne (EPFL).

Odoemelam, S. A., & Ukpe, R. A., (2008). Heavy meal decontamination of polluted soils using *Bryophyllum pinnatumafrican*. *Journal of Biotechnology, 7*(23), 4301–4303.

Oh, K., Li, T., Cheng, H., Hu, X., He, C., Yan, L., & Shinichi, Y., (2013). Development of profitable phytoremediation of contaminated soils with biofuel crops. *Journal of Environmental Protection, 4*, 58–64.

Oyelola, O. T., Babatunde, A. I., & Odunlade, A. K., (2009). Phytoremediation of metals from contaminated soil using *Lvcoperciam escalentam* (tomato) plant. *Int. Jor. P. App. Scs., 3*(2), 44–48.

Parth, V., Murthy, N. N., & Saxena, P. R., (2011). Assessment of heavy metal contamination in soil around hazardous waste disposal sites in Hyderabad city (India): Natural and anthropogenic implications. assessment of heavy metal contamination in soil around hazardous waste disposal sites in Hyderabad City (India): Natural and Anthropogenic implications E3. *Journal of Environmental Research and Management, 2*(2), 027–034.

Paz-Ferreiro, J., Lu, H., Fu, S., Méndez, A., & Gasco, G., (2014). Use of phytoremediation and biochar to remediate heavy metal polluted soils: A review. *Solid Earth, 5*, 65–75.

Prasad, M. N. V., (2007). Phytoremediation in India. In: Willey, N., (ed.), *Phytoremediation: Methods in Biotechnology* (Vol. 23). Humana Press. https://doi.org/10.1007/978-1-59745-098-0_30.

Pulford, I. D., & Watson, C., (2003). Phytoremediation of heavy metals –contaminated land by tree: A review. *Environ. Internat., 29*, 529–540.

Rascio, W., (1977). Metal accumulation by some plants growing on Zn mine deposits. *Oikos, 29*, 250–253.

Robinson, B., Green, S., Mills, T., Clothier, B., Velde, M., Laplane, R., Fung, M., et al., (2003). Phytoremediation: Using plants as biopumps to improve degraded environments. *Aust. J. Soil Res., 41*, 599–611.

Sarma, H., (2011). Metal hyperaccumulation in plants: A review focusing on phytoremediation technology. *Journal of Environmental Science and Technology, 4*(2), 118–138.

Schroder, P., Navarro-Avino, J., Azaizeh, H., Goldhirsh, A., DiGregorio, S., Komives, T., Langergraber, G., et al., (2007). Using phytoremediation technologies to upgrade wastewater in Europe. *Environ. Sci. Pollut. Res., 14*, 490–497.

Schultes, R. E., & Hofmann, A., (1992). *Plants of the Gods* (p. 208). Healing Plant Press Rochester, NY.

Selvarathi, P., & Ramasubramanian, V., (2010). Phytoremedial effect of *Datura metal* L. on paper mill effluent and its impact on physicochemical characteristics of *Lycopersicon esculentum* mill. *J. Biosci. Res., 1*(2), 94–100.

Shanker, A. K., & Ravichandran, V., (2005). *Phytoaccumulation of Chromium by Some Multipurpose-Tree Seedlings, Agroforestry Systems* (Vol. 64, pp. 83–87). Springer.

Sharma, R. K., Agrawal, M., & Marshall, F., (2007). Heavy metal contamination of soil and vegetables in suburban areas of Varanasi, India. *Ecotoxicology and Environmental Safety, 66*(2), 258–266.

Singh, M., Müller, G., & Singh, I. B., (2002). Heavy metals in freshly deposited stream sediments of rivers associated with urbanization of the ganga plain, India. *Water Air and Soil Pollution, 141*, 35–54.

Sukumaran, D., (2013). Phytoremediation of heavy metals from industrial effluent using constructed wetland technology. *Applied Ecology, and Environmental Sciences, 1*(5), 92–97.

Sun, Y. B., Sun, G. H., Zhou, Q. X., Wang, L., Liang, X. F., Sun, Y., & Qing, X., (2011). Induced-phytoextraction of heavy metals from contaminated soil irrigated by industrial wastewater with marvel of Peru (*Mirabilis jalapa* L.), *Plant Soil Environ., 57*, 364–371.

Suresh, B., & Ravishankar, G., (2004). A. phytoremediation a novel and promising approach for environmental clean-up. *Crit. Rev. Biotechnol., 24*, 97–124.

Terry, M., (2003). Phytoremediation of heavy metals from soils. *Advances in Biochemical Engineering/Biotechnology, 78*, 97–123.

Thi My Dung Huynh, Thi My Hanh Diep, Anh-Thu Pham-Thi, & Corinne Rouland-Lefevre, (2011). Enhanced Lead Phytoextraction of *Lantana camara* L. by a Biological Agent, Earthworm Pontoscolex corethrurus. As*ian Journal of Biological Sciences, 4*, 375–382. URL: https://scialert.net/abstract/?doi=ajbs.2011.375.382.

Weis, J. S., & Weis, P., (2004). Metal uptake, transport and release by wetland plants: implications for phytoremediation and restoration. *Environment International, 30*(5), 685–700.

Wilfried, H. O. E., (2006). Evolution of metal tolerance in higher plants. *Snow Landsc. Res., 80*(3), 251–274.

Zhuang, P., Yang, Q., Wang, H., & Shu, W. (2007). Phytoextraction of Heavy Metals by Eight Plant Species in the Field. *Water Air and Soil Pollution, 184*(1–4), 235–242. https://doi.org/10.1007/s11270-007-9412-2.

Ziarati, P., & Alaedini, S., (2014). The phytoremediation technique for cleaning up contaminated soil by *Amaranthus* sp. *Environ. Anal. Toxicol., 4*, 2.

CHAPTER 17

Phytoremediation: A Case Study

17.1 PHYTOREMEDIATION

Heavy metals such as iron (Fe), lead (Pb), zinc (Zn), cadmium (Cd), chromium (Cr), and nickel (Ni) are essential nutrients, but become toxic when their concentration exceeds the recommended levels. Heavy metals such as lead and cadmium are not essential. Even tiny amounts of them are toxic. Such trace elements are deposited in sediments as a result of the direct discharge of contaminants. Plants absorb these elements through their roots and/or shoots. As far as their ability to accumulate elements in different parts, such as roots, stems, and leaves, different species show different behaviors.

The plant organ that absorbs the most heavy metals must be identified. Plant remediation has several advantages and is most important for the study of sublethal levels of bioaccumulated pollutants within the tissues of organisms, which indicate how much pollutant is integrated into tissue over time.

Bio-monitoring of pollutants using plants as accumulator species accumulates fairly large amounts of specific pollutants, even from diluted solutions without exhibiting obvious toxic effects. Passive bio-monitoring of endemic, or native, organisms are referred to as passive monitoring, while active bio-monitoring is referred to as active monitoring.

Hazardous waste sites in an industrial area in Bhopal that require clean-up contain a variety of pollutant types in their soil, groundwater, and wastewater. Consequently, most of the state's major cities and industrial towns, including Indore, Bhopal, Ujjain, Dewas, Nagda, Pithampur, Mandideep, and Ratlam, are located on top of highly toxic sludge underground.

Plants are used in phytoremediation to remove, contain, or render harmless environmental contaminants such as heavy metals, trace elements, organic compounds, and radioactive compounds.

However, industries are necessary for the economic modernization and development of the city of Bhopal. Phytoremediation is beautifully pleasing

Bioremediation and Phytoremediation: Technologies for Toxic Pollution
Ashwini A. Waoo, PhD (Author)

to people who live and work around the area while it is being cleaned up and is supposed as a more natural solution than large amounts of equipment and noisy machinery, the cost is inviting because phytoremediation can be undertaken for less than half the cost of other technologies and in some case, the pollutant (e.g., copper) can be reclaimed and recycled.

Bhopal and the surrounding areas are one of India's most important industrial hubs. Several industries are located in this region, which may cause environmental pollution, which can be harmful to our health. This problem requires an effective solution. Plant remediation can solve this problem.

There is a global problem with environmental pollution caused by toxic chemicals, metals, and xenobiotics. Typical remedial methods are expensive and inhibit soil fertility, which harms the ecosystem.

The purpose of this study was to identify some accumulators that may be associated with important biomass production as well as effective absorption and translocation of heavy metals. This study focuses on the ability of some native plants in Bhopal to accumulate and tolerate excessive quantities of heavy metals, which are often associated with polluted areas.

17.2 OBJECTIVES OF THE RESEARCH WORK

Introduction of invasive, foreign species for phytoremediation purposes in Bhopal may affect the local biodiversity of that area, therefore, identification and selection of native locally grown plant species for phytoremediation research and application is one of the challenges that need to be completed and a pre-requisite for successful phytoremediation research.

Following are the main objectives of this case study

Aim 1: To find out the native species at the contaminated area having phytoremediation Potential and also to detect the heavy metal content in soil of the respected contaminated site.

Aim 2: To optimize rhizosphere conditions and to find out the factors and parameters that enhance a plant's capacity to uptake more contaminants in its above-ground part to develop a successful phytoremediation technology.

17.3 EXPECTED OUTCOME

A primary goal of this research project is to determine which plant species are most suitable for phytoremediation at a specific contaminated site according to the type of heavy metal contamination. Furthermore, to identify the microorganisms present in the rhizosphere of heavy metal accumulator plants.

Scientists and remediation specialists will be able to use the results, interpretations, and methods developed from this research in real-world conditions to achieve a greener environment.

17.4 PUBLIC RELEVANCE

Taking advantage of plants' natural processes makes phytoremediation the most commonly used tool against industrial pollution. Because plants do most of the work for soil remediation, it requires less equipment and labor than other methods. Plants and trees can make a site more attractive. It is possible to clean up the site without removing polluted soil or pumping polluted groundwater. This prevents workers from coming into contact with harmful chemicals. Phytoremediation has been tested in many places and is currently being used at several locations.

So far, most of the phytoremediation experiments have taken place on the laboratory scale, where plants grown in hydroponics settings are fed heavy metal diets, while these results are promising. Thus, efforts are being made at the actual industrial level. The government and industry must be dedicated to the phytoremediation program. It must be a primary requirement to set up or run the industry. The phytoremediation plan for a particular industry must take into account the pollutants it emits. Exercises that promote phytoremediation should be a major criterion for grading the industry.

17.5 MATERIAL AND METHODOLOGY

Plant-based remediation is a cost-effective, eco-friendly, esthetically pleasing approach most suitable for developing countries and urban cities such as Bhopal. A city like Bhopal needs industries to modernize and develop economically.

- ***Study Area***

There are many industries in the Govindpura industrial area of Bhopal, including electroplating, chemical manufacturing, and many others, where wastewater has been used for many years to irrigate plants. The environment is increasingly polluted with heavy metals such as lead, chromium, nickel, cadmium, and copper alloys due to industrial development.

Heavy metal pollution of soil causes plant uptake, which in turn leads to accumulation in plant tissues and eventually phytotoxicity and changes in plant communities. This research was conducted in an area affected by emissions released by electroplating industries. For heavy metal analysis, four to five samples of plants were collected and three samples of soil were collected. A plant's homeostatic mechanisms prevent it from overproducing unnecessary ions and prevent them from damaging cells.

- ***Plan of Work for the Present Investigation***
  1) Qualitative characterization of contaminated sites
  2) Sampling of plant and soil material (20 cm topsoil) from different sites. Total metal content in and soil samples (Cd, Cr, Ni, Pb), pH.
  3) Heavy metals (Cd, Cr, Ni, Pb) concentrations were measured in dried plant material
  4) Preliminary screening of the phytoremediation potential of candidate plant species.
     - o Produce botanical surveys of several contaminated in Bhopal
     - o Assess the phytoremediation potential of candidate plant species.
     - o Growth chamber/in vitro studies.
     - o Elucidate degradation mechanisms.
  5) Manipulation of plant rhizosphere for enhanced phytoremediation.
     - o Manipulation of factors for increased root exudation.
     - o Manipulation of rhizosphere microbial communities.

The analyzes carried out to determine these selected heavy metals like Cd, Cr, Ni, and Pb contents in soil and plants revealed that they absorb and accumulate high amounts of heavy metals in their tissues. The research work presents the spectrophotometric analysis of heavy metals present in polluted soil and accumulated metal content in plants.

17.6 QUALITATIVE CHARACTERIZATION OF CONTAMINATED SITES

To study the heavy metals present in the soil, soil sampling was carried out at 3 different locations from the Govindpura industrial area. The soils were sampled at six layers 0–20 cm, 20–40 cm, 40–60 cm, 60–80 cm, 80–100 cm, and 100–120 cm, at the same time, four to five plants species were selected for the present study, which was represented the major plant species frequently growing in the area of study.

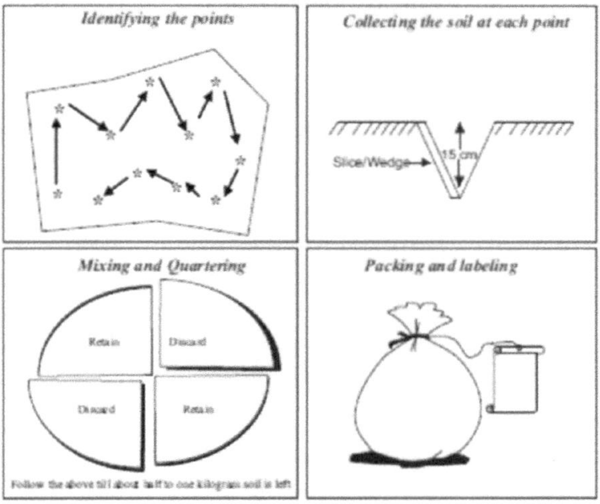

FIGURE 17.1 Random method of soil sampling

At each sampling site, replicate three samples were randomly collected. Approximately, more than 1kg of soil and plants were collected from each layer at each point.

17.6.1 ANALYSIS OF SOIL FOR TRACE AND HEAVY METALS BY AAS

A soil sample was taken from a contaminated site in triplicate and dried at 105°C to a constant weight. Microwave-assisted Kjeldahl digesters were used to acid digest 0.25g sediments. Five ml of concentrated nitric acid, two ml of hydrochloric acid, and one ml of hydrofluoric acid were added to each microwave extraction vessel. Microwave units at 800 watts were used to heat the vessels at 210°C for 20 minutes at a pressure of 40bar. The heavy metals were detected by the atomic absorption spectrophotometer using flame atomization (Figure 17.2). The results are reported on a dry weight basis.

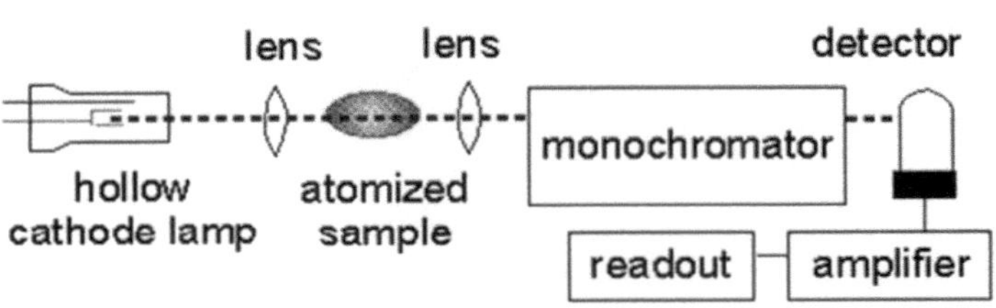

FIGURE 17.2 Schematic diagram of atomic absorption spectrometer.

17.6.2 ANALYSIS OF PLANTS FOR TRACE AND HEAVY METALS BY AAS

The plants frequently grown within the target study area have been identified, and fresh plant parts, including leaves and stems of all five varieties of plants selected for experiments, have been collected by chance from these plants. Seven days of air-drying were conducted on all samples. Using a 60°C oven, the samples were dried and ground to a fine powder. Microwave-assisted Kjeldahl digestion was carried out with three dried samples digested with a mixture of concentrated nitric acid and hydrofluoric acid (3:1). 0.8 g of plant material was added to each microwave extraction vessel along with 6 ml of nitric acid and 2 ml of hydrofluoric acid. The vessels were heated in a microwave unit at 800 watts for 20 minutes at a pressure of 25 bar and a temperature of 190°C. As soon as the samples were diluted to 50 ml, the atomic absorption spectrophotometer with flame atomization was used to analyze the metals. Dry weights were calculated for each component.

- ***Growth Chamber***

The mechanisms of plant defense against industrial pollution are relatively poorly studied. For a long time already, in vitro or growth chamber culture has been used to investigate responses of plants to toxic metal ions. With the use of plant culture in vitro, researchers have selected plants tolerant to some metal ions, including aluminum, copper, manganese, and nickel.

There is evidence that nodal segments of plants in vitro respond similarly to the whole plant, therefore in vitro culture could be used for pre-selection and evaluation of salt and hence for metal toxicity. Even so, the mechanisms involved are little understood, which could be useful for screening for metal tolerance and accumulation during the early stages of development.

Germplasm screening in contaminated fields could be slow, expensive, and labor-intensive. For studies on tolerance in plants, in vitro testing of organs or explants would be a straightforward, easy, and efficient alternative method. Plants have been studied in vitro in response to toxic metal ions. Researchers have used in vitro techniques to select plants that can tolerate aluminum, copper, manganese, and nickel. Several in vitro experiments have examined the effects of high concentrations of heavy metals on the growth of plants tolerant or sensitive to industrial pollution. In vivo and in vitro selection of plants under natural pollution conditions can lead to clones that are toxic metal ion tolerant or accumulators. By selecting plants that are tolerant to toxic ions present in the soil in vitro, it may be possible to develop plants that can accumulate and remediate environmental pollution and can also enable better management of degraded soil, such as in industrially contaminated areas.

Present experiments aimed to analyze the effect of heavy metals, present in the investigated sites such as copper, cadmium, chromium, nickel, and lead, on the growth of those plants grown at the sites, in vitro. This would allow for the selection of plants with higher accumulators of pollution from industry. Likewise, such plants would be valuable research specimens for studying how plants are affected by metal ions.

This study aimed to determine heavy metal tolerance and accumulation of four weed species that commonly grow in the fields of industrial sites.

17.6.3 IN-VITRO METAL TOXICITY STUDIES

Under controlled conditions, in vitro culture involves the isolation, inoculation, and regeneration of plant cells, tissues, and organs in culture bottles containing artificial nutrient medium. As a result of the chemical composition of the medium, as well as controlled environmental conditions such as humidity, aeration, light, temperature, etc., any genomic or phenotypic potentials expressed in the explants are effectively controlled.

A Murashige and Skoog medium supplemented with 3% (m/v) sucrose was used for the culture and maintenance of experimental plants. Murashige and Skoog (1962) modified the nutrient medium by dilution of micro- and macronutrients.

The fresh initiated green young shoots of experimental plants were taken from the investigated sites of Govindpura industrial area, Bhopal, the leaves were removed with the help of a blade, and the nodal region was isolated which is used as explants, were washed three to four times with plain water and then with liquid soap followed by washing with tap water. Further surface sterilization was conducted in a laminar flow chamber. Explants were exposed for surface sterilization. Surface-sterilized explants were inoculated aseptically on MS medium. The cultures were incubated at 24–25°C. After 10–20 days, *in vitro* initiated shoots were used as a source of explants for heavy metal treatment experiments.

17.7 METAL TOLERANCE TREATMENT

A stock solution of heavy metals like Chromium sulfate ($CrSO_4$), Copper sulfate ($CuSO_4$), Nickel sulfate ($NiSO_4$), Cadmium chloride ($CdCl_2$), and Lead Sulfate ($PbSO_4$) were prepared and filter sterilized. After autoclaving the media, a suitable aliquot from the stock solution was added into the medium to attain the final concentration of Cr, Ni, Pb, Cd, and Cu respectively. After the addition of metal salts, the media were stirred thoroughly before distribution in culture tubes. 20 ml medium was poured into each tube.

Plants-initiated shoots were transferred in culture bottles containing medium supplemented with individual heavy metal salts. Medium without metal was used as control. Four to five explants were cultured in each culture jar. Cultures were incubated at 25 ± 2°C under 16 hrs photoperiod in a 32 μ mole, $m^{-2}s^{-1}$ light intensity. The experiment was repeated thrice with twenty replicates in each treatment. One month after growing in one medium with one concentration, the cultures were transferred to a medium containing a high concentration of heavy metals. (20 mg/l to 3o mg/l).

After about 5 weeks, an assessment of culture development was started. The assessment involved percentage of survival, measurements of length and number of shoots, and evaluation of culture quality, i.e., the level of chlorosis and browning of the explant. All experiments were conducted in a randomized block design with four replications of six jars each. This study aimed to assess the ability of these plants in tissue cultures and whole plants to accumulate heavy metals in their tissues and hence use the plant for phytoremediation purposes.

17.8 STUDY OF PLANT RHIZOSPHERE FOR ENHANCED PHYTOREMEDIATION

Heavy metal accumulation in food chains threatens human health by adversely affecting soil microbial activity and crop production. A phytoextraction technique offers the potential for cost-effective remediation of toxic heavy metal contamination of soils, as opposed to conventional chemical and physical remediation technologies which are generally too costly and can harm soil qualities and human health. Plant species that grow in heavy metal-contaminated environments have been identified as hyperaccumulators in recent years. Researchers found that these plant species accumulate unusually high levels of heavy metals without affecting their growth and development. Due to their small biomass and slow growth, the majority of hyperaccumulators identified so far do not meet field phytoremediation requirements.

Heavy metal mobility and availability to plants may be affected by soil microorganisms. The rhizosphere, for example, is the volume of soil adjacent to and influenced by plant roots. Bacteria increased the uptake of Cd in *Brassica napus* and Ni in *Alyssum murale*.

Arbuscular mycorrhizal (AM) fungi are symbiotic associations found in both natural and heavy metal contaminated sites. They may stimulate phytoextraction by essentially improving plant growth and increasing the total metal uptake by the plant (Wang et al., 2007). Mycorrhizal fungi associated with these plants play a great role in the establishment of these plants on polluted soils. Heavy metal-solubilizing microorganisms are therefore a promising approach for increasing heavy metal bioavailability and uptake in soils contaminated with heavy metals. Bacteria that produce siderophores, indole acetic acid, and 1-aminocyclopropane-1-carboxylate deaminase and phosphate-solubilizing bacteria can also stimulate plant growth in contaminated soil. Bacteria that promote plant growth can help plants cope with external stresses, such as toxic metals. Conversely, little is known about the potential of heavy metal-resistant, heavy metal-accumulating, and heavy metal-solubilizing bacteria for phytoextraction of Pb and Cd from heavy metal-contaminated soils.

Over the last few decades, cadmium, nickel, zinc, lead, copper, etc., have become more prevalent due to smelting, mining, manufacturing, agricultural fertilizers, pesticides, municipal waste, traffic emissions, industrial chemicals, and industrial effluents. Many small and large manufacturing industries produce these wastes, such as metallurgical, electroplating, and metal finishing. They are contaminated with extremely high levels of heavy metals.

All major microorganism groups can be found in soil, which is a remarkably diverse habitat. Organic compounds are degraded and synthesized by the soil microbiota. As well as contributing to the cycling of plant nutrients, it also contributed to the weathering of primary minerals. Microflora, such as bacteria and fungi, and soil fauna, such as protozoa and invertebrate groups such as nematodes, mites, and earthworms, are found in soil.

Traditional methods of groundwater cleanup and conventional soil cleanup have been limited by their hazards, which has led to investigations into in-situ bioremediation, which uses microorganisms to destroy or immobilize pollutants in contaminated sites. Microorganisms are nature's original recycling units, converting toxic organic compounds into harmless products, often carbon dioxide and water. Conventional methods of removing heavy metals from wastewater and contaminated soil include chemical methods such as precipitation, neutralization, or physical methods such as membrane separation, electrodialysis, and activated carbon adsorption (Atkinson, 1998). Low concentrations may not be suitable for these processes. Low concentrations may not be suitable for these processes. A low concentration may not be suitable for these processes; in addition, these processes are more expensive and not environmentally friendly.

Researchers have thus developed a viable method to enhance decay and removal by enhancing microbial and associated biota (flora and fauna) within the ecosystems to degrade, accumulate and remove contaminants from industrial sites. To assess the metal tolerance microbes of industrial contaminated sites, we have examined soils. During bioremediation, microorganisms such as fungi and bacteria treat pollutants by converting them into less toxic or non-toxic compounds.

Bioremediation encourages the growth of microorganisms to degrade and remove contaminants by utilizing those as carbon and energy sources. The bioremediation process in use today is based on microorganisms native to the contaminated sites, which are encouraged to work by providing them with the nutrients and other chemicals they require for their metabolism. Researchers have been looking at ways to augment contaminated sites with non-narrative microbes, including genetically engineered microorganisms suited to degrade the contaminants in particular sites. Through the bioaugmentation process, bioremediation may become more effective in the future.

Our goal in this study is to identify, isolate, and characterize microbes that are resistant and able to solubilize heavy metal particles from heavy metal contaminated soils. To determine which strains would be useful in increasing plant biomass production and metal uptake by plants under unfavorable environmental conditions for improving the effectiveness of phytoextraction of heavy metals from polluted soils.

- ***Samples Collection***

This study focuses on isolating microorganisms that survive in soil contaminated with heavy metals. In this study, soil samples were collected from depths well specified at the selected sites, and the composite sample was formed by combining soil from several sites in the study area and mixing them thoroughly before analysis. The presence of microbes was studied in soil samples from three industrial sites. The soil surface (0–5 cm) and 20 cm below the surface was sampled using a sterilized spatula. A sterilized polyethylene bag was used to store the samples.

Until analysis, all samples were stored at 4°C. To remove large debris and vegetation from soil samples, a sieve (2 mm) was used.

Isolation of fungi from the soil samples:

Two methods were used to count soil microfungi, namely the soil dilution and soil plate method using media such as Potato Dextrose Agar and Czapek,s Dox Agar.

- ***Soil dilution plate method***

Microbial suspensions (10^{-1} to 10^{-10}) were prepared by suspending 1 gram of soil sample in 10 ml of double-distilled water. To isolate fungi, 10^{-3} and

10^{-5} dilutions were used. To each dilution, 1 ml of a microbial suspension was added to 15 ml of sterile Potato Dextrose Agar and Czapek's Dox Agar in sterile Petri dishes (in triplicate). Before pouring the medium into Petri plates, 1% streptomycin solution was added to prevent bacteria from growing. Incubation was carried out at 28°C in the dark. After three days, the plates were observed.

- ***Soil plate method***

The soil particles were sprinkled on the bottom of sterile Petri dishes, and then a molten cooled PDA and CZA agar medium (40–45°C) was added, followed by gentle rotation to disperse the particles in the medium. The Petri dishes were incubated at 28 ± 20°C in the dark for three days.

- ***Identification of soil fungi***

Several colonies were identified and counted after incubation. Cultures were identified using morphological characters such as colony morphology, texture, shape, color, diameter, and appearance of colonies, as well as microscopic features such as septations in mycelium, shape, and structure of conidia, and presence of specific reproductive structures. A PDA slant was used to preserve the fungi for future research. Using a compound microscope, conidia, conidiophores, and spore arrangement of fungal colonies dyed with lactophenol cotton blue were observed. The fungi were identified with the help of literature.

17.8.1 ISOLATION AND IDENTIFICATION OF HEAVY METAL RESISTANT BACTERIA

NAM media were used for the selective isolation of bacteria from soil. A distillery water dilution of 10^{-9} was applied to the collected effluent. Initially, microorganisms were isolated by plating serial dilutions of soils on nutrient agar media and incubating at 37–38°C for 48 hours.

Striking individual colonies of bacteria with a variety of shapes and colors on nutrient agar plates allowed them to be selected and purified. Biological isolates were maintained on nutrient agar and subcultured every four weeks. Four bacteria were isolated and purified for further study. Several distinct colonies of bacteria were isolated. Subcultured on the selective nutrient media and obtained as pure cultures, they were identified by their morphology and biochemical characteristics.

This study aims to identify and isolate microorganisms that survive in heavy metal polluted environments. Staining and biochemical tests were conducted to identify the isolated bacterium. Gram staining identifies the gram reaction of an organism.

To identify the motility of the organism, a hanging drop method was used. To identify the organisms, biochemical tests such as indole, methyl red, Voges Proskauer, citrate, Urease test, starch hydrolysis, catalase test, mannitol fermentation, etc. were used.

- ***Indole test (tryptone broth)***

Tryptone broth can be enriched with bacterial culture by adding a loopful. The broth should be incubated for 48 hours. Due to the growth of bacteria, the broth must become turbid after incubation. Testing organisms that grow in a clear broth is impossible. The Indole reagent (tryptone broth) should be poured into the broth culture. Positive results will have a red layer at the top. A negative result will have a yellow or brown layer.

- ***Methyl Red test***

To prepare MRVP broth, add a loopful of bacteria. Incubate for 3 to 5 days. Put 1 ml of broth into a sterile tube. Put three or four drops of methyl red into the original broth. If it is positive, a layer of red will appear on top of the broth. If it is negative, a yellow layer will appear. *Escherichia coli, Salmonella,* and *Proteus* are some examples of gram-negative bacteria that can be differentiated.

- ***Voges-Proskauer test***

The MRVP broth should be inoculated with bacteria. For three to five days, incubate. It is impossible to test the organism if the broth is clear after incubation. Barritt's reagent A (VP A) contains naphthol and Barritt's B (VP B) contains KOH. Test 1 ml of the culture from the MRVP broth. To the 1 ml of broth culture, add the entire contents of the VP A reagent (15 drops) and 5 drops of the VP B reagent. During a positive reaction, the medium will turn pink or red, indicating the presence of acetoin. During a negative reaction, the broth will not change color or will be copper-colored. Bacillus species like Enterobacter and Serratia show positive results.

- ***Citrate test (Simmons's Citrate slant)***

Stripe a loopful of bacteria on a citrate agar slant, but do not stab the butt. The incubation time for Bacillus species should be between 24 and 48 hours.

Keep the cap loose when you incubate. Positive results are indicated by a Prussian blue slant after incubation. Negative slants have no bacteria growth and remain green.

- ***Urease test***

Incubate a tube of urea broth with your test organism for 24 to 48 hours. Yeast extract, urea, and phenol red are the components of urea broth. The pH will rise if urease is present. A positive urease test shows a change from yellow to cerise (a light cherry color; pH 8.1 or greater). A negative test shows no change in color.

- ***Mannitol fermentation***

The majority of bacteria are inhibited by the 7.5% sodium chloride in Mannitol Salt Agar. The bacteria that can grow on this agar can be differentiated based on mannitol fermentation. Mannitol fermentation produces acidic products that change phenol red pH indicator from red to yellow. Incubate the MSA plate at 37°C for two days. If the color changes to yellow, the test is positive. A negative test does not change color.

- ***Starch Hydrolysis***

Test an organism by inoculating a starch plate with it. Allow it to incubate at optimum temperature for at least 48 hours.

Observe results after flooding the plate with iodine. A blue zone indicates no hydrolysis, while a clear zone indicates hydrolysis. B. subtilis showed positive results.

- ***Catalase test***

Take a small amount of the unknown microorganism from the agar slant or broth culture and place it on a glass slide.

Test the organisms with 3% H_2O_2 and check for the appearance of gas bubbles to indicate a positive result. Negative results indicate no bubbles. Bacillus spp. is used as the positive control and Streptococcus lactis is used as the negative control.

- ***Eosin Methylene Blue (EMB) Agar***

Gram-negative bacteria are selectively cultivated in this medium. In Levine's EMB agar, methylene blue inhibits gram-positive bacteria. Enteric differences will distinguish lactose fermenters from non-fermenters. Lactose non-fermenters will have cream-colored colonies after incubation. Lactose fermenters form pink colonies sometimes with dark centers. *E. coli* often has a greenish metallic shine.

A soil survey revealed that industrial areas have rich soil resources. Organic matter content and pH values are around 8. There are mostly granular fragments and debris in the soil, and its color is generally brown-red, while the soil textures are mostly light soil, medium soil, and deep soil.

- ***Concentrations of heavy metals in soil***

Environmental pollution caused by toxic metals is now a concern in the majority of major metropolitan areas. Different depths of heavy metal concentrations showed significant variations, which may be related to the soil properties influencing heavy metal availability at different depths. There was significant enrichment in 0–20 cm soil as compared to other soil layers, and relatively stable enrichment in deeper soil layers; the fundamental reason was the long-term industrial sewage irrigation.

Results show the presence of heavy metals like Copper, Chromium, Lead (Pb), Cadmium, and Nickel from the different sample sites, in comparison with the standard soil sample, Lead and Chromium concentrations surveyed in these areas were above the corresponding values of standard. Cadmium was a nonessential element of human health with high biological toxicity, it mainly accumulated in the surface soil, which entered the body through the digestive system, and the local environmental protection departments should take steps to control the Cadmium pollution.

All environmental crises are concerned with heavy metal pollution today. There are a variety of heavy metals found in soil as free metal ions, soluble metal complexes, as well as organically bound metals precipitated or bound to insoluble compounds, including oxides, hydroxides, and carbonates. In soil, they are produced by repeated use of sewage sludge, municipal waste, and animal waste, smelting and mining effluents, impurities in fertilizers, and the decomposition of air pollutants caused by the burning of fossil fuels. Due to their persistence in soils and slow leaching rate, heavy metals tend to accumulate in soils, making plants susceptible to them.

17.9 ANALYSIS OF PLANTS FOR TRACE AND HEAVY METALS BY ATOMIC ABSORPTION SPECTROPHOTOMETER

The plant species collected from different experimental sites were identified by the higher authorities they are identified as follows:

1. *Acacia arabica*
2. *Lantana camara*
3. *Datura inoxia*

4. *Calotropis procera*
5. *Prosopis*
6. *Thevetia nerifolia (kaner)*
7. *Polygonum*
8. *Ricinus communis*
9. *Ipomea fistulosa,* etc.

Of these, the most common such as *Datura inoxia, Calotropis procera, Polygonum, Ricinus communis,* and *Lantana camara* plants were selected for the present investigation for estimation of heavy metals.

These plants have adapted to stressful conditions induced by excessive levels of heavy metals in soil managing to grow even in such a hostile environment. *Datura inoxia, Calatropis lantana camara,* etc. are the plant species recognized as capable of tolerating high levels of metals from the soil.

17.9.1 HEAVY METALS ACCUMULATION IN PLANTS

Hazardous waste sites can be cleaned up using phytoremediation, an effective and inexpensive method. Our natural water resources can be restored and managed cost-effectively and ecologically by using plants to degrade, assimilate, metabolize, or detoxify contaminants.

In India, it is imperative to use phytoremediation technology to decontaminate polluted soils and water bodies. New technology and resources will be brought to India to solve environmental issues generated by industries.

According to the results, heavy metal concentrations in plants varied from plant to plant and species to species, in addition to soil levels of heavy metals and heavy metal contaminants.

Several studies have shown heavy contamination of agricultural soils and vegetation, with metal concentrations declining steeply with distance from the road, as observed in the present study. Heavy metals deposition to the ground is highly dependent on the presence and nature of vegetation. In the investigated area, metals are deposited on plants about four times faster than on bare soil. According to the present study, the concentration of metals in five cultivated plants was quite higher than the national standard for environmental quality, which is 50 mg/kg. In general, heavy metals in plants depended on the species and collection site. These plants have adapted to stressful conditions induced by excessive levels of heavy metals in soil managing to grow even in such a hostile environment. A heavy metal concentration value of the kinds of plants in different parts was like stem and leaves in the studied plants.

In Figures 17.3 and 17.4, P-1 Castor (*Ricinus* sp.), P-2 *Calatropis procera*, P-3 *Datura inoxia*, P-4 Polygonum, and P-5 *Lantana camara* was recorded.

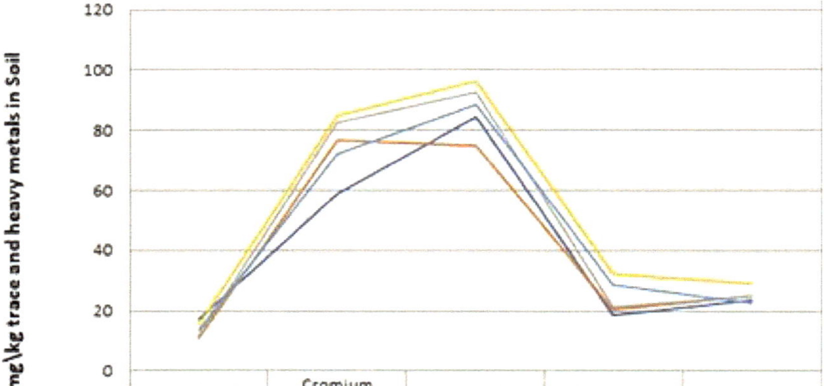

	Copper (mg/kg)	Cromium (mg/kg)	Lead mg/kg	Cadmium mg/kg	Nickel mg/kg
PB-1 castor	17.41	58.92	84.35	18.71	23.58
PB-2calatropis	10.79	76.5	74.71	20.52	25.01
PB-3 dhatura	11.59	82.5	92.48	21.53	25.03
PB-4 polygonum	15.46	84.71	96.12	32.48	29.12
PB -5 Lantana	13.82	72.3	88.4	28.76	22.76

FIGURE 17.3 Status of trace and heavy metals in the stem of plants.

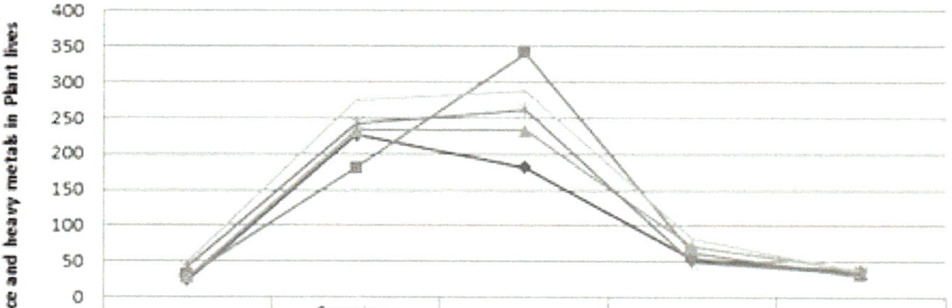

	Copper (mg/kg)	Cromium (mg/kg)	Lead mg/kg	Cadmium mg/kg	Nickel mg/kg
PL-1	23.59	226.5	180.9	53.8	37.36
PL-2	30.11	180.5	340.8	61.9	30.11
PL-3	28.58	232.5	232.8	70.5	38.5
PL-4	49.17	275.4	287.6	81.8	36.84
PL-5	42.56	242.7	262.2	49.4	34.77

FIGURE 17.4 Status of trace and heavy metals in leaves of plants.

Given the foregoing literature, it is programmed to make an attempt to study the nature of phytoremediation technique in the soils of the study area with help of selected terrestrial plants, namely, *Datura inoxia*, *Ricinus communis*, *Calotropis procera,* and *Polygonum persicaria.*

- ***Ricinus communis***

Castor oil plant or *Ricinus* communis is a species of flowering plant in the spurge family, *Euphorbiaceae*. The species belongs to the monotypic genus, *Ricinus*, and subtribe, *Ricininae*. A mixed pollination system favors selfing by geitonogamy, but can also be an outcrosser by anemophily or entomophily.

A castor bean is its seed, which is not a true bean. A native of southern Africa, the Mediterranean Basin, and India, castor is also widely cultivated in other tropical regions.

Castor seed produces castor oil, which has a variety of uses. There is approximately 40% to 60% oil in the seeds, and most of it is ricinolein. Ricine, a water-soluble toxin, is also found in low concentrations throughout the plant, including the seed.

Originally native to the southeastern Mediterranean Basin, eastern Africa, and India, Ricinus communis now thrives throughout tropical regions. Castor is an easily established invasive plant that often grows on wasteland in areas where the climate is suitable. *Ricinus communis* is a plant species also recognized as able to tolerate high levels of heavy metal concerning the lowest values of Copper content in plant parts Significant accumulation of Chromium in the mean value of 58.92 mg/kg was founded on the shoot and 226.5 mg/kg for leaves. Interaction effects of different areas on the accumulation of lead show the highest value of 84.35 mg/kg in plant shoots and 180.9 mg/kg in leaves and Cadmium was significant on a shoot in the value18.71 mg/kg with the difference of the value of 53.8 mg/kg. It seems that accumulated Nickel heavy metal has been scarcely transmitted to Ricinus plant shoot and according to the higher dry matter of leaves in the amount of 23.58 mg/kg that leaves 37.36 mg/kg a significant difference about accumulated nutrients in shoot and root can't be observed.

- ***Calotropis procera***

In ayurvedic formulations such as Arkelavana, *Calotropis procera,* a member of the *Asclepiadaceae* family, is used. Traditional systems of medicine have long known the medicinal potential of *Calotropis procera*, whose leaves are widely used in traditional medicine. Modern pharmaceutical compounds have always been derived from plants, plant extracts, and pure compounds isolated from natural sources. In traditional medicine, this plant is used to treat diarrhea, stomatitis, sinus fistulas, and skin diseases, while the leaf is used to treat jaundice. Numerous reports exist on the geographical distribution, habitat, and morphological characteristics of the plant. Yet no studies have been conducted on the leaves of this plant, which may contain ethnomedicinal drugs. *Calotropis procera* is a dominant and widespread xerophyte that grows in warm and urban environments. This species has a high capacity for taking and accumulating heavy metals into its tissues due to its ability to absorb and tolerate heavy metals without serious physiological damage.

The bioaccumulation decreased according to the order of leaf> stem > root for *C. procera.* The increasing trend of heavy metals for *C. procera* was in leaf, stem and root, Cr > Pb > Ni > Cd > Cu, in stem and in leaves, Pb > Cr > Cd > Ni > Cu. Among the heavy metals studied, Chromium (76.50mg/kg) in stem and lead (340.8 mg/kg) in leaves had the highest concentrations in organs of *C. procera* and Cu was the least bio-accumulated element.

So it was observed that the highest rate of Cr accumulation was in leaves related to the stem of Calotropis sp. Also, it was noticed that the mean of Cr accumulated rate in leaves was twice more than shoot.

The lowest value of copper accumulation was determined in stem < leaves when the plants were grown on soil with the most industrial pollution. The hyperaccumulator properties of some weed species allow them to survive in highly polluted soils and exclude metals from them. When compared with crops, weeds are commonly stress-resilient and can survive in soils polluted with heavy metals as well as adverse water and fertilizer conditions.

- ***Datura inoxia***

In the present investigation, another attempt has been made to exploit *Datura inoxia* for the remediation of industrial waste containing soil. The plant *Datura inoxia* is a suitable candidate for phytoremediation techniques, as it has lower biomass but accumulates more heavy metals from the soil.

Datura inoxia, also known as thorn-apple, downy thorn-apple, and Indian-apple, is a shrub belonging to the family Solanaceae. It is indigenous to Central and South America. Annual shrubs, *Datura inoxia* typically grow to a height of 0.6 to 1.5 meters. The leaves and stems of this plant are covered with tiny and soft hairs that give it a grayish appearance. When crushed or bruised, all parts of the plant emit a foul smell similar to rancid peanut butter, though most people find the scent of the flowers quite pleasant when they bloom at night. This observation was noted with interest. Datura species is a known heavy metal tolerant or heavy metal indicator plant and has been suggested as bioremediation of metal contaminated soils and groundwater. The heavy metal tolerant nature of *D. inoxia,* is one of the species frequently grown on the investigated site.

From the present studies, it is observed that heavy metal accumulation is in higher concentration leaves than stem. The occurrence of the metals in *Datura* leaf follows the accumulation pattern of Pb > Cr > Cd > Ni> Cu. As *Datura inoxia* is a plant species recognized to tolerate high levels of heavy metal, significant accumulation of Chromium in the mean value of 82.50 mg/kg was found on the shoot and 232.5 mg/kg for leaves. Interaction effects of different areas on the accumulation of lead show the highest accumulation value of 232.8 mg/kg in plant leaves and 92.48 mg/kg in shoots. It seems that accumulated Nickel heavy metal has been scarcely transmitted to *Datura* plant shoots and according to the higher dry matter of leaves.

- ***Polygonum persicaria***

The level of soil pollution has also been assessed by measuring heavy metal concentrations (Pb, Cd, Ni, Cr, and Cu) in the wetlands of the investigated sites. A sample of *Polygonum persicaria* plants was chosen because it grows in different areas of sites, in which the levels of pollution vary due to different factors.

Polygonum persicaria of family *Polygonaceae*, commonly known as curl top lady thumb, smartweed, dock-leaf smartweed, is an annual weed plant. Naturally grows in light (sandy), medium (loamy), and heavy (clay) soils. The plant mainly grows in semi-shade (light woodland) or no shade. It prefers moist soil to grow in heavy biomass.

The results of the research studies on the concentration of the present heavy metals from the sites show that the plant *Polygonum species* can be successfully used in plant biomonitoring, primarily because of its favorable ecological characteristics.

Through the assessment, it was observed that the concentrations of heavy metals in the above-ground parts of plant species showed quite a high accumulation. The results show that the concentrations of Pb accumulation are considered very high as compared to other plants. The maximum concentration of Pb (287.6 mg/kg) was measured from a polygonum leaves sample as compared to stem (96.12 mg/kg).

The results show the accumulation of heavy metals in polygonum sp. On the decreasing pattern of metals as Pb > Cr > Cd >Ni > Cu. found in leaves > stem.

Among the four plant species, the mean concentration of each metal accumulation in the vegetative part above the ground (shoot) was found in the order of Pb > Cr > Cd >Ni > Cu. From an industrial contaminated site. The maximum remediation of lead was found in the concentration (340.8 mg/kg) of metals accumulations in the leaves part of plants grown in contaminated sites as in *Calotropis* species and 275.4 mg/kg Chromium in *Polygonum* spp. While for Cadmium, (81.8mg/kg) and Nickel (36.84 mg/kg) accumulation was found in polygonum. The plants that were able to accumulate the lowest amounts of heavy metals in contaminated soil were Ricinus communis and *Calotropis procera.*

This case study analysis approves the use of the grown plants as a tool for the removal of heavy metals from the soil environment. As an alternative to remediation and decontamination, phytoremediation provides an esthetically pleasing result. As a result of these advantages, phytoremediation provides considerable environmental remediation efficiency, especially for industrially polluted areas.

It is observed that these plants are normally slow-growing and produce small biomass but they accumulate heavy metals in their plant parts and tissues lowering the number of metals in soil and controlling the movement of heavy metals through the soil profile.

17.9.2 IN VITRO CULTURE AND METAL TOLERANCE TREATMENT

This study aimed to determine heavy metal tolerance and accumulation of the weed species that commonly grow in industrial fields.

These plants can tolerate and accumulate toxic metal has never been tested scientifically and systematically. In the present study, we used the shoot culture of the plant of these species. *In vitro* cultures are ideal models for analyzing the effect of different metals from their deficiency to toxicity. Nodal explants of *in vitro* cultured shoots have been used in the study of heavy metals stress in *Bacopa monniera* and – *Holarrhena antidyzenterica,* and shoot cultures of jojoba have been used in the study of salt stress. Plant tissue culture provides an excellent tool to assess the natural tolerance and accumulation of plants for various metal salts and to study the mechanism of metal tolerance.

Experimental plants were induced and maintained on Murashige and Skoog, 1962 medium (MS) supplemented with (0.5–1.0) mg/l benzyl adenine to assess the metal accumulation and tolerance. Different concentrations of heavy metals (Cd, Cu, Cr, Pb, or Ni) were added to the culture medium as contamination agents.

- ***Calotropis procera and Polygonum species***

Induction and maintenance of seedlings of *C. procera* and shoot cultures of polygonum seem to favor high levels of BAP and lower levels of NAA. Cytokinin is generally included in a culture medium to stimulate cell growth, induction, and stimulate the growth of shoot apices/ shoot tip cultures and seedlings growth. In the preliminary experiment effect of cytokinins was tested alone and in combination. Maximum (90–80%) number of bud break and initiation of shoots were reported in the medium containing BAP alone (0.5–1.0mg) in the control condition. About 1–2 shoots were developed in every culture. The maximum length of shoots was observed 1–2 cm in the medium containing BAP and NAA. A higher concentration of BAP gives somewhat short and weak shoots with the formation of non-fragile callus.

Treatment of heavy metals (Cr, Cd, Cu, Ni, and Pb) content in different concentrations for seed germination and shoot establishment from seedling derived axillary and apical meristem. The result showed 80% of seed germination and 1–2 shoot induction in every medium supplement with different concentrations of heavy metal. It is reported that there was a progressive decline in the percentage of seed germination in calotropis and shoots induction in polygonum with increasing concentration of metals in the medium. Based on the results above, shoots were induced on MS medium containing a higher concentration of metals, and the evidence suggests there was a difference in heavy metal accumulation between seedlings and shoots derived from contaminated and uncontaminated sources

This study's results showed that the shoots initiated from contaminated sources were tolerant of chromium, cadmium, copper, nickel, and lead despite concentrations of 300 mg/l, for *C. procera* and Figures 17.5–17.9 for *Polygonum. C. porcera* may accumulate heavy metals and be more tolerant of the same metals added to the media, despite their growth inhibition. In the species tolerant or accumulator to toxic metal ions *in vitro* screening of plants tolerant to toxic ions contained in the soil may lead to the production of plants that are better altered to environmental pollution and can enable better management of degraded soil.

- ***Datura inoxia and Ricinus sp.***

The cultures of *Datura inoxia* and *Ricinus* sp. were initiated in the medium MS with 0.5 mg/l BAP. For the study, the heavy metal effect experiment the medium was supplemented with appropriate concentrations of heavy metal salt solutions. Initiated shoots were transferred to a fresh medium containing different concentrations of every heavy metal salt increasingly for each subculture. The cultures were kept in a growth chamber at 22–23°C, with fluorescent mercury illumination (7500 W/m2) for 16 h/day.

After 30 days of growth on media with one concentration, the cultures were transferred to a medium with another concentration. After about 5 weeks, an assessment of culture development was started. The assessment involved measurements of length and number of shoots and evaluation of culture quality. The growth rate of the shoots growing in the presence of the metals was recorded after 30 days.

The cultures show 90% of shoot induction of 1–2 shoots numbers of 2–1 cm length in the medium absence of any metal. It is observed that the cultures of *Datura* and *Ricinus* also show tolerance to chromium, cadmium, copper, nickel, and lead sources even at the concentration of 300 mg/l. The growth of the shoots shows affected by the concentration of heavy metals with the manner of increasing concentrations of all metals supporting 90% growth, inhibition increased with increasing concentration of the metals.

At 300 mg/l concentration of heavy metals, 50–60% of shoots growth was observed from treatment within all five metals. Single node explant has been used in the study of the role of metal in adaptation mechanisms under stress conditions. Such *in vitro* system would also be valuable material for research on mechanisms of metal tolerance in plants sensitive to toxic metal ions. In the present investigation, shoot explants were cultured in media containing various concentrations of Cr, Cu, Ni, Pb, and Cd. The shoots show 80% shoot establishment with one shoot of one cm length in the control medium while the plant can tolerate all the metal at the concentration of 300 mg/l studied. The results show 40–60% of shoot establishments and grow easily and healthy having 0.5–1.0 cm length.

The shoot cultured of *Datura* and *Ricinus* on the medium with heavy metals remained green, even those that did show little further growth. Results show that affected plantlets grown from cultured segments. On medium with 100 to 300 mg/l of metals shoots developed were too small to measure.

Plantlets developing on medium containing the equivalent values for plantlets grown on medium with 300 mg/l metals were 38, 50, and 40%. At concentrations of 200mg/l and 300mg/l, heavy metals strongly inhibited shoot development (number and length of shoots) and reflected an increase in chlorosis and browning. As the concentration of the studied ions in media increased, their toxic influence was enhanced.

High concentrations (100 mg/l – 300 mg/l) of Cd, Ni, were found to be more toxic than those of Cr and Cu and Pb for shoot growth (Tables 5 and 6). The results showed that *Datura* and Ricinus were able to *tolerate* heavy metals and to be more accumulator plant to the same metals added to the media, considering the growth inhibition.

Repeated passaging of these plants to media containing different heavy metals (about 12 months of culture), resulted in the production and

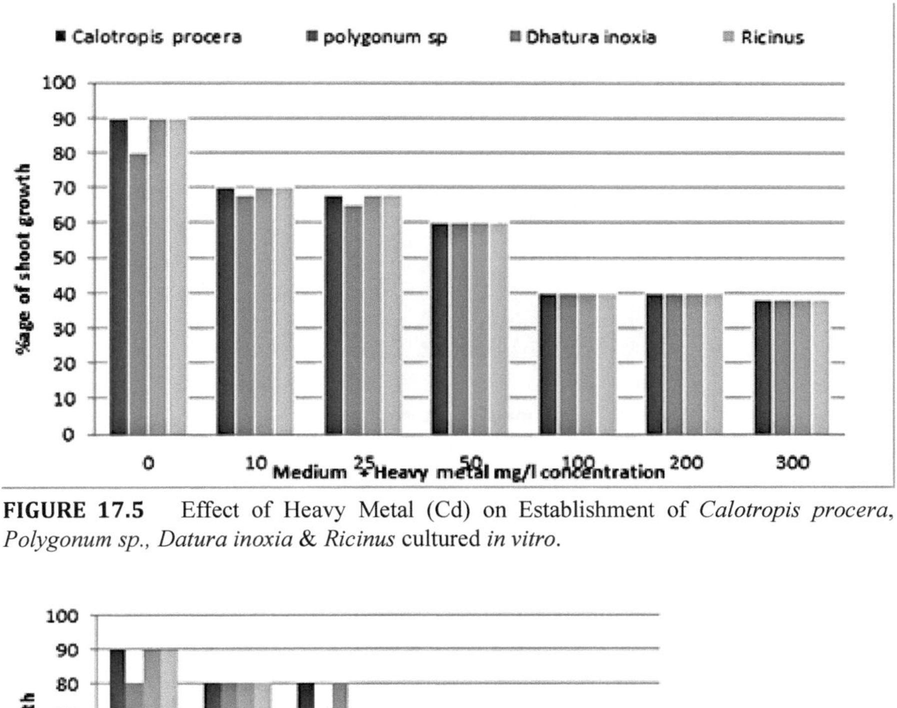

FIGURE 17.5 Effect of Heavy Metal (Cd) on Establishment of *Calotropis procera*, *Polygonum sp.*, *Datura inoxia* & *Ricinus* cultured *in vitro*.

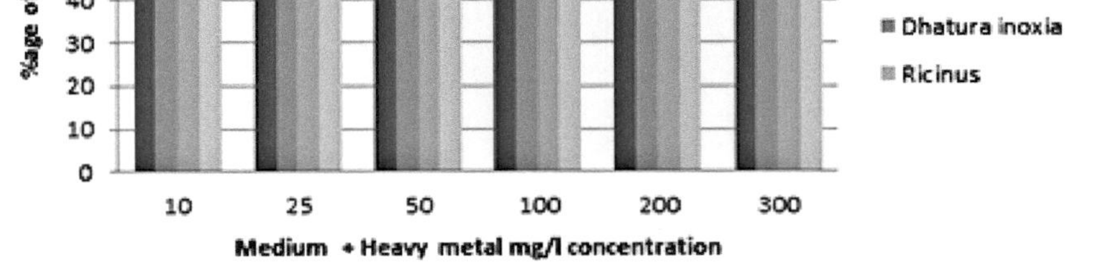

FIGURE 17.6 Effect of Heavy Metal (Cr) on Establishment of *Calotropis procera*, *Polygonum sp.*, *Datura inoxia* & *Ricinus* cultured *in vitro*.

establishment of a line of cultures, which under heavy metal stress conditions (in media with heavy metals) regenerated better than cultures grown earlier on control media. Plants derived from those cultures (grown with metals), were characterized with more rigorous development, particularly in media containing higher concentrations of toxic ions in the case of Ricinus, than control plants (grown earlier without metals). As *Datura* and *Ricinus* plants

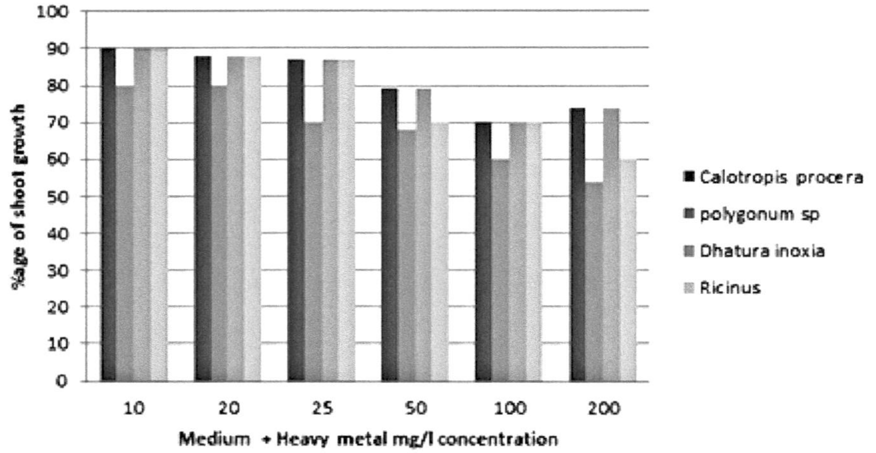

FIGURE 17.7 Effect of Heavy Metal (Pb) on Establishment of *Calotropis procera, Polygonum sp., Datura inoxia* & *Ricinus* cultured *in vitro*.

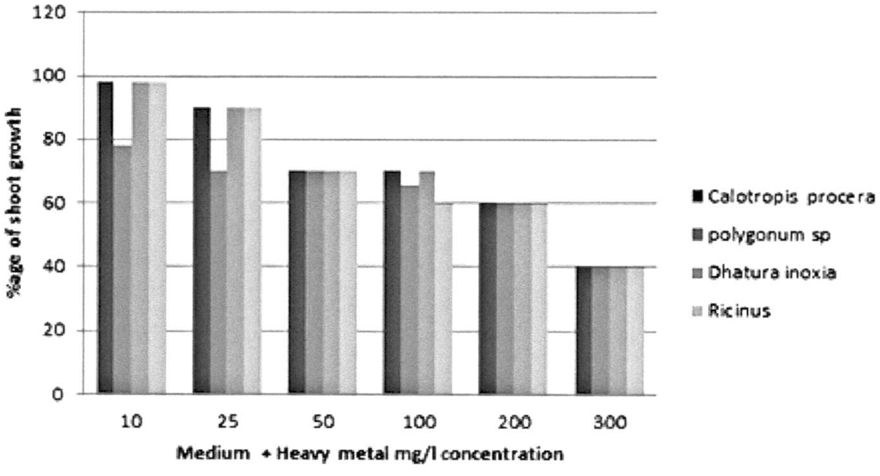

FIGURE 17.8 Effect of Heavy Metal (Ni) on Establishment of *Calotropis procera, Polygonum sp., Datura inoxia* & *Ricinus* cultured *in vitro*.

are known as pollution indicators and also recommended as natural remediators the cultures are tolerant to every metal were present in the polluted sites, also more tolerant to every ion. Thus, induction of tolerance to toxic metal can result in reduced sensitivity to other heavy metals. These types of heavy metal stress-tolerant plants produced *in vitro* preserve a better tolerance to toxic metal ions in the next few generations, also after transfer to natural conditions (Ali, Srivastava, & Iqbal, 1999; Stefani, Arduini, & Ounis, 1991).

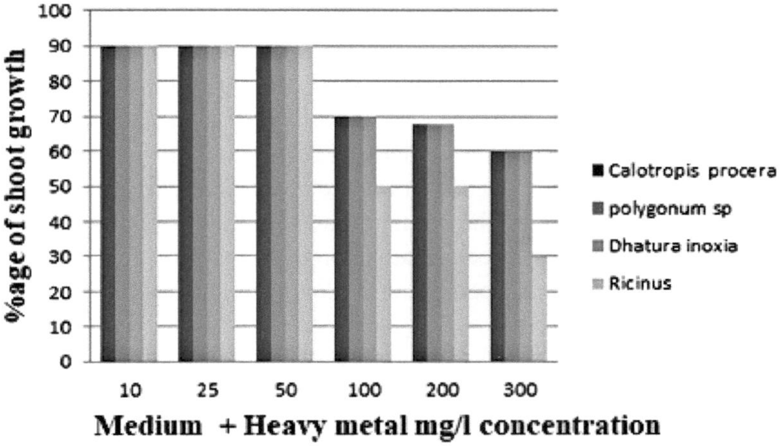

FIGURE 17.9 Effect of Heavy Metal (Cd) on Establishment of *Calotropis procera*, *Polygonum sp.*, *Datura inoxia* & *Ricinus* cultured *in vitro*.

In the present, *in vitro* studies, it is observed that cadmium accumulation in all four plant species shows equal effects. These plants have a potential of Cd tolerance even in the concentration of 300 mg/l. In the case of chromium heavy metal *in vitro* cultures shows a higher concentration of accumulation observed in all four plants except *Ricinus* species while *Calotropis* and *Datura* show the best chromium tolerance potency. *In vitro* culture of Polygonum shows the minimum percentage of establishment in Pb containing medium while *Ricinus* species shows the lowest percentage of growing tolerance in Ni and Cu heavy metal-containing medium. (Figures 17.5–17.9). Studies *in vitro* have shown that shoots produced in cultures on media with different heavy metals proved to show more accumulation than in nature obtained from the soil. Such plants would also be treasured material for studying mechanisms of plant sensitivity to toxic metal ions.

In vitro selection of plants that accumulate toxic ions contained in the soil may lead to the production of plants that are better adapted to environmental pollution and can enable better management of degraded soil (e.g., industrial areas and highways).

There have been numerous in vitro studies examining the effects of high levels of heavy metals on plants that are tolerant of industrial pollution or sensitive to it (El-Aref, Hamada et al., 1998, 1999). The selection of plants in natural conditions of environmental pollution or in vitro may result in the selection of clones adapted to accumulate toxic metal ions.

In the present in-vitro studies, it is concluded that the plant's species grown in the studied polluted areas have a maximum (300 mg/l) tolerance capacity against the heavy metals present in the soil.

From the present study, it was found that the heavy metal accumulation was more in leaves when compared to the stem and roots of four experimental plants. The findings show that the *Datura inoxia* plant has up taken 80–85% of heavy metal from contaminated soil which will have direct use to mobilize toxic metals from the contaminated soil environment. It was noted that *Datura inoxia* is a good candidate plant for phytoremediation implementation. *Datura inoxia* is a weed plant *Datura* Linn. (Family: *Solanaceae). a* genus of poisonous herbs.

Datura is a heavy metal tolerant plant that can row at ease in metal-polluted sites. The hyper accumulative nature of plants has been exploited for the recovery of toxic heavy metals from contaminated ground.

Thus, *in vitro* screening of plants accumulated the toxic ions contained in the soil may lead to the production of plants that are better adapted to environmental pollution for better management of contaminated soil.

Researchers found that seeds and shoots of studied plants collected from studied contaminated sites could accumulate cadmium, chromium, copper, lead, and nickel, and could therefore be used in metalliferous industrial waste revegetation programs.

There has been recent interest in metal tolerant plants as a simple and economical solution to many of the problems encountered in metalliferous polluted lands. Metal concentrations in the medium increased significantly with the presence of metals. On average, the concentrations of heavy metals in shoots grown in tolerant medium reached 300 mg/1. However, in vitro cultured shoots demonstrated high tolerance for all concentrations since the accumulation of heavy metals in the field was high.

17.10 RHIZOSPHERE STUDIES

Remediation of heavy metal-contaminated soils will be improved with microbial populations specifically adapted to high concentrations of heavy metals. The present study suggests that the microorganisms present in the soil investigated may also prove useful for the remediation of heavy metal contaminated soils.

The study aimed to isolate and identify microbes that can survive in heavy metal contaminated environments. In previous samples, the total number of heavy metals was detected. There were isolated bacteria and fungi from this study, which showed positive results for the biochemical tests and the organisms identified (Table 17.1).

17.10.1 IDENTIFICATION OF BACTERIA

Results show that there is the presence of *Staphylococcus aureus* which is gram-positive cocci and shows positive motility test, Coagulase test, and Mannitol fermentation test.

Positive results have been observed for the biochemical tests like Citrate, Voges Proskauer, Urease and Catalase tests and the organism also shows positive reactions in the EMB agar. The organism identified is the *Enterobacter* species (Table 17.2).

Results showed positive for the biochemical tests like citrate urease, catalase tests, motility test, etc. The organism also grows well on Pseudomonas agar. Thus, the microorganism is identified as *Pseudomonas* species (Table 17.4). It is observed that the positive results for the biochemical tests like starch hydrolysis, motility test, and catalase test. The organism thus identified is the *Bacillus* species (Table 17.3).

TABLE 17.1 *Staphylococcus aureus* Showed Positive Results for the Mannitol Fermentation Test and in Motility Test

Biochemical Tests	Result
Organisms	Staphylococcus aureus
Gram staining	(+) coccus
Motility	+
Emb agar	-
Mannitol salt agar	+
Pseudomonas	_

TABLE 17.2 Positive Results for the Motility Test, Voges Proskauer Test, Citrate, Urease Test, and In EMB Agar

Biochemical Tests	Result
Organisms	Enterobacter
Gram staining	(-) rod
Motility	+
Starch	-
Catalase	+
Emb agar	-
Mannitol salt agar	-
Pseudomonas agar	-

TABLE 17.3 Presence of the *Bacillus Subtilis* Showing Positive Tests for Motility Test, Catalase Test, and Starch Hydrolysis

Biochemical Tests	Result
Organisms	*Bacillus subtilis*
Gram staining	(-) rod
Motility	+
Starch	+
CATALASE	+
EMB AGAR	-
MANNITOL SALT AGAR	-
PSEUDOMONAS AGAR	-

TABLE 17.4 Presence of *Pseudomonas aeruginosa* Showing Positive Results for Motility Tests, Citrate, Urease, and Catalase Test and the Animal Grows on *Pseudomonas* Agar

Biochemical Tests	Result
Organisms	*Pseudomonas aeruginosa*
Gram staining	(-) rod
Motility	+
Indole methyl red	-
Voges Proskauer	-
Citrate	-
Urease	+
Starch	+
Catalase	-
EMB agar	+
Mannitol salt agar	-
Pseudomonas agar	+

17.10.2 IDENTIFICATION OF FUNGI

Any soil's fungi diversity is influenced by several factors such as organic content, pH, and moisture. According to the physicochemical analysis of soil, pH ranged from 5.1 to 7.5 while soil textures were related to fungal population and diversity.

Several colonies were identified and counted after incubation. The cultures were identified at the generic level based on macroscopic characters such as colony morphology, texture, color, shape, diameter, and appearance of the colony and microscopic characteristics viz., presence of specific reproductive structures, septation in mycelium, shape, and structure of conidia and presence of sterile mycelium. Various microorganisms were isolated from industrial residues soil in various polluted areas, where the concentration of heavy metals is very high. The present study investigated the effectiveness of isolated fungi from contaminated soil samples and isolated fungi treated with heavy metal at different concentrations. Many different fungal species were isolated from collected soils samples.

It is observed that *Fusarium* sp., *Aspergillus niger,* and *Mucor* sp., were present and dominant in all collected soil samples of drain and selected for resistance analysis. The presence of *Aspergillus niger, Fusarium* sp., *Mucor* sp., have shown that these species were able to resist heavy metals tolerance.

Interactions between fungi, plants, and heavy metals were expatriated in many ways. Because of their resistance and detoxification mechanisms to heavy metals, fungi can leach, absorb and transform heavy metals efficiently. Though heavy metals are very toxic to the fungal population and harmful to its processes, some special fungi have been used for the treatment of industrial waste materials and the remediation of soils polluted by heavy metals. The use of fungal biomass and activity to evaluate the pollution situation and ecological risk of heavy metals in the environment is also significant. A filamentous fungus can also be highly resistant to metals if it can accumulate these elements. This study showed that fungi exhibit high levels of heavy metal resistance, indicating their potential as bioremediation agents, as well as helping the surrounding plants to absorb more metals. The low cost, high efficiency, and environmental friendliness of microbial metal bioremediation make it a very effective and feasible strategy. Metal-microbe interactions have been studied recently for their application to metal accumulation and detoxification. A growing population and anthropogenic sources are responsible for the increased levels of metals. Heavy metals over the normal distribution in the soil raise concerns about the soil's quality.

17.11 SUMMARY

The presence of plants and microorganisms in the rhizosphere appears to enhance contaminant removal. Phytoremediation, even though relatively slow, is eco-friendly, cheap, easy to perform, and does not require soil removal, so it is an in-situ method. The hyperaccumulator can also harvest and extract precious metals, such as gold, zinc, and chromium, as phyto-extraction. An important factor in the success of phytoremediation is the identification of a plant that is tolerant and suitable for each area, one that can accumulate high metal concentrations. Microbes have evolved several mechanisms to tolerate the uptake of heavy metal ions under metal-stressed conditions. Thus, this study highlights the presence of plants, bacteria, and fungi that can survive in a highly heavy metal-polluted environment.

Using weeds as the test plants, we chose bacterial-assisted phytoextraction due to their fast growth and high biomass. In vitro experiments have demonstrated that species facilitate growth, but cannot increase it. In nature, bacteria must be able to quickly colonize the root system during the growing season to promote plant growth.

The microbes that colonized and developed in the soil could, however, promote plant growth in the rhizosphere soil. Several types of research have demonstrated the importance of bacterial inoculation for plant growth and heavy metal accumulation in industrial areas.

Phytoremediation should utilize plants that are fast-growing and capable of producing large amounts of biomass. Unfortunately, many plants that can do phytoremediation lack these characteristics. Therefore, it is a good idea to be able to produce new plants through genetic engineering. The creation of a plant that combines both of these qualities can be achieved by adapting two methods. The first method is to increase the biomass of plants that are already suitable for phytoremediation, while the second method is to enhance the remediation ability of plants with a large amount of biomass, such as populous and corn. As a result of the literature review, it was revealed there has been tremendous progress in this area in recent years. Genomic technology allows us to accomplish more.

Environmental pollution in Bhopal is becoming more severe, and this subject has been extensively explored. In the long run, bioremediation or phytoremediation of the environment is a mandatory task for the government. By combining phytoremediation with other approaches to achieve these objectives, phytoremediation will become a sustainable mechanism and play a crucial role in the remediation of the present environment. Tolerant and accumulator plants would be suitable for remediating soils with high levels of organic and metal pollution. As these plants have extensive root systems, they can reach far deeper and over a larger area of ground. Moreover, it allows for a larger rhizosphere microbial community. The removal of contaminants would also be enhanced if the surrounding soil interacted with it. Additionally, these plants have a high rate of regeneration, which increases their chances of surviving in harsh conditions. This makes them suitable for phytoremediation.

17.12 EXERCISES

Objective Response Questions

1. **Metal is extracted from an ore by: ________.**
   a) Biosorption
   b) Bioextraction
   c) Bioleaching
   d) Microbial extraction

2. **It is not recommended to burn solid waste for the following reasons: __________.**
   a) It is very costly
   b) It requires a lot of space
   c) It causes several environmental issues
   d) It requires modern technologies

3. **__________ is the method of disposing of organic wastes in which solid wastes are converted to residues and gaseous products through combustion.**
   a) Incubation
   b) Incineration
   c) Incarnation
   d) Incarceration

4. **Bioaugmentation involves __________.**
   a) bioventing
   b) eliminating sludge

c) addition of microbes to a cleanup site
d) plants usage for bioremediation

Practice Work

Qu 1. Describe qualitative characterization of contaminated sites for heavy metal detection.

Qu 2. Explain the method of sampling plant and Soil for heavy metal detection.

Qu 3. Explain the principle of AAS.

Qu 4. Describe various in vitro techniques for metal Tolerance treatment.

Qu 5. How will you modify the rhizosphere for phytoremediation?

Qu 6. Explain the techniques of identification and preservation of fungi.

Qu 7. Describe the various biochemical test for rhizosphere microorganisms.

Qu 8. Explain the public relevance of in vitro studies and phytoremediation

Qu 9. Explain the methodology of soil sampling.

Qu 10. Write a short note on the Kjeldahl digester.

REFERENCES

Murashige, T. and Skoog, F. (1962), A Revised Medium for Rapid Growth and Bio Assays with Tobacco Tissue Cultures. Physiologia Plantarum, 15: 473–497. https://doi.org/10.1111/j.1399-3054.1962.tb08052.x

Zhuang, P., Yang, Q., Wang, H., & Shu, W. (2007). Phytoextraction of Heavy Metals by Eight Plant Species in the Field. *Water Air and Soil Pollution, 184*(1–4), 235–242. https://doi.org/10.1007/s11270-007-9412-2.

CHAPTER 18

Implementation of Phytoremediation and Heavy Metal Recovery

18.1 HEAVY METAL RECOVERY

Heavy-metal phytoremediation harvests of hyperaccumulator biomass must be considered hazardous waste that should be contained or treated appropriately before disposal. The leaching of heavy metals from biomass could be used to detoxify the material and convert it to a useful fertilizer or mulch. Most plants reflect their environment in their elemental profile, but some accumulate high concentrations of certain metals. These plants are known collectively as hyperaccumulator plants. The above-ground biomass of hyperaccumulator plants contains high concentrations of metals, such as Ni, Mn, or Zn. These plants can be used for prospecting (finding orebodies), phytoextraction (cleaning soil), phytomining (producing metals, metal products, or catalysts), or agromining (the extension of phytomining into agriculture). A material sometimes referred to as "bio-ore" can be produced by growing hyperaccumulator plants, harvesting them, and then burning the biomass.

Pollution in India is caused primarily by urbanization, extensive industrialization, exploitation of natural resources, and population explosion. Pollution of air, water, and soil in India is caused by toxic wastes and industrial effluents, soil contamination, and degradation of wildlife habitats. The conservation of soil and remediation of polluted soils are important thrust areas in developing countries like India. There are severe issues with soil erosion ranging from loss of topsoil in 130.5 million ha to terrain deformation in 16.4 million ha. A study of soil loss under different land-use options found that trees and grass coexisted to produce the least amount of soil loss. Agricultural and food products are contaminated by pesticide residues. Refineries

Bioremediation and Phytoremediation: Technologies for Toxic Pollution
Ashwini A. Waoo, PhD (Author)

generate a great deal of oily sludge, which is highly harmful to the environment. Heavy metals cannot be degraded like organic compounds, so they must be removed for cleanup. The conventional remedial technologies are very expensive and affect soil fertility, which causes ecosystem imbalances.

The phytoremediation method, on the other hand, is a cost-effective, environmentally friendly, and esthetic approach, which can be used in developing countries like India. The technology of phytoremediation is still not commercially available in India. The phytoremediation approach consists of selecting non-edible, disease-resistant, and tolerant plants with high biomass so that contaminants will not directly enter the food chain. Today, various methods for increasing phytoextraction and utilizing by-products have been developed. As a result of this phytoremediation process, many biomasses need to be disposed of and managed appropriately (Sharma and Pandey, 2014).

Phytoremediation has been studied extensively using plant species that are hyperaccumulators of heavy metals.

18.2 RE-EXTRACTION OF METALS FROM HYPERACCUMULATORS

The major research area in this field is dealing with creating some new techniques and for incineration of accumulator plants and re-extraction of heavy metals from them, which can be used for industrial purposes. Following the recovery of metals from phytoremediation, the remaining biomass is an excellent source of organic matter. But biomass generated through the phytoremediation process contains very high concentrations of heavy metals. Thus, proper heavy metal recovery methods must be adapted and assured for each type of biomass depending on the concentration and types of metal present.

18.2.1 THERMAL TREATMENT/INCINERATION

Thermal treatment refers to the process of destroying organic matter from the hyperaccumulator plant material and recovering metals from ash in the form of oxides. Then metals can be recovered by hydrometallurgical processes and bio-hydrometallurgical processes, ion exchange (Banerjee et al., 2013; Sivrikaya et al., 2011; Mizoguchi et al., 2010; Kır et al., 2006), flotation (Min et al., 2012), magnetic field (Jainae et al., 2010) or electrolysis (Keller et al., 2005), and bioelectrochemical techniques (Modin et al., 2012).

During combustion, plant biomass requires relatively little air due to its high oxygen content. Besides, the relatively low sulfur content of the biomass is an advantage because there are no harmful gases released in the atmosphere during the combustion (Tan and Lagerkvist, 2011). Incineration at very high temperatures (over 600 °C) can be detrimental because some heavy metals, including those that are most toxic, are volatile (Nzihou and Stanmore, 2013). Metal volatilization can be exploited to recover metals since these elements condense in the cooler areas of the oven and are absorbed by fine particles of ash or by filters. A higher metal concentration by mass of ash makes metal recovery more efficient.

A metal's properties during the incineration process are typically determined by two factors: (a) evaporation in combustion areas and condensation in lower temperature zones, (b) physical adsorption on the specific surface. Unburned carbon is used to determine the surface area of ash particles by scanning electron microscopy (SEM). The adsorption of volatilized metals depends on this property. To minimize volatile metal releases into the atmosphere, the incineration system must be equipped with ash particle filtration and retrieval (filters and cyclones). Mercury, selenium, and arsenic are the most volatile metals in the incineration system.

The element mercury is highly volatile and almost entirely vaporizes as HgO and $HgCl_2$.Mercury in its oxidized form is easily collected from the air pollution control system. By comparing pyrolysis with incineration, Keller et al., proved that pyrolysis provided better results by thermal processing the biomass under reducing conditions.

These statements are conditional upon the heavy metal content of ash, which must be below the maximum permissible level if they are used as an amendment. Phytoremediation uses the shoots of Salix trees to recover metals (especially copper and zinc).

18.2.2 HYDROMETALLURGICAL PROCESSES

Ash leaching is an extraction technique for heavy metals that generates precipitated metals that can be quantified. The solutions used for leaching ash must be environmentally friendly, efficient, cheap, and capable of regenerating.

The extraction of heavy metals from leaching depends on such factors as the solubility and availability of the metal. Solubility can be influenced by pH, the chemical form of inorganic species, organic matter, and its reducing properties. Most metals from waste indicated a higher solubility in acid solutions]. As Singer et al., demonstrated, the amount of aluminum extracted by citric acid and temperature were significantly influenced by the acid concentration and the temperature at which the acid was used. A study by Machado et al., is based on the recovery of Ni and Zn in a multicomponent solution by precipitation as alkali metal hydroxide (Meawad et al., 2010).

To increase the solubility of metals from ores, Hoque and Philip proposed the use of microbes to convert insoluble metal sulfides into soluble metal sulfates. A microbe oxidized copper sulfate by oxidizing copper sulfide, for example. Solid residues were removed from the aqueous phase while metal ions were concentrated. Using the same technology, solid metals can be converted to water-soluble forms by microorganisms. This technology, called bio-oxidation, is used for the microbiological oxidation of metal-containing minerals to be extracted.

18.2.3 ELECTROCHEMICAL PROCESSES

The advantage of electrochemical processes for recovering heavy metals is that recovery depends on the reduction potential of the metal. To attain a high degree of purity for every metal, it is necessary to optimize the factors that affect the deposition of metal on the electrode (pH, electrolyte concentration, electrolyte bath temperature, and metal species).

Electrolysis is a common method for the extraction of metals in metallurgy, such as zinc, copper, nickel, etc. Every metal has a specific ion discharge potential that corresponds to the minimum potential at which an ion electrode can discharge continuously and visually (substance discharge). A substance with more ions undergoes successive discharges as the ions reach the potential of each download.

As Machado et al., in 2011 point out, since Ni and Zn have only a 0.25 V difference in deposition potential, separate recovery by electrolysis by these metals may be compromised, with the risk of metal co-deposition.

The electrolyte was then used to test whether the two metals could be removed by precipitation. As a result. Kirkelund et al., conducted the study to remove Cd from residual plant material by electrochemical methods, based on the principle of electromigration of ions in solution. Researchers optimized the process by using membranes to exchange anions and cations. Another method for optimizing results for metals extraction was developed by Modin et al., who used a bio-electrochemical process to recover copper, lead, and zinc from dilution solutions. In the anode chamber, the microorganisms from sewage sludge were inoculated and a nutrient solution was circulated. This method reduces the energy consumption for the metal discharge and improves the selectivity of metal extraction (Won et al., 2013).

18.3 SCOPE OF PHYTOREMEDIATION TECHNOLOGIES

The phytoremediation process is slower than mechanical or chemical clean-up methods and is also confined to the depths within the reach of the roots. Phytoremediation is only effective in low to medium pollutant concentration areas, and contamination in the shallow upper layer of soils where the roots of plants can easily access the pollutant. Phyto-remediation can also be used as a 'finishing' or 'polishing' step in combination with other clean-up technologies for better results. Phytoremediation technologies have been applied for remediation through metals, explosives, pesticides, solvents, polyaromatic hydrocarbons, and crude oil. Generally, Plants can also be used to remediate pollutants in water streams as well as groundwater. Extensive research has been done for finding the trees that allow absorption of contamination at greater depths. In the case of groundwater firstly pumping the water out of the ground by mechanical means and then using surface plants to treat the contamination of that water.

Further research must be conducted on the effects of bioaccumulation and biomagnification in the food chain, for example, if insects or small rodents consume plants containing absorbed pollutants that are then consumed by larger mammals. There is also a need for studies to determine if it is released at the time of leaf fall in the autumn or during the time when firewood or mulch from trees is used.

Biotechnology advancement strategies for standardizing phytoremediation led to the development of molecular mechanisms for the process. The research focused on analyzing various plants concerning phytoremediation potential. Use of bioinformatics tools in the compilation of all the aspects related to phytoremediation, such as genes responsible for phytoremediation, related products, and their mode of action. Industrially contaminated soil areas were assessed for their metal tolerating microbiome and metal removal efficiency of plants from polluted soils. For survival under metal-stressed conditions, microbes and plants have evolved various metabolic pathways to accumulate, tolerate and uptake the heavy metal ions. Bioinformatics study highlighted the presence of plants and microorganisms that survived in a highly heavy metal-polluted environment.

Since the industrial revolution, the amount of toxic heavy metals on the globe has increased dramatically. Primary sources of this pollution are the mining of metalliferous ores, municipal wastes, fertilizers, and pesticides. In addition to sites polluted by human activity, natural mineral deposits with huge quantities of heavy metals are present in many regions of the biosphere.

All living system requires special mechanisms for transport and handling toxic metals to keep them free from them. Severe toxicity and health concerns occur in humans due to heavy metal pollution by soil and water contamination and due to industrial exposure. Some heavy metals are essential to us in trace quantities but are toxic at their higher concentrations than that of the threshold value. To recover from environmental pollution, many remediation technologies have been already there to treat polluted soil, mainly mechanical or physio-chemical-based remediation methods.

Bioremediation processes are very much cheap and feasible as compared with physicochemical methods such as ion exchange, electrochemical treatment, precipitation, evaporation, and reverse osmosis for toxic metal removal techniques since they can have a lower cost and low efficiency at low metal concentrations. Several biological organisms can be used to remove metal from wastewater, such as yeasts, bacteria, and seaweeds. Microorganisms that are capable of growing in the presence of heavy metals could be used in the treatment of wastewater, where microorganisms directly affect the treatment of wastewater, and their growth is affected by heavy metal toxic effects. Major metal resistance mechanisms in microbes include precipitation of heavy metals as their salts in the form of carbonates, phosphates, or sulfides, cysteine-rich proteins, volatilization via methylation or ethylation, etc.

The mechanisms of plant defense against industrial pollution are relatively poorly studied. For a long time already, in vitro or growth chamber culture has been used to investigate responses of plants to toxic metal ions. Researchers have selected plants tolerant to some heavy metals through plant culture in vitro. Studies in vitro have shown that nodal culture of poplar and birch, produced in cultures on media containing aluminum, proved to be more tolerant than standard culture media without aluminum. Heavy metal accumulators are mainly divided into three groups based on their tendency to accumulate different metals. Approaches for phytoremediation focused on the identification of natural hyperaccumulators or developing genetically engineered plants having that capacity. Introducing metal-binding proteins and peptides into plants is a compelling strategy.

Heavy metal mobility and availability to plants can be affected by soil microorganisms. It depends on the nature of the available rhizosphere, i.e., the volume of soil adjacent to and influenced by the plant roots. Microorganisms like bacteria increased the uptake of Cadmium in *Brassica napus* and Nickel in *Alyssum murale*.

Arbuscular mycorrhizal (AM) fungi are symbiotic associations found in natural and industrially contaminated sites also. They may stimulate phytoextraction by essentially improving plant growth and development and also increased uptake of heavy metals. Mycorrhizal fungi associated with plants play an important role in the survival of these plants on polluted soils. Therefore, the use of heavy metal-solubilizing microorganisms is a promising approach for increasing heavy metal bioavailability and uptake in heavy metal contaminated soils. Also, bacteria producing siderophores, and phosphate-solubilizing bacteria are capable of stimulating plant growth in the contaminated area. Plant growth-promoting bacteria could improve the response of the plant to pollutants stress such as toxic metal content. The potential for accumulation and oxidization of heavy metals by bacteria is little understood. Chemicals and heavy metals such as Cd, Ni, Zn, Pb, and Cu have significantly polluted the biosphere in recent decades. Hundreds of small and large manufacturing industries generate effluents and industrial chemical pollutants that contain considerable amounts of heavy metals at elevated concentrations.

Soil is a species-rich habitat containing millions of microbial species. The soil microflora is involved in the degradation and synthesis of organic compounds. It was also contributed to the cycling of plant nutrients and also in the weathering of minerals. Soil consists of microflora such as bacteria, fungi, and the soil fauna like protozoa and invertebrate groups. Now a day, the effects of plants on soil microbiomes have greatly advanced. A limited amount of research has been done on the rhizosphere communities of aged-contaminated industrial soils.

Through the accumulation of heavy metals in food chains, elevated levels of heavy metals not only decrease soil microbial activity but also threaten human health. As an emerging technique for the remediation of toxic heavy metal-contaminated soils, phytoextraction appears to be more cost-effective than conventional chemical and physical treatments, which are generally expensive and harmful to soil properties and human health. Several plant species have been identified as hyperaccumulators growing in heavy metal-contaminated soils in recent years. These plant species have been found with the ability to accumulate unusually high concentrations of heavy metals in their parts without impacting normal growth and physiology. However, most hyperaccumulators identified so far are not applicable for actual field phytoremediation implementation due to their small biomass and slow growth.

The limitations of traditional cleanup technologies methods have encouraged investigations into *in-situ* bioremediation, which is the use of microorganisms to destroy or immobilize pollutants in contaminated places. Microorganisms represent nature's original recycling units, which convert toxic organic compounds into harmless products, often carbon dioxide and water. Conventional techniques such as chemical methods, precipitation, neutralization, or physical methods include ion exchange, membrane separation, and electrodialysis. These methods are commonly applied to remove heavy metals from wastewater and contaminated soil. These processes may be non-suitable at low concentrations. Also, these processes are much expensive and not eco-friendly.

Thus, the researcher develops a viable method to enhance the process of decay and removal by boosting the microbial growth in the biosphere, to degrade, accumulate and remove the contaminants from the industrial sites. After metal extraction plant material must be removed from the contaminated site without extensive excavation, traditional remediation practices involve disposal cost top layer of soil. For better land restoration or remediation. Genetic tools can aid in the discovery of genes influencing key traits, allowing them to be altered. Microarray chips for DNA and RNA provide systematic genome mapping, sequencing, and functioning, based on studies about mutations and polymorphisms, allowing a deeper understanding of structure-function interactions in complex genomes under high concentrations of heavy metals.

Multiple Sequence Alignment will be used to analyze and compare DNA sequences from different hyperaccumulator plants, and their efficacy against metal concentration can also be assessed. Multi-Sequence Alignments (MSAs) are sequence alignments that compare DNA, RNA, or proteins of various biological sequences. There is an evolutionary relationship between the input sequence sets, as the sequences are descended from a common ancestor. Multiple sequence alignment is commonly performed with T-Coffee. Using T-Coffee, all pair-wise alignments between sequences are pre-processed. This provides us with a library of alignment information. Combining local and global pairwise alignments generate the library.

Heavy metal accumulation has a bad impact on the environment, plants, and human health. Metal degrading genes for phytoremediation purposes can be further enhanced by modifying their metal binding property to increase their affinity to specific metal ions. In this study, by using the Metal Detector web server, the metal-binding ability of protein metallothionein has been examined.

18.4 AGRONOMIC ENHANCEMENT OF PHYTOREMEDIATION

Phytoremediation technology must be enhanced by the proper use of soil amendments and incorporating concern agronomic practices. A soil amendment can increase or decrease the biological availability of a contaminant for plant uptake depending on the soil composition. Some plants absorb metal ions directly from the soil solution. Amendments to soil can greatly enhance the soil's fertility and increase plant growth and establishment. Many metals can be made bioavailable and plant-available by lowering soil pH, adding fertilizers, adding chelating agents, and altering the soil ion composition. The most successful amendment is to add chelates such. as EDTA and hydroxyethyl ethylene diamine-tri acetic acid, which can increase ambient soil solution levels of certain heavy metals greater than 1000-fold for easy absorption and simultaneously modify root/shoot partitioning in a wide variety of crop plants.

18.5 GENETIC ENGINEERING APPROACH FOR PHYTOREMEDIATION

To enhance the efficacy of phytoremediation, various factors involved in the process of contaminant uptake and restoration can be optimized using genetic engineering techniques. Here plants were genetically engineered for valued-based parameters such as what they absorb, sequester, destroy, and tolerate, etc. genetic engineering can alter several aspects of plant root structure, such as root depth, penetration into anaerobic zones, and root density. Deeper roots would increase the depth from which a contaminant could be retrieved for phytoremediation. For biodegradable organic contaminants, improved penetration into anaerobic zones would allow increased phytoremediation. Increasing the quantity of plant degradative enzymes, e.g., peroxidases, oxygenases within the root tissue and into the soil can increase their utility. Extraction is more efficient with increased root densities. Extensive root proliferation is caused by *Agrobacterium rhizogenes*, which is usually an undesirable characteristic, but it gains appreciation in the phytoremediation approach. Genetic analysis of mutants could be a promising start to understanding mechanisms concerned with metal biology. For example, chemical mutagenesis has produced a large variety of mutants. In *Pisum sativum,* a recessive mutation that causes 10- to 100-fold higher accumulation of Fe and shoots Pb accumulation is also increased by 50% over that of the parent genotype in recent hydroponic conditions. Varying degrees of enhanced tolerance was reported in the resulting transgenic lines; however, metal uptake levels were not dramatically enhanced. It is not yet clear whether metal hyperaccumulation can be achieved through this approach (Scott and David, 1996).

Bioremediation refers to the use of living ones such as plants and microorganisms to degrade eliminate and transform contaminants in soils, sediments, and water and render them in non-toxic forms. The technology has been widely accepted worldwide in recent years because of its potential to remediate organic and inorganic pollutants from the environment. It gives effective treatment under *in situ* and ex-situ conditions in an eco-friendly manner. The potential of microorganisms and plants has been exploited to achieve maximum remediation of air water and soil.

Phytoremediation is mainly suggested for low contaminated soils and waters at surface level depth. Plants roles in the decontamination of soils are as follows:

- Uptake of contaminant from soil particles into their roots;
- Chemical or physical binding of the contaminant into their root tissue
- Transportation of the contaminant from their roots and accumulation into shoots and leaves.

Moreover, the plants used in phytoremediation should not only accumulate, degrade the contaminants but also grow fast at the contaminated site and prepare themselves for easy harvesting. Plants in that area that have died will return contaminants to the soil. Thus, the cutting and incinerating of the plants are necessary to remove all contaminants from the site. Some examples of plants used in phytoremediation are the following: water hyacinths (*Eichornia crassipes*); poplar trees (*Populus* spp.); alfalfa (*Medicago sativa*); coontail (*Ceratophyllum demersum* L.); Kentucky bluegrass (*Poa pratensis*); American pondweed (*Potamogeton nodosus*); and the emergent common arrowhead (*Sagittaria latifolia*) amongst others (Melinda et al., 2013).

Green plants act as weapons against pollutants and they provide one of the most economical and eco-friendly techniques to remove the organic and inorganic contaminants from water, soil, and air. The major challenge is to find and use exactly the right plants for specific pollutants. Plants having specific properties such as deep roots, special adsorptive properties, luxuriant leaves, and the associated rhizosphere flora, hyperaccumulation of pollutants, metabolize and/or degrade them. According to various functional aspects of plants, phytoremediation is characterized into, phytoextraction, phytostabilization, phytotransformation, rhizoremediation, and phytovolatilization. Many plant species have been successfully applied in the removal of organic and inorganic pollutants from water and soil. For example, poplars remove and metabolize polychlorinated biphenyls (PCBs), 2,4,6-trinitrotoluene (TNT), trichloroethylene (TCE), and hexahydro-1,3,5-trinitro-1,3,5-triazine (RDX), Brassica family (Indian Mustard & Broccoli) take-up heavy metals, such as cadmium, chromium, nickel, and zinc (Zhai, 2011).

Microbiome represents complexity with diversified microorganisms. There are tremendously complex interactions that occur between plants and microorganisms. The diversified networks of microorganisms and their associations with the plant affect the resilience capacity of plants against abiotic stresses, plant growth, and productivity, thus these associations also affect phytoremediation mechanisms in plants. Extensive research work has been made to trace out these molecular interactions between plants and their associate microorganisms, still, there is a lack of understanding of the complex and multi-trophic interactions. Rhizosphere supports a much higher density of microbial communities as compared to the bulk soil, according to the 'rhizosphere effect' and That plant select microorganisms based on available pollutant concentration (Andria, Reichenauer, and Sessitsch, 2009; Kawasaki Bell et al., 2014,). Biodegradation can occur by a group of microorganisms and not by a single taxon. In a microbial community, microorganisms can detoxify inorganic and organic contaminants, according to a review of literature:

To obtain a considerable effect, contaminant degraders must present in high densities

When contaminant concentrations are low, plants may favor or selection of particular growth-promoting microorganisms but opportunistic microorganisms not contributing to phytoremediation may also compete for plant-produced compounds.

For contaminant degradation and other functions, microorganisms are capable of interacting additively or synergistically with each other.

In some cases, the growth of plants is not affected by the biodegradation of microorganisms.

Plants and their microbiome interact to determine the outcome of phytoremediation. Phytoremediation research on a proper framework of ecological and biochemical studies is one of the thrust areas of research to better manage the plant-microbiome relationship today. There have been hundreds to ten thousand species of prokaryotic organisms found in the rhizosphere, and plants are associated with up to 1011 microbial cells per gram root (Mendes, Garbeva, and Raaijmakers, 2013). Rhizospheres still contains a vast number of uncharacterized organisms, their genes, and enzymes involved in associations. Culture-independent techniques are therefore indispensable for estimating microbial diversity and examining the function of microbial systems relevant to plant growth, contaminant degradation, and the environment. Next-generation analytical techniques include 16S rRNA gene sequencing, stable isotope probing with shotgun metagenomics, functional gene microarrays, and metatranscriptomics. The goal of future research is to characterize enzymes, regulatory pathways, interaction networks, and new microbial species. Using these methods, it is possible to study single cells in complex communities and determine the exact mechanism of phytoremediation in whole living systems in their natural context. Microorganisms are intrinsically related to plants, making it a challenging task to boost the bioavailability of trace elements. The effects of thousands of microbial taxa on trace element bioavailability and sequestration are still under investigation. Using metabolomics and proteomics, Li et al., (2014) evaluated the effect of metal-resistant bacteria on copper uptake by maize plants grown in soil contaminated with trace elements. There is a need for more studies in which meta-organism-based approaches can provide useful insights. Phytoextraction efficiency has recently been improved with whole-plant omics analyzes and powerful imaging techniques, which will enable better plant selection and management of plant microbiomes. Researchers are trying to identify gene loci that are related to trace element uptake, translocation, and tolerance, i.e., phytoremediation. Plant breeding programs cannot use current knowledge of specific phenotypic traits and root exudates in situ to optimize phytoextraction. There is a need for integrated plant-microbiome studies to identify novel gene targets that can be exploited to enhance beneficial interactions between microbes and plants.

18.6 PUBLIC RELEVANCE

The most powerful tool against industrial pollution is phytoremediation, which is based on natural plant processes. Plants do the majority of the work, requiring less equipment and labor. Workers avoid contact with harmful chemicals because plants do the majority of the work. Many sites have successfully tested phytoremediation, and it is currently being used on several others. Plants in hydroponic settings are fed heavy metal diets, and most phytoremediation experiments have been conducted on the lab scale, with promising results. Thus, efforts are being made at the industrial scale.

It is important to keep the public informed and involved in the management of pollution issues. More importantly, effective management of industrial pollution requires a community-based effort. Thus, a policy to manage industrial pollution should encompass control; including the provision of information about the pollutant, its hazardous effects, plants that can be used for phytoremediation purposes. This research work raises awareness towards industrial pollution and the importance of phytoremediation for its solution.

Phytoremediation must be supported by industry and the government. It must be an essential requirement for the establishment or operation of the industry. According to the type of pollutants that an industry emits, the phytoremediation plan must be followed. In grading the industry, physical rehabilitation workouts should be a major factor.

By utilizing natural plant physiological processes, phytoremediation can decontaminate agricultural soil, industrial sites, brownfields, sediments, and water containing inorganic and organic pollutants or improve food chain safety through phytostabilization of toxic elements. The technology is low-cost and environmentally friendly. It targets the removal, degradation, or immobilization of contaminants.

Emerging phytoremediation companies need to share two characteristics. Those standards will be developed after extensive research by university scientists, and every pollutant producing company will be required to abide by them. A phytoremediation company develops expertise in a particular contamination or plant process. Partnerships with environmental engineering and consulting firms are the most effective way to market their services.

18.7 SUMMARY

Heavy metals can be recovered from plants used in the phytoremediation of polluted soils by treatment, leaching, and electrolysis. Reduced soil toxicity due to the removal of these elements contributes to environmental protection. An evaluation of phytoremediation processes and heavy metal extraction indicated that ash leaching and electrolysis were effective methods for removing metals from phytoremediation wastes – the plant biomass. The recovery of saleable heavy metals through phytoremediation and the creation of low-concentration heavy metal wastes without toxicity risks were the results of this process.

Phytoremediation can create biomass that is highly toxic due to its application to polluted soils containing heavy metals. Biomass is considered waste and requires controlled, responsible disposal because of its potential effects on the environment and food chain. During decomposition, metals are washed by rain and returned to the soil. The phytoremediation process must end with quantitative recovery of heavy metals if it is to have effective outcomes in terms of decreasing heavy metal levels in the environment more than just moving them between areas. In addition, heavy metal recovery increases the economic value of the phytoremediation process as it transforms this into a financially self-supporting approach for environmental remediation. Further research was conducted to identify methods of recovering heavy metals from ash formed during the incineration of biomass. The recovery of heavy metals in polluted soil and ash can be achieved through phytoremediation. The polluted soil and ash can be used for fertilization because they are low in heavy metals. The industry can recuperate heavy metals as financial support for phytoremediation.

18.8 EXERCISES

Objective Response Questions

1. What can affect the microbial population?

a) Green algae
b) Kitchen waste
c) Heavy metals
d) Oxygen

2. Which of the following terms has nothing to do with affecting microbial populations?

a) Reduction in abundance
b) Diversity of species
c) Color of species
d) Resistant population

3. Tolerant microbes can grow better in metal stress.

a) True
b) False

4. **Which of the following soil pollution is not responsible for the reduction in bacteria?**
   a) Copper
   b) Cadmium
   c) Zinc
   d) Glass

5. **What can dissolve metals?**
   a) Biogeochemical cycle
   a) Microbial metabolic secretions
   b) Afforestation
   c) Deforestation

6. **Which of the following methods of phytoextraction removes metals?**
   a) Absorption by plant roots
   b) Absorption by water
   a) Absorption by soil
   b) Absorption by weathered rock

7. **Heavy metals cannot be removed by plants.**
   a) True
   b) False

8. **What are the benefits of bacteria in metal extraction?**
   a) Oxidation of ore
   a) Reduction of ore
   b) Assimilation
   c) Accumulation

9. **The following method does not remove heavy metals from water.**

10. **Adsorption**
   a) Chemical precipitation
   b) Biogas
   c) Reverse osmosis

11. **Copper is extracted from contaminated soil by phytomining.**
   a) True
   b) False

Practice Work

Qu 1. Describe the process of re-extraction of metals from hyperaccumulators.

Qu 2. Explain the scope of phytoremediation technologies.

Qu 3. Write a short note on the agronomic enhancement of phytoremediation.

Qu 4. How genetic engineering approach is useful for phytoremediation.

Qu 5. Write a short note on thermal treatment.

Qu 6. Describe hydrometallurgical processes in detail.

Qu 7. Explain the electrochemical processes of extraction of heavy metals.

Qu 8. What is electrolysis?

Qu 9. Write a short note on bioaccumulation and biomagnification in the food chain and its health consequences.

Qu 10. Explain the role of mycorrhizal fungi in heavy metal removal.

REFERENCES

Banerjee, D., Rao, M. A., & Wattal, P. K., (2013). Separation and recovery of Cs from high active waste simulant using resorcinol formaldehyde polycondensate resin: Batch and column studies. *Separation Science and Technology, 48*(1), 133–139.

Blaylock, M. J., Salt, D. E., Dushenkov, S., Zakharova, O., Gussman, C., Kapulnik, Y., Ensley, B. D., & Raskin, I., (1997). Enhanced accumulation of Pb in Indian mustard by soil-applied chelating agents. *Environ. Sci. Technol., 31*, 860–865.

Clístenes, W. A. N., Amarasiriwardena, D., & Xing, B., (2006). Comparison of natural organic acids and synthetic chelates at enhancing phytoextraction of metals from a multi-metal contaminated soil. *Environmental Pollution, 140*(1), 114–123. ISSN 0269-7491, https://doi.org/10.1016/j.envpol.2005.06.017.

Cunningham, S. D., & Ow, D. W., (1996). Promises and prospects of phytoremediation. *Plant Physiol., 110*(3), 715–719. doi: 10.1104/pp.110.3.715.

Grcman, H., Vodnik, D., Velikonja-Bolta, S., & Lestan, D., (2003). Ethylenediamine dissuccinate as a new chelate for environmentally safe phytoextraction. *Journal of Environmental Quality, 32*(2), 500–506. doi: 10.2134/jeq2003.5000.

Hoque, Md. E., & Philip, O. J., (2011). Biotechnological recovery of heavy metals from secondary sources: An overview. *Materials Science and Engineering C, 31*, 57–66.

Jainae, K., Sanuwong, K., Nuangjamnong, J., Sukpirom, N., & Unob, F., (2010). Extraction and recovery of precious metal ions in wastewater by polystyrene-coated magnetic particles functionalized with 2-(3-(2-aminoethylthio) propylthio)ethanamine. *Chemical Engineering Journal, 160,* 586–593.

Keller, C., Ludwig, C., Davoli, F., & Wochele, J., (2005). Thermal treatment of metal-enriched biomass produced from heavy metal phytoextraction. *Environ. Sci. Technol., 39*(9), 3359–3367.

Kirkelund, G. M., Damoe, A. J., & Ottosen, L. M., (2013). Electrodialytic removal of Cd from biomass combustion fly ash suspensions. *Journal of Hazardous Materials, 250, 251*, 212–219.

Kır, E., Çengeloğlu, Y., & Ersoz, M., (2006). Influence of chelating agents on the recovery of Al (III), Fe (III), Ti(IV), and Na(I) from red mud by cation exchange membranes. *Separation Science and Technology, 41*(5), 961–977.

Machado, M. D., Soares, E. V., & Soares, H. M. V. M., (2010). Selective recovery of copper, nickel, and zinc from ashes produced from *Saccharomyces cerevisiae* contaminated biomass used in the treatment of real electroplating effluents. *Journal of Hazardous Materials, 184*, 357–363.

Machado, M. D., Soares, E. V., & Soares, H. M. V. M., (2011). Selective recovery of chromium, copper, nickel, and zinc from an acid solution using an environmentally friendly process. *Environ. Sci. Pollut. Res., 18*, 1279–1285.

Meawad, A. S., Bojinova, D. Y., & Pelovski, Y. G., (2010). An overview of metals recovery from thermal power plant solid wastes. *Waste Management, 30*, 2548–2559.

Min, X., Yuan, C., Liang, Y., Chai, L., & Ke, Y., (2012). Metal recovery from sludge through the combination of hydrothermal sulfidation and flotation. *Procedia Environmental Sciences, 16*, 401–408.

Mizoguchi, K., Ida, J., Matsuyama, T., & Yamamoto, H., (2010). Straight-chained thermo-responsive polymer with high chelating group content for heavy metal ion recovery. *Separation and Purification Technology, 75*, 69–75.

Modin, O., Wang, X., Xue, W., Rauch, S., & Fedje, K. K., (2012). Bioelectrochemical recovery of Cu, Pb, Cd, and Zn from dilute solutions. *Journal of Hazardous Materials, 235, 236*, 291–297.

Nzihou, A., & Stanmore, B., (2013). The fate of heavy metals during combustion and gasification of contaminated biomass: A brief review. *Journal of Hazardous Materials, 256, 257*, 56–66.

Sharma, P., & Pandey, S., (2014). Status of phytoremediation in world scenario. *International Journal of Environmental Bioremediation & Biodegradation, 2*(4), 178–191. doi:10.12691/ijebb-2-4-5.

Singer, A., Navrot, J., & Shapira, R., (1982). Extraction of aluminum from fly ash by commercial and microbiologically produced citric acid. *Applied Microbiology and Biotechnology, 16*, 228–230.

Sivrikaya, S., Altundag, H., Zengin, M., & Imamoglu, M., (2011). Separation, preconcentration, and recovery of Pd (II) ions using newly modified silica gel with bis (3-aminoprop- yl) amine. *Separation Science and Technology, 46*, 2032–2040.

Tan, Z., & Lagerkvist, A., (2011). Phosphorus recovery from the biomass ash: A review. *Renewable and Sustainable Energy Reviews, 15*, 3588–3602.

Won, S. W., Lim, A., & Yun, Y. S., (2013). Recover y of high-purity metallic Pd from Pd (II)-sorbed biosorbents by incineration. *Bioresour. Technol., 137*, 400–403.

Zhai, G., (2011). Phytoremediation: Right plants for right pollutants. *J. Bioremed Biodegrad., 2*, 102e. doi: 10.4172/2155-6199.1000102e.

Index

D

H

I

Q

R